Medicine in Iran

Medicine in Iran

Profession, Practice, and Politics, 1800–1925

Hormoz Ebrahimnejad

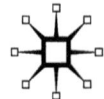

MEDICINE IN IRAN
Copyright © Hormoz Ebrahimnejad, 2014.

All rights reserved.

First published in 2014 by
PALGRAVE MACMILLAN®
in the United States—a division of St. Martin's Press LLC,
175 Fifth Avenue, New York, NY 10010.

Where this book is distributed in the UK, Europe and the rest of the World,
this is by Palgrave Macmillan, a division of Macmillan Publishers Limited,
registered in England, company number 785998, of Houndmills,
Basingstoke, Hampshire RG21 6XS.

Palgrave Macmillan is the global academic imprint of the above
companies and has companies and representatives throughout the world.

Palgrave® and Macmillan® are registered trademarks in the United
States, the United Kingdom, Europe and other countries.

ISBN: 978–0–230–34102–9

Library of Congress Cataloging-in-Publication Data

Ebrahimnejad, Hormoz.
  Medicine in Iran : profession, practice, and politics,
  1800–1925 / Hormoz Ebrahimnejad.
    pages cm
  ISBN 978–0–230–34102–9 (alk. paper)
  1. Medicine—Iran—History—19th century. 2. Medicine—
Iran—History—20th century. 3. Medicine, Persian—History—
19th century. 4. Medicine, Persian—History—20th century.
  I. Title.
  R631.E27 2013
  610.955—dc23   2013014602

A catalogue record of the book is available from the British Library.

Design by Integra Software Services

First edition: January 2014

10 9 8 7 6 5 4 3 2 1

Transferred to Digital Printing in 2014

To my dear daughter, Darya

Contents

List of Illustrations	ix
Preface	xi
Introduction	1
1 The State of Medical Theory and Practice in Nineteenth-Century Iran	15
2 The Physicians and Their Encounter with Western Medicine	49
3 The Reform Movement and Medical Institutionalization	87
4 Medical Transition under the Constitution	121
Conclusion	163
Notes	167
Bibliography	219
Author Index	241
Subject Index	251

List of Illustrations

Figures

1.1	Body and its diseases caused by foul air	30
2.1	Mirzâ Mohammad-Taqi Shirâzi (Malek al-Atebbâ) (Aqâ-Bâbâ Rashti)	68
2.2	Malek al-Atebbâ and Tholozan, examiners at the Dâr al-Fonun	69
3.1	Mirzâ Mohammad Khân-e Sepahsâlâr	109
3.2	Mirzâ Abolfazl-e Tabib-e Kâshâni	115
4.1	Kholâsat al-hekma (late nineteenth to early twentieth century), a modern medical text used by traditional physicians	141
4.2	Prescription according to Kholâsat al-hekma and a sample prescription in 1332 (1953)	143
4.3	Traditional and modern educations	143
4.4	Pharmacist Octav Le Comte	148
4.5	Certificate issued by the American hospital	149
4.6	Source for the salary of Mirzâ ᶜAbdollah Loqmân (*Asnâd*)	151
4.7	Source for the salary of Mirzâ Yahyâ ᶜEmâd al-hokamâ (*Asnâd*)	151

Map

3.1	Tehran's map in 1859	111

Preface

In one of his speeches, Ayatollah Ruhollah Khomeini (1902–1989), the founder of Iran's Islamic regime, contended that traditional medicine is being neglected for the benefit of modern medicine, and its revival is necessary for overall Islamic revival. Subsequently, associations were formed with the help of "Islamist" physicians to create institutions of traditional medicine (*tebb-e* sonnati). Considering that these institutions responded to Khomeini's call, we can presume that by traditional medicine, Khomeini meant Galenico-Islamic medicine or Galenic medicine as assimilated in Islam. Humoral medicine's association with Islam, as suggested by Khomeini, goes a long way to explain its assimilation by Islam throughout several centuries, thanks in part to the works of scholars like Ibn Qayyim al-Jawziyya (1292–1350). Initially, the humoral medicine of Hippocrates and Galen, even after being translated into Arabic, was not considered to be in harmony with the Koranic teaching and the Prophet's traditions.[1] In fact, Jawziyya's task was to bring these two in agreement. In the modern period, however, with the rise of Islamism, the dichotomy between Galenic medicine and Islamic/Koranic teaching on health resurfaced. With the state-sponsored revival of traditional medicine, the Islamization of medicine made headway when other associations were created, such as the *tebb-e qor'âni* (Koranic medicine), strictly based on the teaching of the Koran and rejecting Galenico-Avicennian medicine.[2] Yet, when Khomeini was diagnosed with the fatal stomach cancer, chemotherapy was done for his treatment with the assistance of state-of-the-art medical technology rather than the medicine of the Prophet, Koranic medicine, or even humoral medicine of Hippocrates and Galen. In 2002, his successor, Ayatollah ʿAli Khamenei, officially authorized embryo stem cell (ESC) research and therapy.

This maneuvering between traditional and modern medicine obviously indicates that the Islamic regime has used the revival of traditional medicine for its Islamic identity and modern medicine for its practicality, and also for enabling the regime to be viable in an age when modern science and technology has the last word. The application of modern medicine, a symbol of Western influence, by an Islamic power whose ideology is based on fighting Western influence and values indicates how far modern technology has penetrated all the interstices of the society. But at the same time, such shifts seem to be a requisite in a fast developing and changing world. While in an Islamic revolution, it appears normal to revive traditional medicine alongside other Islamic values, it also appears normal that a newly established clerical regime uses modern technology to strengthen its power and therefore its legitimacy. Today, the Islamic regime boasts of having mastered the most advanced technologies in various fields, including biomedicine. Such a discourse of "medical modernity" is so entrenched socially and politically that today the most conservative individuals employ all legal means, including *feqh,* jurisprudence, and *ejtehâd* (independent reasoning) to justify the use of modern medicine as if humoral medicine has never existed. Probably it is due to this omission that historians have failed to explain how this change in medical paradigm happened.

Focusing on the medical transformation in Iran for more than a decade, I have endeavored to fill this gap. My first attempt to cover this topic did not result in a full monograph as half of the book was devoted to the editing and publication of a document that illustrated the state of medical knowledge and practice in Qâjâr Iran (1898–1925).[3] Furthermore, it focused mainly on public health, while the development of medicine in theory and practice was only touched upon. By taking up this latter question and examining it at length, the present volume completes my initial research project on medical transition in Qâjâr Iran. How far has modern medicine been the product of pure scientific and conceptual transformation? And to what extent did this transformation come in response to sociopolitical developments? Intrinsically interlinked but rhetorically distinguished for analytical purposes, these are the two major questions that inform the structure of this work. The intellectual and theoretical analysis of medicine is based on the examination of different types of medical literature in the nineteenth century, and the information related to social, political, and institutional contexts within which medicine has transformed have been drawn from contemporary journals and archives.

Traditional and modern medicines are examined in relation to, and in dialogue with, each other. Can subjects like "traditional" and "modern" be looked at in isolation in a given historical context? Before the nineteenth century, when Persia came into increasing contact with the West, no such distinction between tradition and modernity was made. Both traditional and modern medicine found their identities as a result of encounters with each other, and their study makes sense only when examined in relation to each other. Such identification does not signify that they were static phenomena. Traditional medicine, unlike received wisdom, underwent changes during the nineteenth century, particularly in Iran, when it came into contact with Western medicine.[4] Although both traditional and modern medicine underwent changes, we will still employ these terms, representing two different entities, for the purpose of analysis.

By traditional medicine, I mean learned medicine based on the Galenic system, and not folk or magic/religious healing, even though the latter had integrated elements of humoral medicine. This is not because folk and religious medicines did not occupy significant place in practice, but because it was learned medicine that was involved in the transmission and development of modern medicine in Iran, through the collaboration of traditional physicians within modern institutions such as the Dâr al-Fonun and the sanitary councils.[5]

Some explanation is required for the terms "Iran" and "Persia," which have been alternately used throughout the text. The country was always called Persia by the Europeans but never by the Iranians themselves. All historical and literary sources in Iran, including diplomatic correspondences, used the term "Iran." However, since many of our European sources, archives, and travel accounts use the term "Persia," we employ both Persia and Iran.[6]

For the transliteration of Arabic and Persian names and terms, I have followed a simplified diacritical style, with "â" for long "a" (such as in fall, social) and "ᶜ" for *ayn* (ع) in Arabic such as in *aᶜlam* (اعلم) and "e" for *ezâfeh* in Persian, instead of "i" that is often used, also by the *Encyclopaedia Iranica*, but is confusing. Thus, I use Eᶜtezâd al-Saltaneh and not Iᶜtizâd al-Saltaneh, and Karim Khân-e Zand and not Karim Khân-i Zand.

This volume is the most recent piece of work that is the result of years of research on the history of medicine generously funded by the Wellcome Trust since 1998. I take this opportunity to once more acknowledge this invaluable support. During my research at the Wellcome Trust Centre for the History of Medicine in London, I had a marvelous time in the company of wonderful colleagues, including Harold Cook, Lawrence Conrad, Vivian Nutton, Nikolaj Serikoff, Nigel Allen, Cornelius O'Boyle, Henrietta Bruun, Tony Woods, Sanjoy Bhattacharya, Roger Cooter, Fiona McDonald, and of course the late Roy Porter; I benefited hugely from both their expertise and humanity. I extend my heartfelt thanks to all of them.

The preparation of this book began with a research project on the military hospitals in Qâjâr Iran, funded by the Wellcome Trust and hosted by the Wellcome Trust Unit at Oxford University. I am indebted to Mark Harrison, the Head of the Wellcome Unit, for his support and the valuable advice he was always ready to give during my time there. In the dynamic Wellcome Unit at Oxford, I benefited from weekly seminars, and discussions with colleagues and fellows, particularly Roy MacLeod and John Manton. My acquaintance with Roy MacLeod was extremely beneficial for the present volume as he very generously accepted to read and comment on the manuscript, providing me with invaluable feedback. I owe a special debt of gratitude to Anne Marie Moulin and Seref Etker for reading the earlier drafts of my manuscript and providing me with valuable information. I must also thank the University of Southampton for granting me a semester's sabbatical leave that enabled me to complete this monograph on time. I would also like to thank Chris Chappell and Sarah Whalen at Palgrave for taking care of the manuscript during the production stage. I am particularly obliged to the anonymous reviewers whose remarks and comments were extremely helpful in reducing the mistakes or inaccuracies of my manuscript during the final revision. However, any deficiencies in the content or style that one might find in this volume are only mine. The librarians and staff at various libraries in Tehran, including the Central Library of Tehran University, Malek Library, Majles Library, the Sepahsâlâr Library, and the National Library, were always cooperative and professional. I feel grateful to all of them, particularly Ali Kassai, Zahra Taheri in the manuscript department at the National Library, M. Khalili in the Public Relations department at the Majles Library, and Rasul-e Ja'fariyân, the then head of the Majles Library, who were extremely helpful in facilitating my access to sources and obtaining copies. Last, but not least, I owe special thanks to my brother Bozorgmehr-e Ebrahimnejad, who has always been generous in providing me with accommodation in Tehran during my research trip to Iran.

Introduction

The history of the relationship between Iran and the West has been studied from a socio-political point of view, but the development of modern science as part of this relationship has not yet been the object of serious and sustained examination.[1] Most of the literature covering the modern history of science in non-Western countries has focused on science in Empire within the framework of the relationship between the metropolis and its colonies.[2] However, the development of science in countries that were not formally colonized (such as Iran) has yet to be explored.[3] Such a study can shed new light on the nature of what is called "colonial science" and on the relationships between local and Western knowledge. Although the "orientalist" discourse of modernization in the "Orient" does not seem to discriminate between colonized and non-colonized countries, physicians in the latter countries had a different relationship with their Western counterparts as they were not bound to consider them their master or superior, whether socially or intellectually.

The perception of medical modernization is shaped by a model highlighting the introduction of concepts, methods, institutions, and drugs by Western physicians. As a result, the changes that occurred within local medicine, whether in isolation from or in contact with modern medicine, have not been seriously examined even though some scholars have highlighted the importance of the interaction and dialectical relationship or cross-fertilization between Western and non-Western medicines.[4] This is due, at least partly, to the "Orientalist" or "Eurocentric" conception of modernity that equates "traditional" with "static," "unchanging," and "unhistorical."[5] According to Tavakoli-Targhi a considerable body of persianate literature that engaged with modern science in Iran and India at least since the sixteenth century has been ignored in modern historiography. As a consequence, the author maintains, modernity, and even "Orientalism," was initially the fruit of European–non-European or Western–Eastern collaboration.⁶ Even when Persian literature is referred to or cited by contemporary or modern historiography, its content is not appropriately scrutinized. A detailed study of medical literature produced in the nineteenth century helps us to find that medical modernization in Iran followed a pattern of transformation from Hippocratic medicine to biomedicine quite similar to that in the West, albeit at a different level. This pattern also has been ignored in modern historiography given the presumed intrinsic antagonism between "traditional/Iranian" and "modern/Western medicines." Quite paradoxically, while "orientalizing" traditional medicine, in Saidian terms,⁷ could imply a Western-style examination of its history and consequently lead to the built-in evolution in both theory and practice, it has presumed traditional medicine devoid of any capacity for change. Consequently, mapping the transformation of traditional medicine even as a result of its encounter with Western medicine in the nineteenth century has never been on the agenda of modern historiography.

The intellectual or theoretical transformation in which traditional/local medicine was involved was underpinned or generated by historical and sociopolitical contexts, which will also be examined in this volume. Even though the state and Western missions played a crucial role in medical modernization, they operated in relationship with a traditional medical establishment that they could not avoid. This situation also was at work in nineteenth-century India, to the extent that physicians, such as John Tytler and Alloy Sprenger, were of the opinion that "Muslims could not be educated in Western medicine 'without a knowledge of their own literature' [in Arabic and Persian]."⁸ When modern medicine was first introduced, both its theoretical and practical applications went through a long process of negotiation and dialogue with the local political and medical establishments. The political context and the relationship between the court and learned traditional medicine (in particular the political and social movements for modernization in the second half of the nineteenth century) informed the medical transition. From this perspective, modern medicine is not seen as a system of knowledge transplanted and developed in line with the West, but rather as a development deeply imprinted by the local context.⁹ It is due to different local conditions that Galenico-Islamic medicine reacted to modern Western medicine differently in different countries.¹⁰

Medical modernization was a multilayered process that involved not only Western influence and the introduction of modern techniques and concepts, but also an evolution in local medicine and the local medical profession. This book is an account of this evolution—an account of the socio-political reforms that framed the acceptance of modern ideas, the political power that managed the establishment and development of modern institutions, and, finally, the international relationships peculiar to the Qâjâr state that informed the pace of the introduction of modern medicine and decided the form of "modern" medical institutions.

The employment of Western experts in the military or in other fields had been a custom since the Safavids and continued under the Qâjârs.[11] Western medicine was thus not perceived as "colonial," and its assimilation did not pose a "nationalistic" problem.[12] However, the introduction of Western science was not the initiative of the state only. Prior to modernization by the Qâjâr state, elements of modern medicine (mainly Paracelsian iatrochemistry) were introduced because of the interest and initiatives of individuals.[13] Furthermore, before the eighteenth and nineteenth centuries, when the encounter with Western medicine took place, traditional learned physicians in Iran were already accustomed to different opinions via the commentaries (*sharh*) and biographies of physicians (*tazkerat al-atebbâ*) of various regions and epochs.[14] In the modern period, Western physicians were incorporated in these biographies. To my knowledge, the first and probably the last literature of this kind in medicine is the *Matrah al-Anzâr* of ʿAbdol-Hossein-e Zenuzi-ye Tabrizi (known as Filsuf al-Dowleh, 1866–1941).[15] Written in 1324/1906, during the Constitutional Revolution, this book was influenced by modernist ideas.[16] Filsuf al-Dowleh was a "hybrid," a traditional physician but with an inclination for modern medicine. There were, however, also traditional physicians who referred to modern Western ideas in their writings.[17] This literary context played an important role in the assimilation of new medical concepts in the nineteenth century, if not earlier.

This suggests that medicine was seen by physicians in the second half of the nineteenth century as an evolving and/or plural knowledge encompassing both old and new, or Iranian and Western, ideas. However, for a long time, the conceptual overlaps between Galenico-Islamic medicine and nineteenth-century Western medicine were overshadowed by a cultural and regional Iranian-Western dichotomy, to the extent that in Qâjâr Persia, the Neo-Hippocratic medicine of Grisolle and the nascent biomedicine of Pasteur were identified with "Western" or "modern" medicine in spite of their conceptual differences. To some extent, such denomination was also used by Western instructors in Iran. They identified the medicine they taught at the Dâr al-Fonun with the medicine that was in full-fledged development in Europe; however, the anatomical pathology introduced by Drs. Polak and Tholozan was substantially imbued with (neo-) Hippocratic theories, lagging behind the anatomical pathology taught and practiced in the West. In the preface to a book written by Nasrollâh Mirzâ Qâjâr, Tholozan stated that this "is an exact account of our medical knowledge and its use is very worthwhile in Persia."[18] In the book, health and disease were explained according to Galen; fevers were given the same classification found in traditional texts: they were either continuous (*motassel*), in which the four "states" and "durations" of the disease were linked, or intermittent (*monfasel*). In the 1870s, when this book was published, medicine in Europe was distancing itself from such Hippocratic principles. In the same vein, no discrimination was made between Persian traditional physicians who, as we will see in Chapter 2, differed in their opinions on diseases and their treatments.

Nevertheless, the strong dose of neo-Hippocratism in Dâr al-Fonun's anatomical pathology benefited medical transformation in nineteenth-century Iran. Familiar to some extent with the language of these fields, traditional medicine entered into dialogue with them and, inspired by the new ideas and methods of modern medicine, experienced a kind of self-reflection and revision. The neo-Hippocratism, as taught at the Dâr al-Fonun, shed new light on humoral theory and fine-tuned its terminology or at least enriched it, furthering the clarity of its concepts. For example, an expression like *homâ-ye lâzemeh* (literally "necessary fever") found a clearer translation in modern medicine, where it was called as "continuous fever." Similarly, *hommâ-ye yawm* (fever lasting one day) was translated as "ephemeral fever" and *hommâ-ye dâyerh* (literally circular fever) and *nowbeh* (recurrent) were translated as "intermittent fever."[19] Neo-Hippocratism and anatomy gave meaning to the age-old humoral literature on the inner body, which had been, until then, obscure.

In relation to content, the borderline between old and new, Iranian and Western, was defined in terms of its clarity and efficiency. For the critics or the "modernists," a mastery of humoral (or modern) medical literature was understood to mean ʿ*elm* (theoretical knowledge). What became prominent in the nineteenth century was ʿ*amal* (practice). ʿ*Elm* without ʿ*amal* was no longer valued. Traditional medicine, according to this view, did not appreciate ʿ*amal* (manual medicine). Without practical use, such theoretical knowledge was but futile. A similar movement was at work in literature. The pedantic and meaningless style of prose, which included redundant terms for the sake of rhyme, or poems for embellishment, was deemed useless. The modernists were seeking concision and meaning rather than empty wordings.[20] The superiority of Western medicine over humoral medicine in the nineteenth century resided in surgery, which was based on new anatomical knowledge. So, it was not surprising that modern medicine equated to manual medicine although some modern trained doctors who had no practical skills were also considered inefficient and only those skilled in surgery were appreciated by the local population.[21] However, this is not to separate the purely intellectual interest of local physicians in modern medicine from their interest in its practical use for the benefit of the state or the society. It was the new understanding of the functions of the body, veins, muscles, and nerves in the light of practical dissection that led physicians to become efficient surgeons. In this respect, it would be inaccurate to identify "colonial" or "semicolonial" countries (for example, India and Iran in the nineteenth century) by the fact that these countries privileged applied science to the detriment of research.[22] If in nineteenth-century India and Iran there was an imbalance between application of modern techniques on one hand and research on the other, it was mainly because they lacked institutional and material infrastructure to organize and implement research and not because of the lack of interest in research.

The inherent link between intellectual interest in modern medicine and its practical aspect and efficiency can be seen in the efforts made to bridge theoretical gap between traditional and modern medicines. Throughout the period, in most non-Western or Islamic countries, in North Africa, Iran, and the Indian subcontinent, we find two categories of physicians: those who remained attached to humoral medicine and those who adhered to modern medicine. Modern trained physicians made the effort to find common ground between traditional and modern medicine as a way to justify modern medicine and also as a necessary theoretical step toward understanding and assimilating modern medicine. However, the intellectual and institutional mechanisms of the introduction and development of modern medicine varied in different countries. For example, in the political and colonial context of the nineteenth century the encounter with modern medicine and the discourse of modernity that ensued raised the question of identity (Western science versus local knowledge). The relationship between the European powers and Iran and between the British empire and India was that of dominant–dominated, but the type of relationship these countries had with the West informed the discourse of identity. In Qâjâr Iran, the local intelligentsia adopted modern medicine for the improvement of medical knowledge and public health. The fact that it was introduced by imperialist countries was overlooked. Those who introduced it, the Shah and his ministers, as well as those who taught it, were hailed as noble founders of modern science in the country. In this way, modern science was given a national identity. In India, on the other hand, the major concern in the transformation of modern medicine was to preserve the identity of Ayurveda and Unani medicine by reviving them in the face of Western (colonial) medicine. At the same time, the Unani hakims endeavored to justify the "scientificity" of their traditional medicine by reminding people of "the points of confluence between tibb and biomedicine, or between germ and poisonous matter."[23] In modern Arab countries, with yet different historical contexts to those of Iran and India, "the common roots shared by Western medicine and 'Islamic medicine' are often mentioned to reinforce the legitimacy of modern medicine,"[24] which is at odds with some aspects of

Islamic customs. At the same time, "the wealthy rulers of the Persian Gulf region employ state-of-the-art medical knowledge and technology as a foundation for their own legitimacy."[25] At work in all these strategies was medical transformation, no matter what the discourse in which they were expressed.

The development of science in the modern period has always been affected by power relationships; cultural, religious, and racial differences; and conflicts at national and international scales. The countries that create new technologies endeavor to protect them from others. Both military and civil technology is often developed in such a way that it cannot be duplicated, unless its use would not threaten the interests of its creator nation. Most state-of-the-art science and technology is in the possession of the superpowers, the successors of the nineteenth-century imperial and colonial countries. Despite the universal character of science and technology—one cannot talk of Western and Eastern space science for example—the discourse on indigenous versus foreign or Western knowledge still dominates in an India scarred by nineteenth-century colonialism. Although India has largely benefited from cooperation with the West in the development of its space technology, its elites like to remind the world that India is now self-sufficient and able to construct missiles and satellites "entirely indigenously."[26] A similar imperial relationship between Europe and China in the nineteenth century has affected Chinese scientific developments too, especially in the military and in space science. In 2011, in proportion to their gross national income (GNI) per capita, China (GNI $8,390) and India (GNI $3,590) spent far more on scientific research than the United States (GNI $48,820) and the United Kingdom (GNI $36,010), as a means of asserting their power against ex-colonial or imperial masters and present rivals.[27]

It is hard to establish how genuine the claims by the clerical government in Iran concerning achievements in space science are. However, progress has effectively been made in nuclear technology and medical research, which can better be understood within the framework of the power relationships with the West, particularly since the 1979 Revolution. The cloning of animals has been taking place in the Islamic Republic since 2006, and Iran is one of the few countries at the forefront of embryo stem cell (ESC) research. Research on embryos is forbidden by conservative Christians who believe that human life is created at conception and research on an embryo destroys it. In Islam, the question of when human life begins is subject to debate among physicians and religious scholars because "the references in the Qur'ân and the Sunna [the tradition of the Prophet] are indecisive concerning the exact starting point of human life."[28] The authorities of the clerical regime in Iran, on the other hand, claim that this question has been settled through *ejtehâd* that takes into consideration primary rules required by *shariʿa* and secondary rules dictated by special circumstances.²⁹ The account given by Bâqer-e Larijani, a physician close to Ayatollah Khamenei, on this question indicates that all medical institutions and parliament that regulated the practice of ESC research and cloning followed the "religio-political" decision of the leader.³⁰ It was therefore decided that the life of the embryo begins with ensoulment (when the soul enters the fetus), which occurs between 40 and 120 days after conception.³¹ Thus, embryonic stem cell research and therapeutic cloning are "permissible only in the pre-ensoulment stages of fetus development, i.e. up to 120 days after conception."³² In a similar vein, gestational surrogacy has been allowed despite the fact that in principle the sperm of a man is transferred to the womb of the surrogate mother, either directly (artificial insemination) or via in vitro fertilization (IVF). In Sunni Islam this is not permitted because "it involves introducing of a sperm of a man into the uterus of a woman to whom he is not married."³³ It is true that the theory of *ijtihâd* allows the Shiite clerics to (re)interpret or revise the *shariʿa* according to new conditions but significantly such a principle can be affected by historical contexts: the Shiite clerics, who in the nineteenth century opposed the introduction of railways and modern schooling system, now advocate the construction of satellites and the cloning of the human body. Ayatollah Khamenei's endorsement for ESC research correlates with his support for all branches of science and technology "in an attempt to enhance the country's global status."³⁴ Such a dramatic position from a conservative Shiite cleric would not have taken place had the clerical regime not been in conflict with the West. However, adoption of foreign science and reinterpretation of Islamic principles to make this possible had a long history, beginning with the translation movement in early Islam. Ibn Qayyim al-Jawziyya introduced principles of Hippocratic and Galenic medicine in the *Medicine of the Prophet* whenever they did not contradict the teaching of the Prophet.³⁵ The necessity of negotiation and harmonization between Islam and science has increasingly been felt with the encroachment of modern science in everyday life, to such an extent that whenever "scriptural texts contradicts scientific knowledge, their apparent meaning (*al-zâhir*) should be abandoned and preference should be given to their metaphorical interpretation (*taʾwil*)."³⁶

Change, reform, or "modernization" does not occur in a vacuum. Medical modernization in nineteenth-century Iran probably would not have occurred without interaction with the West even if we take into account its internal dynamics. This principle also holds true for medical transformation in the West. Numerous instances of the development of medicine since the early modern period, due to contact of the West with non-Western countries, has now been well-documented.³⁷ As Tavakoli-Targhi pointed out, the historiographical discourse on modernity, nurtured by Western

"Orientalism," has ignored the heterogeneous nature of modernization.[38] However, in the same way that the West in the development of its science, commerce, and economy was conditioned by its encounter with other countries, any change or development in non-Western countries was informed by their relationship, within colonial context or not, with the West. The perspective of the present volume is not to study medical modernization *in comparison to* modern medicine in Europe but *in relation to* it and within the context of Western influence, not least because it was a compelling historical phenomenon. Such an analytical approach echoes the harmonious relationship between Western and Persian physicians despite contrasts or conflict of interests between them. It was within the framework of this relationship that medical transformation took place.

The Qâjâr medical establishment did not initiate transformation, but underwent change from within as a result of its contact with the West and the prevalence of epidemic diseases. Chapter 1 provides a glimpse of the major waves of epidemics and their impact on medical knowledge to contextualize the intellectual transition of traditional medicine. This is because epidemics constituted the most striking medical and social phenomenon that physicians had ever experienced. The dynamics at work within Galenico-Islamic medicine, described in this chapter, were to a large extent due to external factors but this was made possible, firstly, because they occurred in an institutional environment, namely, the state, and, secondly, because, theoretically, it was receptive to the neo-Hippocratic version of modern medicine with which it first came into contact. The Islamic world is not unfamiliar with integrating non-Islamic science and knowledge, as it directly benefited from Greek science and philosophy in developing what is called Islamic science, that is, astronomy, medicine, and mathematics. Although epistemological gaps appeared between the original Hippocratic and Galenic medicine and its Islamic application, the continuation of the format and curriculum of Galenic medicine in Islam secured the continuity of Hippocratic (medicine of observation) and Galenic (anatomy) discourses. Even though the discourse that animated the literary format, or curriculum, had lost some of its Hippocratic and Galenic content, it preserved a formal identity with the content that favored the understanding and assimilation of neo-Hippocratic and anatomical-pathological medicine in the context of the nineteenth century.

Through a narrative of the encounter with Western medicine, Chapter 2 analyzes the historical developments by means of which Iran tried to counterbalance the imperial domination of Russia and Britain by seeking an association with other countries, notably France. The different European powers were not seen through the same lens, and the presence of the West in the field of medicine was regulated and attuned according to the different relationships Iran had with these powers. This scheme was quite different from that of a binary colonial/colonized relationship. It had an important effect on the attitude of the physicians toward Western medicine. Not only were the traditional physicians familiar with the language of neo-Hippocratic and early anatomical pathology or at least studied such literature with sympathy, they did not find them representative of a foreign imperial or colonial domination. This situation did not provoke the traditional physicians to self-defend by trying to revive traditional medicine or create an institution of their own to oppose modern medical institutions. Rather, they reacted to modern medicine on an individual basis and with different attitudes toward it. Dividing physicians into three categories, the orthodox, the hybrids, and the modernists, Chapter 2 examines the intellectual transition of medicine within the framework of the state institutions and the encounter with the West. Most physicians tried to combine traditional (*qadim*) and modern (*jadid*) medicines, a combination that is explicit in several treatises.[39] Others, like Mirzâ ᶜAli-Akbar-e Hamadâni, taught both traditional and modern medicine in parallel, but did not combine them in their writings. This combination is due to educational and institutional factors. Pedagogically, the grammar or conceptual framework provided by the existing knowledge is necessarily used for, or affects, the understanding of a new science. Institutionally, Western and local physicians were working within the same bodies, the court, hospitals, and sanitary councils. Thus the conciliatory attitude of Dr. Tholozan and his good relationship with the traditional court physicians should not be seen merely as a tactic for ingratiation and self-promotion;[40] it also reflects a genuine intellectual and sociological affinity.

By linking medical modernization to the Qâjâr reform movement, which was characterized by its growing dependency on the West, Chapter 3 describes the nature of education in general and the social mix of the modern medical institutions, including the polytechnic School of Dâr al-Fonun. The School itself was not exclusively devoted to modern science nor were its students solely studying modern science, even though it was the major and the first institution for the introduction of modern science and technology. The mixed nature and composition of the Dâr al-Fonun and its role in the medical transformation has not been closely examined. The Qâjâr government in this project aimed not only at modernization but also at controlling the educational system no matter if it was not successful due to lack of appropriate economic organization to create and finance new schools. This model of modernization was initiated by Mirzâ Taqi Khân-e Amir-Kabir (1807–1852), the first Grand-Vizier of Nâsser al-Din Shah, who accorded fundamental importance to state centralization. Amir-Kabir's political legacy survived after his death, and, as we will see in Chapter 3, his projects of reforms in public health and hospital continued under his successors albeit selectively or discontinuously, while Nâsser al-Din Shah himself, as Abbas Amanat underlined, was influenced by him and bore marks of this influence during the rest of his reign with respect to state centralization, relationship with the West, and modernity.[41] As a legacy of Amir-Kabir, modernization was thus not limited to the introduction of modern science and technologies but was also meant to expand and reinforce state authority and for this particular aim the government attempted to incorporate traditional institutions within the state or modern institutions. This was not exclusive to Qâjâr Iran but if the Qâjârs did not fully succeed it was because of the weakness of their administration and a lack of consistency in their reforms.[42]

In this period, a combination of political and intellectual factors led the Iranian modernist elites to form a community of intelligentsia, coalescing around the ideology of progress as its intellectual framework and freemasonry as its social and professional network. In the patrimonial system of the Qâjâr state the creation of an independent intelligentsia seemed unlikely but such attempts allowed the elite to create an informal network that played a crucial role in reforms.[43] The key personalities linked to such associations were involved in the reform of the state, and the creation of newspapers and civil associations like sanitary councils, schools, and hospitals. Significantly, these institutions created new spaces of activity for both modernists and traditionalists particularly in medicine.

Chapter 4 examines how, under the Constitutional Revolution, the ideas of *tajaddon* (modernity) and *taraqi* (progress) that were dominant among the elite were crucial in completing the integration of modern medicine by ending the hybrid function of modern institutions like the Dâr al-Fonun. The terms modernity and progress had become the leitmotivs of the modernists' discourse that provided the ideological platform for the assimilation of what was called *tebb-e jadid* (modern medicine) in the nineteenth century and biomedicine under the constitution and the aftermath. The ideas of both modernity and modern medicine in nineteenth- and twentieth-century Iran should be qualified in relation to the historical context, to the men who perceived them, and the institutions and strategies through which they were adopted and implemented. Even when we discuss the adoption of biomedicine we need to contextualize it and try to describe it as it was meant and practiced in Iran by practitioners and patients at the turn of the twentieth century. Even the theories, concepts, and "universals" were perceived differently in different contexts, as they are associated with "time, place, and circumstances, . . . transient fashions and temporary opinions," to use the terms of Clifford Geertz.[44] The transition from Galenico-Islamic medicine to biomedicine through conceptual transformations does not therefore mean that biomedicine was finally assimilated in its Western version.

From the late nineteenth century on, a movement for modern education emerged. Under the constitution, there was a focus on illiteracy and comparisons were made to the literacy rates of other countries. For the journal *Tarbiyat* (education), created by Zakâ ol-Molk-e Forughi, education meant the acquisition of modern science. Most of its articles were on modern sciences, as traditional education and science were not deemed to be capable of putting the country on the path to development. It was evident to the editors of this weekly journal that for several centuries Europe had advanced, in comparison to other continents, thanks to modern science. Traditional education was deemed responsible for Iran's stagnation. While Ptolemy's geocentric theory was still dominant in Iran in 1896, an article said: "Copernicus dismantled Ptolemy's theory and Kepler improved Copernicus's . . . but the change did not stop there, Galileo and Newton continued to shed new light on astronomy. Compared to the large number of such stars in the West, in Iran one finds nothing."[45] As a result of this growing consciousness, new journals, like the *Tarbiyat* (December 1896 onward), in contrast to the state journal *Waqâyeᶜ-e Ettefâqiyeh* (1852 onward), which published articles on both traditional and modern medicine, devoted their medical articles to modern medicine. The military school, according to the state periodical, *Iran*, "is the first school that is not in line with the existing traditional system, but organised along the standards of the reputable schools of the advanced countries and its instructors and teachers are [selected] from the graduates of European universities."[46]

Modern science was introduced as a means for colonial or imperial expansion. Scientific explorations by eminent naturalists such as Charles Darwin, Joseph Banks, and others were undertaken within an imperialist framework.[47] However, these generated intellectual interest among both Western and native scientists, independent of the imperial or political agendas. Moreover, one may, for the purpose of analysis, distinguish intellectual interest from financial or other material forms of interest but in reality they are not distinct but interrelated. For instance, one can hardly deny that Dr. Cochran's medical service, analyzed in Chapter 4, was entirely disinterested and that he did not support the political or imperial aims of his government. However, it would be equally hard to exclude any humanitarian, intellectual, or religious and pious motivation behind Cochran's enthusiasm for running a hospital and treating patients in hostile conditions. The inherent and organic link between various forms of interest that motivates provision of medical service, education, or undertaking research could lead to different interpretations or perceptions according to sociopolitical contexts or circumstances. Under a colonial context, any service could be understood as driven by colonial and imperialist aims. In Qâjâr Iran that was not formally colonized, at worst difference was made between imperial and intellectual goals and at best purely scientific aim was considered as the major aim of Western physicians (despite the political aims of their governments), although this could change from person to person and from one period to the other. In the second part of the nineteenth century attitude toward modern medicine or Western physicians was quite mixed while under the Constitutional Revolution even the traditional physicians endeavored to learn modern medicine. Even the skeptics discriminated between the scientific and intellectual interests of their Europeans colleagues, on the one hand, and the "colonial" agendas of their governments, on the other. From this perspective we can better understand why in a period of increasing imperialism in Iran, the Iranian constitutionalists continued their attachment to modernity and modern medicine, and abolished the teaching of "traditional medicine" at the Dâr al-Fonun.[48]

Alongside the imbalance of power that subjected the Qâjârs to Western influence, the universal triumph of science persuaded the elite in Iran to adopt modern science, as they came to believe that it was key to development and progress. For several decades, intellectual, literary, and political discourse was dominated by belief in the efficiency of modern science. Well before the relationship between science and development was defined by colonialists like Lord Curzon and J. Chamberlain in the late nineteenth century,[49] it was transformed into an "ideology" by the Iranian elites. In other words, the "ideological alliance" between science and development, using the terms of J. Hodge, was perceived by Iranian intelligentsia since at least the mid-nineteenth century.[50] Toward the end of the nineteenth century and particularly under the Constitutional Revolution, this "ideology" became the driving force behind social and political change. No matter how modernity and progress, the major ingredients of constitutionalism, were perceived by the elite or their followers in the *anjomans* (constitutional societies), they created a new authority and, subsequently, legitimacy, so that even those who were anticonstitutionalists at heart feigned support.[51] The swift increase in constitutional societies and newspapers recognizing freedom of speech and criticism in a country that had never before experienced either was due to the ideology of progress that gained momentum in the early twentieth century. Since Qâjâr Iran did not experience formal colonial domination, the belief in modernity and modern medicine was easily established without the need to indigenize modern science or renovate traditional medicine so that it was in line with anatomical pathology or biomedicine.

Integral to the sociopolitical movement that led to the Constitutional Revolution, the modernization of science, including medicine, was associated with a movement against the Old Regime. This echoed what Cabanis referred to in 1804 as "semantic contamination between political revolutions and scientific reforms."[52] From this vantage point, the modernization of medicine was seen as part and parcel of sociopolitical changes more than an imported Western product. This was in contrast to the Indian experience, where "modernity was presented as a colonial import and not something intrinsic to man's rational nature."[53] A comparison between the Iranian and Indian experiences allows us to wonder whether colonialism, while promoting modern science, did not create a paradox: by leading the "colonized" country, the colonial power triggered the question of the identity of knowledge, allowing India to seek its intellectual and cultural independence. In Iran, on the other hand, Western influence and introduction of modern science encouraged the view that the development of science had been the work of universal participation or a transnational phenomenon[54] that could result in acculturation or the assimilation of Western values as it was desired by Western observers.[55]

CHAPTER 1

The State of Medical Theory and Practice in Nineteenth-Century Iran

Background and Origin

Before the introduction of modern schools in the nineteenth century, medical education in Iran had not changed since the Middle Ages. The sources students read included Hippocrates' *Aphorisms* (*Fosul Buqrât*); *Summaria Alexandrinorum* (*Jawâme ͨeskandarâniyeen*, the summary of Galen's works); Râzi's *The Continent* (*al-Hâwi*); Majusi's *The Kâmel al-Senâ ͨa* (the perfect art of medicine); Avicenna's *Canon of Medicine* (*al-Qânûn fi'l Tibb*), and various commentaries on the Canon, such as those by Ibn Nafis and Qutb al-Din-e Shirâzi; Abu-Sahl-e Masihi's *The Mâ'a*; Samarqandi's *Asbâb* and (Qazvini's) al-*Hâwi al-Saghir* (the Lesser Continent); Seyyed Esmâ ͨil-e Jorjâni's *Zakhira* (Treasure); and Hakim Mo'men's *Tohfa*, the last two written in Persian.[1]

There was no institution of higher education in medicine before 1851, when the first state-funded school of Dâr al-Fonun (Polytechnic) was created in Tehran. Elements of humoral theory were taught at *madrasa* (Islamic college) alongside the other natural sciences, but medicine as a speciality at a higher level could be studied only by motivated individuals, either as a fellow to a private master or independently.[2] Most of the physicians we will refer to in this book, for instance, undertook their studies privately, with a master (often a relative) or independently, following self-taught masters like Ebn-e Sina (Avicenna).[3] This situation continued even after the establishment of the Dâr al-Fonun, not least because this school was not extended beyond Tehran and Tabriz. Medical graduates were too few and their knowledge of modern medicine too imperfect to change the landscape. ͨAbdollâh Mostowfi called them "semidoctors" (*doctor-che*) as opposed to doctors who had completed their studies in Europe. According to Mostowfi, these "semidoctors," with the exception of those skilled in surgery, were not popular or respected, and the most famous doctors were the traditional *hakimbâshi*, whose knowledge was based on Hakim Mo'men's *Tohfa*, Avicenna's *Qânun*, and Râzi's *Bor' al-sâ ʿa* (recovery in one hour), and who, in addition, had drugs from the *Makhzan al-advieh* (of ʿAqili).[4] John Gilmour, who reported on the state of medicine in 1924, a period when traditional education was not yet a forgotten practice, stated:

> Formerly it was the custom for the older and more renowned Persian doctors to take an assistant or assistants..... Later, the assistants were allowed to be present while the master examined and treated their patients. They copied the prescriptions ordered for the different maladies. They learned the signs and symptoms of diseases. After two or three years of this training, the teacher presented his pupils with certificates stating that they were competent to practice medicine.[5]

The *ejâzeh-nâmeh* (or license) issued by Galenico-Islamic physicians specified the book or text, such as Chaghmini's *Commentary* (on Avicenna's *Canon*), or al-Majusi's *Kâmel al-Senâʿa*, that a student had read and studied with his master. With the introduction of a modern curriculum, the certificate mentioned only the names of the masters and the subject matter studied, such as physiology, surgery, ophthalmology, and *materia medica*. In a letter dated 17 Jamâdi II 1326 Hejira (July 17, 1908), an anonymous author states that, after studying medicine under Mirzâ Mohammad ʿAli Tabib (one of the distinguished physicians of Kâshân), Mirzâ Hossein Khân-e Kâshâni "has also been present in my classes, and has assimilated a high level of medical knowledge with enthusiasm... and therefore he should be trusted as a skilled physician by patients."[6] With the establishment of modern hospitals, hospital practice was also added to indicate practical skills.[7]

With few exceptions, most medieval sources used by physicians in the nineteenth century were in Arabic. On his arrival in Iran in 1893, Dr. Schneider reported that the principal medical texts, translated into Persian and used by physicians, were those of Râzi.[8] It is not clear to what extent Arabic sources were used, but it was recommended that skilled doctors should read these sources in Arabic rather than in translation.[9] Writing medical treatises in Arabic or claiming to be conversant with Arabic was commonplace among the learned physicians in the Qâjâr period. Mirzâ Mohammad-Taqi Shirâzi Malek al-Atebbâ (d. ca.1873) wrote four books in Arabic.[10] Mohammad-Kâzem-e Rashti Filsuf al-Dowleh (d.1905) used to show off his Arabic knowledge, although he did not write in Arabic.[11] Reading or writing in Arabic was a sign of a high level of theoretical knowledge, which distinguished learned physicians from the rank and file. However, contrary to Gilmour's claim that "medical works were always written in Arabic,"[12] it does not seem that Arabic medical texts constituted the bulk of medical literature in nineteenth-century Iran. Although the ideal language remained Arabic, even the physicians who were able to read and write in Arabic wrote mostly in Persian. A number of treatises were prompted by the occurrence of epidemics, such as *wabâ* (cholera) and *tâ'un*, while others included advice on preserving health, the qualities of food and drugs (based on humoral theories), and the Six Non-naturals (*Setta Zaruriya*).[13] It would be surprising had medical literature on such popular topics been written in Arabic, as it would have, otherwise, limited its audience by excluding those who were able to read Persian but not Arabic.[14]

Following Galen's theory that "only he is a perfect physician, who is at the same time a philosopher,"[15] physicians in Iran at the end of the nineteenth century liked to flaunt their "qualification in philosophy, which they deemed indistinguishable from that of medicine"[16] so as to distinguish themselves as "professional." These learned philosopher-doctors (*hakim*) were rare, because few could afford, either intellectually or materially, to pursue the study of both medicine and allied sciences such as philosophy, astrology, mathematics, geography, and geometry.[17] The "rank and file" doctors to whom patients had recourse were many, however, whether surgeons, ophthalmologists (*kahhâl*), female healers, barbers, or dervish (mendicant Islamic monks),[18] as they were cheaper and more accessible. Nevertheless, no clear-cut theoretical distinction between "learned" medicine and folk and faith healing can be drawn. We find learned physicians advising magic and faith healing in their books.[19] This attitude was similar to that of nineteenth-century Western physicians, who, as Jacques Leonard noted, observed modesty toward magic medicine (cure through pilgrimage) because the popularity of this practice expressed the impotence of their own therapies.[20] Criticisms by learned physicians targeted quacks or utterly unskilled doctors but not faith healers.[21] This combination of science and religion and/or magic was also observed in practice. During the cholera epidemic of 1904, when modern medicine was already well established, all public health measures were taken and medical instructions for the prevention and treatment of cholera were distributed, while at the same time, the population gathered in the mosques, under the guidance of the ulama, who after some discussion on the conditions of health and treatment, prayed to God for the healing of the sick and for the epidemic to cease. The population was receptive to modern scientific solutions as well as to religion and faith healing.[22]

In the same way that faith healing and magic could be found in learned (or rational) medical texts, humoral theories based on Greek rational medicine percolated through all forms of healing methods and knowledge. In a treatise on plague (*Mehnat al-tâ'un*), the author refers to classical authors who set out four causes for the occurrence of cholera and plague: (a) the gas contained in the earth that bursts out through cracks, fouls the air, and the foul air produces choleric diseases; (b) the state of planetary constellations; (c) corrupted matter and water in the environment that pollutes the air—in support of this third cause, the source quotes Galen as saying that in India, following a great fight, the disintegration of cadavers and their subsequent putrefaction produced plague and cholera; (d) God's will to punish his creatures who have committed sin. What proves this to be divine punishment is that physicians are unable to prevent or cure it unless God takes pity on his creatures and sends a remedy.[23] The author acknowledges the importance of divergent opinions and theories without concern for their contradictions, apparently seeing them as complementary. Furthermore, following the theory that "in the face of Destiny a physician is helpless" nineteenth-century physicians in Iran recognized their inability[24] and saw their recourse to irrational medicines as entirely legitimate as well as practical.

At the end of the nineteenth century, most physicians dressed like the *ulama*, indicating that they were knowledgeable in divine science as well as in medicine.[25] Likewise, religious scholars (*mojtahed*s, literally "those who practiced *ejtehâd*," interpreting the Koran and the *sunna* (the traditions of the Prophet) to reach legal decisions on various aspects of everyday life) also studied medicine to enable them to issue opinions on health and medical cases related to the practice of religion. In addition to learned *mojtahed*s, dervishes, who were more available in the market, practiced freely the dogma that all diseases were divided into hot, cold, humid, and dry.[26] If the study or practice of medicine by religious scholars was justified for legal reasons, the study of almost all other "sciences," including jurisprudence and religion, was recommended for medical purposes. A physician needed to master ten "necessary" sciences and these were jurisprudence and the traditions of the Prophet (*feqh va hadith*), so that doubt or satanic inspiration did not cause him to digress from the right way; ethics (*akhlâq*); philosophy (*hekmat*); logic (*manteq*); natural science; geometry (to understand the shape of the organs in surgery); and astronomy and astrology (to know the nature of the seasons, the situation of the planets, and their temperaments and constellations). The physician had to know that when the moon was in the constellation of Scorpio, bloodletting was not recommended unless it was vital due to the severity of the disease. In general, each organ corresponded to a different constellation. When the Moon was in the constellation relating to the organ to be scarified, cupping was not permitted. The ninth science was arithmetic and the tenth, the science of divination and insight (*kehânat va farâsat*).[27]

After the collapse of the Caliphate (1256 AD), medical books were no longer written in the style of compendiums such as Avicenna's *al-Qânûn* and Râzi's *al-Hâwi*. The literature of the post-Caliphate period often mixed irrational and rational medicine. Such a combination was not absent in classical Greek medicine, as it was borrowed by Islam, but the presence of irrational elements in the learned medical literature of the post-Mongol period is more pronounced and contrasts with the rational dimension of medical knowledge during the translation period (eighth to tenth century) right up until the fourteenth century. This development was partly due to the institution of the madrasa, where pre-Islamic and Islamic sciences or rational (*maʿqul*) and religious (*manqul*) sciences were combined, but one can also assume that the Medicine of the Prophet, elaborated since the fourteenth century, reinforced the association between religion and medicine. The Safavids, who imposed *shi'ism* as the state religion, might have strengthened this trend. There were several high-ranking physicians in the Qâjâr period who were particularly close to the religious establishment. Seyf al-Din Astarâbâdi, a cleric doctor, claimed that he had studied medicine in the 1860s with Hâj Mirzâ Khalil in Najaf and Hâj Mirzâ Bâbâ (Malek al-Atebbâ), as well as Mirzâ Mohammad-e Golpâyegâni in Karbalâ.[28] The professional continuum between physicians and the ulama, which became well established under the Qâjârs, finds its origin in this development since the fourteenth century.

The medicine attributed to the Prophet (*tebb al-nabi*) and the Imams (*tebb al-aʾemma*) originated in the development of Islamic schools of law that informed the creation and development of the madrasa. The authors of this medical literature were jurists like al-Bukhâri, who was the major source for later authors like Ibn Qayyim al Jawziyya (1292–1350) and al-Dhahabi (d. 748/1348). Some of the founders of the schools of law were protagonists of medicine based on Greek humoral theory as it was translated and carried into Islam by earlier authors.[29] One of those who handed down the medical and dietary traditions of the Prophet, Ibn Qayyim al-Jawziyya, who compiled *Medicine of the Prophet* in the fourteenth century, is said to have organized and arranged the sayings of the Prophet in health and medicine, or to have applied existing medical knowledge (i.e., Galenico-Avicennian medicine) to the teaching of the Prophet.[30] As a literary source, *Medicine of the Prophet* illustrates the influence of Greek medicine in societies that were conquered by Islam. By encompassing both Hippocratic theories and contemporary folk medicine, *Medicine of the Prophet* was a conscious attempt to integrate medicine, classified as non-Islamic science, into Islam. This is exemplified in one of the widely quoted *hadith*s of the Prophet that says: *al-ʿelm*

ʿalmân, ʿelm al-abdân va ʿelm al-adyân (science is twofold: science of body [medicine] and science of religion). The *Tebb al-nabi* played an important role in the incorporation of faith healing in learned medicine, which continued to characterize medical theory and practice in modern Iran. So it comes as no surprise that physicians either advised, alongside medicine based on herbs and humoral pathology, spiritual or faith healing or at least did not oppose such practices. This state of knowledge and theory resulted in the development of an occupational group that included a wide range of practitioners rather than a well-defined medical profession attached to specific institutions. Nevertheless, the holders of knowledge in "learned medicine," regardless of its content, enjoyed a different status from ordinary physicians or empirical practitioners. This privileged status of learned physicians in Iran was always due to state patronage rather than independent medical institutions.

Medical Institutionalization

In the first part of the nineteenth century, we do not find medical bodies, such as hospitals, or schools, where professional physicians could practice, or to which they could refer as their institutional home. There were semimedical institutions such as the *dâr al-shafâ* (a type of hospital-madrasa) usually attached to mosques or mausoleums, but with the exception of the *dâr al-shafâ* of Mashhad, no record of other *dâr al-shafâ*s providing medical service under the Qâjârs has been found. It is more likely that the Qâjârs inherited this situation from previous centuries, where we find some of the learned physicians teaching medicine at madrasas (Islamic schools) rather than *dâr al-shafâ*s. Among many examples, we can cite Fakhr al-Din-e Kâzeruni, who was appointed to teach at Sadriyeh madrasa in Maragheh in 1266 AD, and Qotb al-Din-e Shirâzi (d. January 1311), the author of the *Tohfat al-saʿdiyyah*, a commentary on Avicenna's *Canon*, who taught in a madrasa in Tabriz.[31] Another example would be the case of Mohammad-Javâd-e Hakim, one of the Qâjâr physicians (d. 1914), who taught and sometimes received patients in the mosque of Aqâ Qâsem.[32]

Learned physicians were often attached to the court or the nobility.[33] Princes were in constant need of doctors, not only for medical advice and for treatment when they fell ill, but also because of their knowledge of food and poisons. For example, Nâsser al-Din Shah (1848–1896) was often accompanied by more than one doctor at the dining table, or when he took a bath.[34] The other function of court doctors was to apply or supervise punishments, such as cutting off ears, hands, or noses, for example, on the order of the prince. Once in the service of a court, physicians became close to the Shah and some of them became involved in politics or business. ʿEmâd al-Din

Mahmud-e Shirâzi noted that some court doctors did not find enough time to concentrate on medicine or to write books, while others, having secured a comfortable material life, devoted their time to medicine.[35]

Under the Qâjârs, physicians attached to the court were not always endowed with property or high salaries and their fortune depended on their relationship with the Shah or the support of influential princes and ministers. The first Western-educated *Hakimbâshi* (chief physician) of the Qâjârs, Mirzâ Bâbâ Afshâr (d. ca. 1842), studied medicine in England between 1811 and 1819. He was a favorite of Prince ʿAbbas Mirzâ, who gave him one of the best villages near Tabriz in Azerbaijan, and through its revenues, he became a rich man.[36] But after the death of ʿAbbas Mirzâ in 1833, he seems to have lost his privileged position. Considering his likely attachment to Britain and the hostility of Prime Minister Hâji Mirzâ Aghassi toward the British mission in Tehran, it is unlikely that he was paid very much.[37] In fact, about three years later, the prime minister nearly succeeded in dismissing Mirzâ Bâbâ as court physician.[38] On the other hand, Mohammad Kâzem-e Rashti, one of the famous court doctors who served under three successive Qâjâr Shahs, received a large amount of property. When he became the personal physician of the Queen Mother (*Mahd-e ʿOlyâ*) in 1846, he received the *khâlesseh* village of Mahmud-âbâd, which was crown property, as *tuyul*.[39] It appears that this *hakimbâshi* was not in need of a state salary (*mavâjeb*), and as reported in 1865, he did not receive a salary, but within the context of court etiquette, or perhaps out of greed, he requested that he should also be paid as the other court physicians.[40] Mirzâ Mohammad Kâzem-e Rashti accumulated a large fortune and shortly before his death built a *madrasa* called *Madrassa-ye filsuf* (named after his title *Filsuf al-Dowleh*—The Philosopher of the State), and endowed it with the revenue of his properties.[41] When he died in 1905, his title, together with its allowance and benefits (*ravâteb va marsumât*), were transferred to his son, Loqmân al-Molk, who was the court doctor to Mohammad ʿAli Mirzâ, the heir apparent and governor of Azerbaijan.[42] The less fortunate doctors, according to their status and relationship with the court, received salaries regularly or occasionally. Salary was not, however, the only important aspect of court employment. The status attached to such employment seems to have been even more precious, especially in the absence of any other appropriate institutions under the Qâjârs.

From antiquity, higher education in Iran was sponsored and supported by the state, whereas elementary and secondary education was assured privately, or through *waqf* endowments.[43] The Qâjâr state was not involved in the modernization of primary education, which was initiated by private individuals toward the end of the nineteenth century, but sent students to Europe and created a modern school of higher education in the middle of the nineteenth century. As a result, the Qâjâr intelligentsia, acquainted with modern Western science and education, was the product of state support. The authors of most books written in the second half of the nineteenth century acknowledged that they had been fostered (*parvardeh*) by the court system and were at its service.[44] The medical establishment created and sponsored by the state or the court can thus easily be identified, whereas there is no institutional or corporate identity for ordinary doctors or surgeons.

In the aftermath of eighteenth-century civil war, the early nineteenth century did not see court medicine developing. There were reportedly two physicians attending Agha Mohammad-Khân's court, namely, Mirzâ Masih-e Tehrâni and Mirzâ Ahmad-e Tabib-e Esfahâni, who in 1791 rescued Agha Mohammad-Khân from what seems to have been a stroke. We find Mirzâ Ahmad-e Esfahâni, with the title *Hakimbâshi* (chief physician), treating Mirzâ Shafiᶜ, the prime minister of FathᶜAli-Shah, in 1819 and Prince ᶜAbbas Mirzâ in 1824.[45] The relatively long reign of FathᶜAli-Shah and the development of his court saw an increase in court physicians. Among them was Mirzâ Hossein-e Hakimbâshi, who in 1836 accompanied the army of Mohammad Shah (1834–1848) in Fars to subdue Prince Farmânfarmâ, the half-brother of the new Shah, who refused to pledge allegiance to his sovereignty.[46] Mollâ Mohammad-e Hakimbâshi-ye Qoboli, the translator of several Western medical sources into Persian, attended the court of Nâsser al-Din Mirzâ (1837), the heir apparent in Tabriz, and later became one of his Hakimbâshis in Tehran.[47] Mirzâ Nazar-ᶜAli Hakimbâshi-ye Qazvini, a rival of Prime Minister Hâji Mirzâ Âghâssî (1836–1848), was the physician of Mohammad Shah. Qazvini was more involved in political affairs than in medicine. He aspired for prime ministership, and perhaps for this reason was in favor of British influence in Iran.[48] In 1261/1845 he was sent into exile in Qom by the order of Mohammad Shah, but after the Shah's death in 1848, he joined the new Shah's army in Qazvin at the head of 600 horsemen, with the intention of taking a state position under Nâsser al-Din Shah. However, he was sent back to Qom by the order of the new Prime Minister, Mirzâ Taqi Khân.[49] As part of the inner court, some of the hakimbâshis were involved in politics to the extent that their political role overshadowed their medical duties. At times they carried out diplomatic missions, as in the case of Mirzâ Bâbâ Afshâr, a graduate of Oxford and Edinburgh, who was sent to St Petersburg in 1828 and also to England to negotiate with the government in London.

Court patronage under the Qâjârs played a crucial role in the development of medical institutions in Iran, although this support was not framed in any "court" or "royal" medical institution but happened in an ad hoc manner. With the exception of their routine responsibilities, such as providing medical services and advice to the Shah and his court and statesmen, physicians were not bound to an agenda set by an institution. This situation left the door open to physicians of all persuasions or qualities. Some of these physicians published relatively prolifically, such as Mirzâ Mohammad-Taqi Shirâzi. Others, like Mirzâ Ahmad-e Tabib-e Tonekâboni, physician of FathᶜAli-Shah, authored fewer works and others, such as Mirzâ Bâbâ Shirâzi, left no book or treatise.[50] In the first three decades of the nineteenth century, we do not find any important medical work authored by court physicians. Even in the later period, only a few of the court doctors wrote books. ᶜAlinaqi Hakim al Mamâlek, for instance, was part of the inner court and accompanied Nâsser al-Din Shah on his recreational activities or official travels while his medical duties took a secondary role. The only book written by Hakim al-Mamâlek is a travel account of a journey to Khorâssân when he accompanied the Shah.[51] Nevertheless, in the absence of other funding, the court played an important role in the promotion of medical knowledge. With the consolidation of Qâjâr power and the increasing number of court physicians, we see medical literature developing rapidly in the following decades.

This institutional transition is a key phase in the history of medicine in Iran. In the nineteenth century, the medical profession was to a large extent sustained by the law of inheritance. This system of inheritance prevented social mobility. Initially, court patronage reinforced this system insofar as the modern institutions inaugurated by the creation of the Dâr al-Fonun in 1851 primarily recruited from the traditional establishment. Mohammad Hossein-e Afshar (son of Mirzâ Ahmad-e Hakimbâshi, a traditional court doctor), who was one of the first medical students at the Dâr al-Fonun, is just one example. In the twentieth century, we still find modern educated physicians who were sons of illustrious traditional doctors. However, the modern school of Dâr al-Fonun also provided opportunities for young and ambitious students, with no familial medical connection, to pursue their studies and become eminent court doctors. Mirzâ ᶜAli Hamadâni was one of those individuals and studied medicine as a vocation. He became a professor of both traditional medicine and modern anatomy at the Dâr al-Fonun and was the personal physician of Prince Nâyeb al-Saltaneh, the governor of Tehran.[52] At the same time, the Dâr al-Fonun, as the only higher education institution, provided a home not only for modern medicine but also for traditional physicians,[53] and provided the most significant context for encounters between the two.

While court patronage constituted the foundation of the traditional medical establishment, it also contributed to its disintegration owing to its protection and sponsorship of modern medicine and the employment of Western doctors. Due to both diplomatic relationships and medical needs, the Qâjârs sought Western doctors while the Western medical presence closely followed the development of diplomatic relationships. The absence of Russian and British doctors and instructors at the Dâr al-Fonun created by Amir-Kamir was a reaction to their intrusive influence, and led in 1886 to the Qâjâr government's attempts to entice the Americans to exploit the natural resources of the country by giving concessions.[54] After opening doors to the United States, Nâsser al-Din Shah showed special favor to Drs. Holmes and Varmeman, the physicians of the American Missionaries in Hamadan.[55]

This special attention to American doctors was politically motivated, but at the same time, the prince, being central to the destiny of the state, called on all medical resources available. As Anne-Marie Moulin has pointed out, across ages and cultures, princes and sovereigns have welcomed foreign physicians, sometimes at considerable costs.[56] Putting aside the project of modernization, such preferences were among the main factors in state sponsorship of Western medicine.

However, the medical needs of the prince or court cannot alone explain the relationship between the state and medicine. There was an element of public health in this relationship as well. Concern for the health of subjects is illustrated in the appointment of surgeons to the army, where their service was of more use than the service of physicians. In the early nineteenth century, we frequently find surgeons and chief surgeons (*jarrâhbâshis*) accompanying the army of Fathᶜ Ali Shah or Mohammad Shah. Thus, medical service beyond the court but sponsored by the court for public service was limited to surgery for the army. In the nineteenth century, with the spread of epidemics, this medical service became more the duty of physicians rather than of surgeons, and the need for the state to support physicians beyond the court was growingly felt. In the introduction to his treaties on cholera in 1852–1853, Mirzâ Musâ Sâvaji stressed that the preservation of the health of the population depended on security in the country and this was not attainable without a strong state. But a strong state could not be created without the protection of its subjects from the invasion of (epidemic) diseases, and such protection could be ensured only by the formation and employment of physicians through state support.[57]

However, this recommendation was based on an ideal picture that a court physician, committed to maintaining the health of the population, had in mind regarding the relationship between state, medicine, and physicians. It did not reflect the actual state of medical organization and welfare in Qâjâr Iran.[58] According to the ideal picture, the state stood as the only social agent potentially able to institutionalize medicine. The task of protecting public health became incumbent on the state because the Shah was considered not only the possessor of the life and property of his subjects, but also their protector against invasion by outsiders and epidemics.[59]

Epidemics

An Overview

The role of the state in the institutionalization of medicine was furthered by the emergence and spread of epidemics in general, and pandemics in particular. Due to increasing trade and travel routes, epidemics introduced from neighboring countries spread across the country. The epidemics that afflicted most in the nineteenth century were cholera and plague, which found their way into Iran along populated regions and trade and pilgrimage routes: Herat and Mashhad in the northeast and the Persian Gulf in the south. The deserts of Beluchestan and central Iran were a barrier to their spread. According to Tholozan's study of epidemics of (bubonic) plague over a period of 300 years (1600–1800), Azerbaijan suffered three plagues across the whole province and one confined to a small locality; the northwest of the country had three epidemics, Guilân one, Khorâsan one, Qazvin one, Damâvand one, Gorgân one, and Mazandaran one. In these epidemics, no region of central Iran (Qom, Kâshân, Isfahan, and Shiraz) was affected, while the south was attacked only once.[60] In 1838, plague occurred in parts of Azerbaijan, particularly Maragha and Ardebil,[61] but not in central and southern parts of the country. The north and west were thus often visited by plague but not the dry and hot areas of central Iran. In terms of temperature, plague appeared in mild humid seasons and disappeared in hot dry seasons.[62] The warm climate, on the other hand, was more conducive to cholera, whereas cold slowed its advance.[63] Increasing communication with neighboring countries integrated Iran into the global epidemic geography, and thereafter the epidemics of cholera and plague intensified their activity in Iran. Only the central state could take countrywide preventative measures, such as the organization of quarantines. The role of the state became still more crucial during pandemics, as the solution was "less in the measures taken by individual states independently than by an international agreement between the states."[64]

The nineteenth-century epidemics were one of the major factors that triggered the medical transformation of Iran. Prior to the nineteenth century, chroniclers as well as medical tracts inform us about sporadic diseases, whose descriptions indicate their epidemic character. But the force and frequency of epidemics in the nineteenth century gave known epidemics a new dimension, motivating physicians to focus on them. According to Mohammad Râzi Kani Fakhr al-Atebbâ:

> This disease, i.e., *wabâ*, occurred also in the past but not frequently enough for the physicians to find it worth investigating by further examination and experimentation. In the past, this epidemic reappeared unexpectedly after several

years of break and lasted from 40 days to two or three months and very rarely up to seven months, while the patients infected by the epidemic often died within two or three days. Therefore it did not give enough time for physicians to test different treatments. Lack of experimentation was also due to the fact that physicians avoided the sick and even fled the area infected because they considered it contagious. For this reason they only touched upon the subject briefly in their treatises..., whereas in our time this disease occurs far more often and each time it invades relentlessly for three or four years allowing physicians enough time to implement or test different treatments and experimentations.[65]

What Kani indicates is significant, as it was not epidemics themselves but the change in their frequency that ushered in modern epidemiology. This change was particularly striking in Iran, where, according to most Western travelers, the climate, with the exception of coastal areas (the Caspian Sea and Persian Gulf), was one of the most healthy in the region. This was because of the dryness of large parts of the country, as compared to most tropical or semitropical countries under colonial administration.[66]

Due to the lack of adequate administration and the absence of regular records, it is hard to accurately map epidemics in nineteenth-century Iran. After the creation of the formal Sanitary Council in 1868, Tholozan collected information from officials on epidemics across the country and intended to complete the information by using his historical data and other contemporary testimonies, and to publish it.[67] However, this document was never compiled or published owing to Qâjâr administrative and financial deficiency.

Furthermore, since the state of medical knowledge did not permit specific definition, so these diseases are known under the generic term of *wabâ*. *Wabâ* did not refer to particular diseases but to "public mortality" (*marg-e ʿâm*).[68] The use of *wabâ* stems from the fact that cholera caused the highest mortality rate in nineteenth-century Iran. The term *wabâ* was so well entrenched in the medical vocabulary that even Western physicians used it to refer to an "epidemic." Dr. Polak describes dysentery as a "*maraz-e wabâyee*" (epidemic disease).[69] Iran certainly experienced cholera in its endemic form before the nineteenth century, but the first recorded cholera epidemic was that of 1821, which arrived to the country via the Persian Gulf and the Ottoman territories; 12,000 to 17,000 people perished in Basra, about 500 at Bushehr, and 1,500 in Shiraz. The cold weather in December that year arrested the progress of cholera, but by then about 8,000 were dead.[70] The epidemic subsided with the cold but resurfaced the following year.[71] According to the chronicler Sepehr, writing in about 1842, the disease, which had not been seen in Iran for a long time, was introduced from China and India via Shiraz, Yazd, Esfahan, Kashan, Qazvin, and the western part of the country including

Azerbaijan, and took the lives of more than 100,000 people. The epidemic spread among the troops of Prince ʿAbbas Mirzâ, who had returned from a campaign in Erzurum in eastern Turkey. On 25 *Ziqaʿdeh* 1237 (August 13, 1822), Mirzâ Bozorg, the prime minister who accompanied ʿAbbas Mirzâ in Azerbaijan, died of cholera.[72]

The 1822 cholera also subsided with the arrival of winter,[73] but a *wabâ* occurred again in 1826 during the war between Iran and Russia in Azerbaijan.[74] Since at this time plague developed in the Armenian plateau, it is possible that the *wabâ* in question was in fact plague, transported by troops to Azerbaijan.[75] As to cholera, after a few years of respite, it swept back in 1828. *Wabâ* was reported by Sepehr in August 1829 in Tehran, where nearly 10,000 perished. In the following year, the same *wabâ*, which in this case was probably cholera, revisited the country.[76] Mirzâ Mohammad-Taqi Shirâzi's *Tâʿuniya* (on plague) obviously examined the epidemic of plague, which spread during 1829–1830 in the north of the country and particularly in Rasht, the capital city of Guilân. This was confirmed by Sepehr, who reported that in 1829, plague in Guilân killed more than 10,000. According to Sepehr, the plague had been raging for several years in Kermânshâh, Hamadan, and Borujerd in the west of Iran and in Azerbaijan, where it killed nearly half the population, while the cities of Kâshân, Yazd, Esfahan, and Qom in central Iran were preserved.[77] According to Tholozan, this epidemic originated in Turkish Armenia and penetrated Mesopotamia via Persia.[78] Polak confirmed this information, adding that the plague decimated the population of Guilan to the extent that parts of the Caspian coasts were devoid of inhabitants.[79] These waves were repeated in the following years. Plague appeared in Baghdad in the middle of the winter of 1831 and broke out among the crew of the boat that was carrying John McNeill, physician and chief assistant to the British mission in Tehran, and his family on his way from Baghdad to Basra.[80] Sepehr noted that, in 1832, plague was accompanied by excessive cold and snow across the country.[81]

Epidemics revisited Iran again in 1833. Sepehr reported that *tâʿun* (plague) spread in Mâzandarân, Brujerd, Erâq, Semnân, and Khwâr, while intermittent fever (*tab-e nobeh* and *tab-e larz*) killed 20,000 in Tehran and 200,000 across the country.[82] This epidemic is not clearly identified in the reports of the British envoy John Campbell in Tehran, who in a letter of October 10, 1833, noted that Dr. Cormick, sent by the Shah to treat Prince ʿAbbas Mirzâ from dropsy of the lower joints, died of the "prevailing epidemic on the 23rd Ultimo."[83] According to Sepehr, Dr. Cormick died of the *tab-e nowba* (intermittent fever).[84] He mentions that this illness also affected Hakim Davoud Khân, who, as a result, was unable to accomplish his mission in replacing Dr. Cormick.[85] Nevertheless, expounding on the reason of the fatal trip of FathᶜAli-Shah in 1834 to collect the taxes of Fars, Sepehr refers to the tax arrears of this province caused by the draught, famine, plague, and cholera that waged a continuous onslaught in the previous years.[86] In June 1835, according to Sepehr, "excessive heat generated putrefaction of the air and *wabâ* visited the population in Tehran, and the court of Mohammad Shah left the city in order to avoid the epidemic."[87] Mirzâ Mohammad-Taqi Shirâzi wrote his first treaty on cholera as a reaction to this epidemic.[88]

A wave of *"wabâ"* is reported to have taken place in 1836 at the beginning of the summer. A royal victim was Prince Hosseinᶜali Mirzâ Farmânfarmâ, governor of Fars (in the south of Iran), who died on July 11, 1836.[89] The epidemic was also reported in Khorâssân the same year.[90] Considering the size of these provinces, it seems that the whole country was affected by cholera in 1836, with 50,000 dead.[91] The cholera of 1846 in Tehran under Mohammad Shah, and in 1853, is reported in various medical treatises.[92] The first case of cholera in 1846 occurred in July causing the death of about 12,000 of the 30,000 population of Tehran.[93] The epidemic of 1853 was called by Sepehr *wabâ*, which was likely to be cholera.[94]

Owing to improved communication and trade in the second half of the century, epidemics occurred more frequently and covered a wider geographical area. In the spring of 1867, a severe form of cholera raged across Khorâssân,[95] disrupting ordinary life and causing famine, which in turn increased predisposition to the disease. This is illustrated in the famine and cholera that went hand in hand four years later, in 1871, in Khorâssân.[96] According to Dr. Basil, an Armenian trained in England, malnutrition predisposed a person to cholera.[97] From 1877 onward, waves of plague raged in the northern and western parts of the country, and quarantines proved to be inefficient as the shallow burial of the dead around the mosques or shrines helped spread the disease.[98] In 1885, plagues afflicted the north and northwest of Khorâssân and in May 1892 a form of cholera that had originated in India and was introduced to Khorâssân via Herat and Mashhad took such a severe turn that most inhabitants of the city left their homes and took refuge in the higher and cooler areas around Mashhad.[99] In the autumn of the following year, 1893, cholera revisited Khorâssân and four years later, in 1897, typhoid appeared in Khorâssân.[100]

The relentless waves of epidemics in the nineteenth century made cholera and plague major subjects of medical literature. While epidemics awakened a spirit of inquiry, they also introduced a new vision of the relationship between medicine and the state and between patients and physicians via the issue of the epidemic or contagious nature of diseases. The following section will examine the impact of epidemics on the evolution of traditional (i.e., Galenico-Islamic) medicine in Iran.

Epidemic Diseases According to Humoral Theory

Diseases were divided into particular (*khâssa*) and general (*ᶜâmma*) categories. The particular diseases included different kinds of *sarsâm* (meningitis or phrenitis), *zât al-janb* (pleurisy), and *zât al-riyeh* (pneumonia).[101] Sarsâm literally means that the head is inflamed or swollen (*sar* being "head" and *sâm* "inflammation" and, by extension, ache), which corroborates the definition of phrenitis, or inflammation of the diaphragm (phrena). That is why an early treatment for a headache was to fasten the head strongly with a handkerchief in order to contain inflammation. *Zât al-janb*, literally disease of the side (or diaphragmitis pleurisy), is Arabic for the Persian term *barsâm*, meaning inflammation of the internal membrane (pleura) of the chest (*bar* being the side). Similarly, *Zât al-riyeh* means disease (*zât*) or inflammation of the lungs (*riyeh*).[102]

The most important of the general diseases were fevers, including the epidemics. According to Galenico-Islamic medicine, the pathology of epidemic fevers did not differ from that of other fevers, which were divided into three categories: *tab* or *hommâ-ye yawm*, also called *tab-e ᶜozwi* (local or organic); *hommâ*, or *tab-e khelt* (humoral fever); and *hommâ*, or *tab-e deqq* (*tab-e ruhi*) (hectic/pneumatic fever). These three fevers related to the three components of the body: the organs (*aᶜzâ*), the humors (*akhlât*), and the spirits (*arvâh*). The relationship of these components to each other was like the relationship between the three components of a bathhouse (*hammâm*): the walls, the water, and the vapors, respectively.[103]

Diseases were caused by the combination of two factors: the environment and physiological predisposition. The first factor that triggered an epidemic was the putrefaction of the air, which could be caused by the rising of putrid vapors from the earth, which due to the particular constellation of the planets and the weakness of the sun did not ascend to the point of being transformed into water. For this reason, the lack of rainfall was one of the signs of an epidemic. Putrid matter remained mixed with and dispersed into the air.[104] If the body was weak, and this corrupted air entered the body via respiration, disease resulted.

Figure 1.1 is a schematic representation of the relationship between the body, its illnesses and the environment (air). Once in the body, the putrid air entered the heart and caused unnatural heat (*harârat-e ghariba*). This unnatural heat was then transferred to other parts of the body and fever appeared.[105] If the unnatural heat affected the organs, it produced *hommâ-ye*

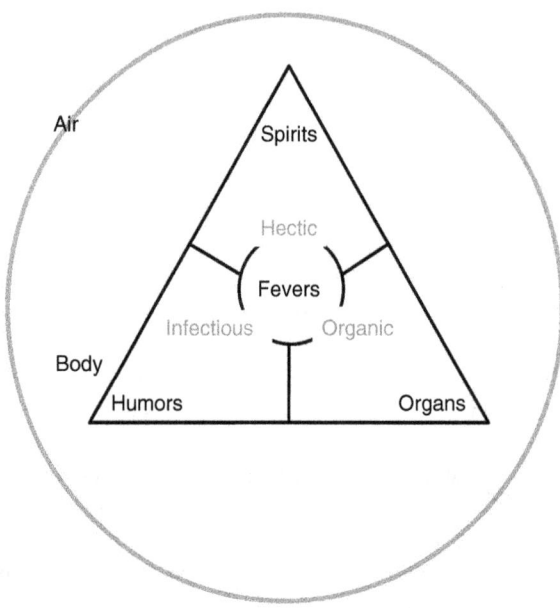

Figure 1.1 Body and its diseases caused by foul air

yowm, also called light fever.[106] If the unnatural heat attacked the humors (*akhlât*), it produced infectious fever (*hommâ-ye khelt*), and if it affected the pneumata (*arvâh*), the result would be *hommâ-ye deqq* (hectic fever, or consumption). Epidemic fever is in fact a *hommâ-ye ʿafan* or *khelt*—infectious fever.[107] Since the infectious humors caused epidemic fever, the first and foremost prophylactic measure was the reduction of humors via bloodletting or other purging.[108]

Unnatural heat (*harârat-e gharîba*), as opposed to natural, or innate, heat (*harârat-e gharîzî*) of the body, was caused by external factors. As a result of unnatural heat, the faculties (*quwâ*) and spirits (*arvâh*) were weakened or impaired, which caused indigestion, which in turn further reduced the natural heat. The diminution of natural heat was caused by the fact that in this process it was pushed into the deepest parts of the body, and as a result the surface of the body, especially the extremities, became cold.[109] The reduction of natural heat diluted the blood, and therefore the color of the skin turned pale. The yellow bile increased at the expense of other humors and heated up the heart with unnatural warmth and infected the humors, giving them more fluidity, and this furthered their penetration even into the narrowest channels in the body and caused diarrhea. The choleric air targeted weak constitutions but did not cause choleric disease in healthy bodies, although it might slightly have altered their temperament (*mezâj*). The changes that the epidemic air created in healthy bodies varied according to the weakness or strength of the affected organs. If the stomach was weak, it created blowing, indigestion, or flatulence; if the heart was weak, fear and anxiety, while a weak-minded person was subject to melancholia and unsound ideas.[110] Both Shirâzi and Mirzâ Musâ Sâvaji, whose treatises are summarized above, claimed they wrote of their experiences during the cholera epidemics. Sâvaji specified that during the cholera of 1846, he drafted a treatise on prevention and treatment and distributed it for public use. He then complemented that treatise in 1852, when another epidemic was looming. Both documents were completed by order of the commander of the army, Amir al-omarâ Amir Asadollâh Khân.[111]

In humoral theory, the fevers were divided and subdivided, but the kinship, or genealogical relationship between them, was mostly cognatic; in other words, many fevers were cousins rather than direct descendants. For example, *hommâ-ye dâ'em* (continuous), such as *mohreqa* and *mothbeqa*, and *nowba* (intermittent, or literally, a turn), like malaria, constituted other branches of "general fevers" (*amrâz-e ʿâm*). Before the nineteenth century and the appearance of cholera, three fevers were considered the most important: *mohreqa*, *mothbeqa*, and *nowba*. *Mohreqa* consisted of the corruption of yellow and black bile in the veins, *mothbeqa* was the corruption of blood in the veins, and *nowba* consisted of corruption of the humors outside the blood vessels. All these fevers were caused by putrefied humors.[112] These fevers can be considered the siblings of most epidemic fevers that were called *hommây-e khelt* (humoral) or *ʿafan* (putrid), such as cholera and plague.

Although the epidemic or pestilential fevers were known in classical works, owing to the epidemics of the nineteenth century, physicians were able to further refine their definitions of the new epidemics. Shirâzi divided putrid fevers (*tabhâ-ye ʿufuni*) into epidemic (*wabâyee*) and nonepidemic. In the nonepidemic fevers, the unnatural heat began in a given organ and was then transferred to the heart and from there it was distributed to all parts of the body, while in the pestilential fever the heat was first produced in the heart and from there was conveyed by blood to other organs. In this fever, the unnatural heat essentially originated in the heart and for this reason in the pestilential fevers the death toll was high, while in nonpestilential fevers symptoms appeared only in the heart.[113]

The principle that the origin of all fevers was corrupted humors was used to explain why cholera, *mothbeqa*, *mohreqa*, *nowba*, and seemingly other epidemic fevers all had the same pathological origin and could transform into each other. This principle of transformation rendered the task of identifying the pathological characteristics of cholera and all the others difficult. Attempts were made to overcome this problem. Dr. Polak, for example, argued that *mothbeqa* was antagonistic to *nowba* (intermittent fever), which could occur every third (tertian fever) or fourth (quartan fever) day, and these two could not transform into each other, because the former was the corruption of blood in the veins, while the latter was the putrefaction of other humors outside the vein.[114]

Nevertheless, even for Polak, different fevers represented different degrees of the same disease. He identified typhus, which corresponded to the description of *hasbeh* by the indigenous doctors in Iran,[115] with both *mothbeqa* and *mohreqa*. "If typhus is mild, it is called *febris mucosa* (*mothbeqa*) and if it is malign, it was named *febris septica* (*mohreqa*)."[116] Polak was of the opinion that during a cholera epidemic, different kinds of fevers (*nowba*) or even simple diarrhea are transformed into cholera, and "physicians have to treat the *nowba* as a matter of urgency."[117] By the same token, Sâvaji believed that in its fourth day, or the crisis point, choleric fever is transformed into *mothbeqa*, whose symptoms are the reappearance of heat and blood in the surface of the body, a sign that heralds the end of the illness.[118] Likewise, for T. E. Aubert, typhoid fever was a reaction to the morbid and lethal typhus. The occurrence of typhoid (an internal fever) signaled the end of the illness and disappearance of typhus.[119] In other words, the transformation of typhus to typhoid signaled that the patient survived the lethal typhus. Polak, on the other hand, maintained that *mohreqa* and *mothbeqa* were one and the same contagious disease that, like smallpox, occurred only once in the life of an individual. Its symptoms were high fever, headache, typhoid (*hasbeh*), and mental disturbance. According to European physicians, Polak observed, the matter of *mohreqa* and *mothbeqa* was one and the same, and the difference of nomination related to different degrees of the strength of this disease, so that if, for example, fever, headache, or anxiety were severe, they called it *mohreqa*, and if these symptoms were mild, they called it *mothbeqa*.[120] All of this echoes the fact that in the nineteenth century, "typhus" (or *mothbeqa* and *mohreqa* in traditional medicine) was a generic term for acute infectious diseases.[121]

In Galenico-Islamic medicine, diseases were considered as ontological realities independent of their symptoms, but they could overlap or coincide. ʿAqili, for instance, writing toward the end of the eighteenth century, observed that whenever a disease and its symptoms coincide, physicians should first treat the disease. But if the symptoms are too critical, such as a severe pain of the colon or pain caused in a serious injury, the pain should first be reduced before treating the disease.[122] Similarly, in the early nineteenth century, contemporary Western physicians distinguished between diseases and their symptoms, but the order of their occurrence could vary or coincide. Augustin Grisolle, one of the authors whose works were translated into Persian and studied by students of "modern" medicine in Iran, stated that (unlike anatomical pathology), fever was not a symptom, but the disease itself, because it could occur before the appearance of lesions or inflammation of the organs or tissue, or it could even cause them.[123] However, according to the physiology of François Joseph Broussais, which was based on anatomical pathology, fever actually represented tissue inflammation. So there was a continuum between states of health and disease. Illness was a lower degree of health, an idea that rejected the ontological existence of "disease," or "fevers," to which the Galenico-Islamic medicine reduced all diseases.[124] The "modern educated" Iranian physicians adhered to Grisolle's ideas (as Mirzâ Rezâ Doktor and Mirzâ ᶜAli Hamadâni, translated his books into Persian) but not to Broussais's anatomical physiology. None of Broussais's works was translated into Persian at that time. Western physicians, such as Polak and Tholozan, who taught at the Dâr al-Fonun, do not seem to have encouraged the translation of Broussais's work, whereas in Egypt, Dr. Barthélémy Clot was a staunch follower of Broussais.[125] The difference might be explained by the strong presence of humoral medicine at the Dâr al-Fonun and also by the fact that in Egypt, under the aegis of the ruler Mohammad Ali and his overarching reforms, complete postmortem dissections were performed at the Qasr al-ᶜAini hospital in Cairo, while in Qâjâr Iran this was not allowed.[126]

Despite these fundamental differences between Galenico-Islamic medicine (or the new reading of Hippocratic medicine) and Western pathology, humoral physiology contained elements of disease localization. Although the *Kholâsat al-hekmat* of ᶜAqili was published in India, it was one of the major Persian medical sources in the nineteenth century. ᶜAqili believed that in fevers resulting (*ᶜârez*) from inflammations or lesions, when the latter are healed, the fevers disappear.[127] This observation predated the physiology of Broussais by several decades. Likewise, the accumulation of putrid blood and/or other humors in and around the stomach in a choleric patient in fact indicated, as in anatomical pathology, the location of the infected organ in and around the stomach. One of the clinical signs of *mothbeqa* and *mohreqa*, according to Polak, was inflation of the stomach,[128] which echoed the idea of Galenico-Islamic medicine that *mothbeqa* was caused by the putrefaction of blood vessels surrounding the stomach or the *cardia* (the opening of the esophagus into the stomach [derived from "cardia": the heart]). Bloodletting was advised to expel putrid humors. In the physiological medicine of Broussais, bloodletting was applied in order to reduce the concentration of blood in the stomach, thus reducing inflammation of the affected organ. This similarity of treatment in Iran and the West could reflect similar clinical experience by traditional physicians in Iran and their anatomo-pathologist counterparts in the West. The treatment that traditional physicians proposed corroborated the way they described the occurrence of disease. Thus, what informed diagnosis and treatment was not only humoral theory, but also clinical observation. Today, modern treatment for cholera prescribes water in order to fight dehydration. For traditional physicians in the nineteenth century, water was given to reduce the (unnatural) heat that was produced in a choleric patient. In other words, the clinical observation prompted by epidemic outbreaks resulted more or less in the same conclusions that Western physicians reached on the basis of anatomical pathology.

The Pathology of Epidemics

In Hippocratic writings, epidemics did not refer to epidemic diseases as understood today, but to any illness found among (*epi*) the people (*demos*). Conditions of air and water, seasons and temperature, were always considered to cause epidemic diseases. Most of the diseases described in detail and based on clinical observation are attributed to different weathers. In Hippocrates's chapter on "Airs, Waters and Places," for example, it is the environmental conditions that cause various illnesses "among the people." A wet and mild winter or a dry and cold winter, for example, produced different kinds of illnesses that changed according to gender, age, and temperament.[129] Hippocrates's emphasis was on unseasonable winter, summer, or excessive temperatures, and not particularly on miasma as a factor in generating epidemic diseases. Cholera was one of these diseases, in which the bile (*chole* in Greek) escaped through vomiting or diarrhea. According to H. Scoutetten, Hippocrates attributed cholera not to miasma but to imprudence, excess in drinking, eating, and sometimes to the misadministration of drugs.[130]

Hippocrates described diseases clinically by referring to specific cases. Islamic physicians, by contrast, tended to interpret and describe what they called pestilential fevers (*humma-ye wabâyee*) in rather vague and inclusive terms. Avicenna's description of *humma-ye wabâyee* (febris pestilentialis), as translated by Tholozan from the *Canon*, reads:

> In pestilential fever, the surface of the body is calm, while the inner body is agitated. Quite often this malady is fatal. One feels that the organism is overheated and that inside there is inflammation. The respiration is long, high and fast and often with difficulty and expiration is fetid. There is a pungent thirst and dryness of tongue, with sometimes nausea or anorexia.... There is ache in the epigastria, swollen spleen, pronounced anxiousness, and agitation. Sometimes other symptoms are added to these: dry cough, a tendency to syncope,

delirium, hypochondria, insomnia, and complete prostration of faculties. The eruption of red or reddish-white pustules also occurs. Sometimes they erupt suddenly and disappear fast. Ulcers and aphtha are formed in the mouth. The pulse becomes more frequent, small, and fever intensifies at night. Tendency to hydropisy as well as flux of bile or other matters in the belly appear. Stools are soft, fetid, and sometimes black and often foamy; they are formed by the liquefaction of certain parts of the body. The urine is aqueous, yellow or black. There is black but quite often yellow vomiting. Sweating is fetid.

This fever starts with these symptoms and when it is severe it leads to syncope and to the cooling of the extremities, lethargy and spasms and convulsions. In some of the pestilential fevers the patient loses the faculty of perception. Those who approach them would not notice any increase of temperature or alteration of pulse and urine. Such symptoms are entirely fatal.... The majority of the patients whose respirations are fetid die promptly because they have putrefaction in the heart.[131]

This pathological description of *hummâ-ye wabâyee* would apply to any kind of disease. As Tholozan remarks, "it is not based on clinical observation, but represents an idealtype of morbidity, which has no specific pathological representation."[132]

For centuries, *wabâ*, literally "containing" or "including," was the generic term to indicate any kind of epidemic disease that affected a large number of people at the same time, caused by putrid air covering an extensive geographical area.[133] The common characteristic of sudden spread and subsidence, the high casualties they caused, and the obvious common cause, namely, noxious miasma, befuddled their pathological identification.[134] A contemporary doctor writing in the mid-1850s informs us that "two years ago, in Tehran, the *khâdeʿa* fever that people call *wabâ* emerged and after a month it disappeared."[135] According to Seyyed ʿAli Tabrizi, after cholera, plague often followed.[136] In his *Diary,* Mohammad-Hassan-e Khân Eʿtemâd al-Saltaneh, the minister of Press, relates that in the "plague year," Fathʿ Ali Shah stayed about 40 days in Suhânak (near Tehran) to escape the bad air of the capital. The "plague year" can be any year from 1829 to 1833, during which both cholera and plague spread alternately or simultaneously.[137]

During the nineteenth century, the term *wabâ* became increasingly synonymous with cholera, not least because cholera was the most widespread epidemic in that period. Following an outbreak in the summer of 1835, Shirâzi wrote *Wabâiyeh kabireh* (*Greater Treatise on Cholera*), which was based on "his experience in the four previous cholera epidemics." In the second section of this work, he was very specific about the symptoms of cholera, and reported a myriad of them observed among choleric patients:

> Fever is a major sign of cholera but it is weakly felt on the surface of the body of some patients and in others the surface is also very hot. In some people all the body is cold except around the navel and the heart, while in others the extremities are cold and the inner body is warm, accompanied by vomiting and passing stools of watery phlegm or yellow bile, or a rice-like water... In some, thirst is strong and in others, thirst subsides but fever (*eltehâb*, literally burning, inflammation) remains; in most patients, urine retention appears after one, two or more days and they become very skinny, with sinking eyes.[138]

The phenomenal spread of epidemics prompted physicians to renew their reflections on *wabâyee* (epidemic) diseases. The change that characterized the epidemiology of traditional medicine was not only limited to the awareness of the epidemic, as opposed to the sporadic, nature of the epidemics. It also related to the perception of the disease itself, though always within the realms of humoral theory.

Nineteenth-century authors writing on epidemics referred to Galen or Avicenna, and were attached to the theoretical frame and literary format used by these authors. But at the same time, they sought further clarification and classification to distinguish epidemic from nonepidemic diseases. For example, pestilential fever (*hummâ-ye wabâyee*) was, according to Sâvaji, a putrid fever (*hummâ-ye khelt*) that was ignited in the heart and from there transmitted to all over the body. Its symptoms were joint pain and backache, extreme heat in the chest, dryness of the tongue and halitosis (ᶜ*ofunat-e dahan*), vomiting yellow bile, and diarrhea of yellow bile. All of these conditions were called *vâfeda* (literally epidemic) if they were not fatal, and *mowtân* (dying in mass) if they were lethal. In both cases they affected a large number of people and for this reason they were called *wabâyee* (encompassing).[139] This distinction between deadly epidemics (*mowtân*) and ordinary epidemics echoed the differences between what were known in Europe as cholera and *cholerin* (referring to the intensity of vomiting and diarrhea, because *cholerin* was a benign disease but could turn into severe cholera.)

Similarly, the principle of clinical observation that was, at least theoretically, a key element of Galenico-Islamic medicine favored a different perception of disease. Sometimes such observation contradicted either the theories or the advice and opinions of the classics so dear to physicians and therefore created a puzzle. The main method of explaining the contrast between theory and observation was the art of interpretation. In epidemics, for instance, bloodletting (*fasd*) was advised by classical authors. However, physicians' experience of cholera showed that this practice could be fatal. How could a traditional physician like Sâvaji combine this clinical observation with "authorities" such as Hippocrates, Avicenna, Majusi, and others who prescribed bloodletting?

Sâvaji seems to have followed the method of the Shiite *mujtaheds*, that is, the intellectual effort of (re)interpreting a classical text in order to uncover its secret meanings, thus helping to understand a contemporary issue.[140] On the other hand, he also justified clinical observation in terms of humoral theory: the bloodletting of a choleric person was not allowed because blood was the vector of spirits, pneumata, and natural forces (*qowâ*), which were necessary for the heart's operation. Therefore, if blood was expelled from the body in order to reduce the putrid humors, it could also expel *arwâh* and *qowâ*, which were vital. In order to overcome the contrast between this idea and the word of the classics, after "much intellectual effort," Sâvaji found out that if Hippocrates and others advised bloodletting, it was at the stage when the fever had subsided or transformed into another, like *mothbeqa*, in which the heat resurfaces and boils (*bothurât*) appear on the body, a feature that occurs usually on the fourth day, at the peak of the crisis, when the patient has overcome the illness.[141] Only then should *fasd* (bloodletting) of the basilica vein (*bâsîlîqâ*) (the "royal vein" on the forearm), be applied, once on the fourth and again on the seventh day of the fever.[142]

We can distinguish two layers of change in outlook with regard to the physiology of disease. The first relates to the application of miasma theory in the physiological alteration in the body, and the second to the distinction of symptoms in the "epidemic" as opposed to "ordinary" diseases.

Effects of the Foul Air on the Body

In Hippocratic medicine, different airs could cause different illnesses. This theory was maintained centuries before epidemics at pandemic levels brought about an awareness of the nature of "epidemic" diseases. Then, physicians pointed to specific kinds of weather—the corrupted air—as a cause. They furthered investigation by examining the effects of putrid air on the body and humors. Foul air corrupted the humors via respiration. Corrupted or altered air was against nature. Choleric air was of two kinds: in the first, the essence (*jawhar*) of the air was altered and in this case it was named corruption or infection (*ʿofunat*). The corruption of the air was like the corruption of water. In the second, change was due to the degree of cold, hot, dryness, and moisture. If one of these qualities became abnormally strong or weak, or if they transformed into one another, the air was not considered "infected" or "corrupted," but "unseasoned." In other words, it became hotter, colder, or more humid than normal and such unseasonal air also generated epidemic diseases.

The epidemics were thus caused in two ways: by the corruption of air, caused by terrestrial factors, such as smog emanating from marshes, swamps, rice fields, and other warm and humid areas, as well as the smell from dead bodies and infected water, and by the alteration of air, as a result of celestial factors, or the particular constellation of the planets. In the first case, the increase of foul matter infested the air and produced epidemics. In the case of the alteration in the quality of air, cholera often occurred at the end of summer or the beginning of autumn, because there was either excessive heat that went against the season's temperament, or an abrupt fall of temperature in the early autumn or at the end of spring following a snowfall.[143]

The major characteristics of epidemic diseases were thus diagnosed via external factors. No attention was given to identifying or differentiating between each disease. All such diseases were classified as *wabâyee* (epidemic) after their common cause, the foul "air." Hence one of the signs of the occurrence of *wabâ* consisted of a change of the weather.[144] Pathogens were external to the body. Foul air entered the heart via respiration and corrupted the spirits (pneumata), which in turn through retraction and expansion of the arteries was carried throughout the body and affected all organs (*aʿzâ'*) and humors (*akhlât*). As for the air that could be corrupted because of its compound nature—for if the air were pure or composed of simple elements, it would not absorb any other matter—humors also have the capacity of being infected because of their mixed compositions; the term *akhlât* (humors) is taken from the root, خلط which means mixture or combination.

Symptoms

Theories were used as guidelines for an observer's description and diagnosis. They informed diagnosis and served as a thread to relate and organize elements of clinical observation in order to reach a meaningful conclusion and achieve an appropriate treatment. However, when various cases were observed, new descriptions allowed new ideas, and this meant that diagnosis or therapy based on theory gradually became secondary to a diagnosis based on observation. It was the observation of symptoms that helped negotiate theory. Miasma theory was used by both Galenico-Islamic medicine in Iran and anatomical-pathological medicine in Europe. For both, the cause of cholera was noxious miasma, with the difference that medicine based on anatomical pathology localized the effects of foul air on the body via dissection, whereas humoral medicine continued speculating on the way putrid air affected humors and the organs via veins and arteries in an opaque body. While undertaking autopsies after the epidemics of 1831 and 1848 in Egypt, Clot Bey found that the lungs, liver, and spleen were saturated with blood and the peritoneal side of the stomach and the mucous membrane were congested and ecchymotic (having an accumulation of blood in subcutaneous tissue caused by ruptured blood vessels). He concluded that this explained why blood did not reach the surface of the body, particularly the extremities, which subsequently became cold, a symptom that in Islamic medicine was called *bard al-atrâf* (coldness of the extremities). The most rational treatment of cholera was to prevent the concentration of the blood in the viscera by conducting it to the surface of the body through bloodletting. Clot Bey insisted that by bloodletting he saved almost all of his choleric patients.[145] In anatomical pathology, the concentration of blood in certain organs resulted in a scarcity of blood at the surface of the body. In humoral pathology, on the other hand, putrid humors reduced the innate temperature of the body. This explained the cooling of the surface of the body. In other words, anatomical pathology and humoral medicine came to a similar conclusion for different reasons. In Iran, bloodletting was considered a logical way to expel the putrid matter that caused the unnatural heat of the body.[146] But it was probably the observation of symptoms that elicited the following physiological explanation: the matter of epidemic fever was the putrid yellow bile that circulated faster than other humors in the body, and for this reason reached the *cardia* (aperture) and the bottom of the stomach quicker, causing vomiting and diarrhea. But this putrid humor could also prevent the expulsion of the putrid matters, the therapy for which was to administer cold water to reduce the force of the putrid yellow bile (*safrâ*).[147]

For nineteenth-century physicians in Iran, the general symptom of choleric fever was severe vomiting and/or diarrhea, with the passing of watery stools and cold hands and feet. A toxic humor was held responsible for excessive dehydration. Sâvaji observed that a choleric patient dramatically loses weight in two hours; the eyes are sunken and reduced in size as small as a small coral deep in the face. These are signs that the patient will die within 12 to 24 hours.[148] Reference by Sâvaji to what he called *ʿezam-e tehâl*[149] (enlargement of the spleen or splenomegaly), probably echoes the observation made by Clot when dissecting choleric corpses.[150] However, Sâvaji also referred to other symptoms that he attributed to a "different kind of *wabâ*, in which the illness is accompanied with dropsy and a dry cough, while in another kind, the patient's weight loss is not accompanied with vomiting or diarrhea, but fingers, lips and the skin around the eyes and ears of the patient turn black and death occurs within ten hours."[151] Different diseases were thus recognized and distinguished by Persian physicians through their symptoms. This differentiation and identification of various diseases under the common denomination of *wabâ* was a far cry from the inclusive description of pestilential fever that referred to all kinds of epidemic diseases without clear distinction.[152]

Epidemics and Contagion

In Islamic medicine, miasma that engulfed a large number of people was deemed to be the origin of *wabâ* (epidemic). This concept, however, could not explain why miasma did not affect everyone. Classical Galenic medicine explained this with the idea of predisposition to diseases, which made some individuals more vulnerable than others. Nevertheless, clinical observation revealed that, in some cases, diseases were transmitted through contact of the sick with the healthy, thus creating the notion of "contagion." But this observation did not find theoretical support in humoral medicine, where the idea of transmission from one to another could not be explained with miasma theory. Miasma explained the way cholera was spread, with a sudden surge and sudden subsidence, whereas the spread of disease by contagion could not be fast, each individual contamination needed a contact and a period of incubation. This did not easily explain how a large number of people who were not in contact with each other were affected by the disease over a short period of time. We therefore find both in Europe, where anatomical pathology made its first forays, and in non-European countries, where medicine was dominated by humoral theory, a lively debate on the "contagious" or "epidemic" nature of diseases such as cholera, plague, and measles.

The Islamic physicians' belief in miasma led them to negate contagion. Shirâzi opened his treatise on plague (*Tâʿuniya*) by criticizing physicians who avoided patients for fear of contagion. He subtitled this treatise '*Mosakken al-fu'âd min khawf seraya* (*soothing of the heart away from the fear of contagion*).[153] The anticontagionism of many contemporary Western physicians was also based on miasma theory.[154] There were also those who found the question baffling. Seyyed ʿAli Tabrizi in 1853 stated: "no one knows whether cholera (or plague) is contagious, as there have been many observed cases where everyone perished except one and in other places one person died and others escaped the disease; or the person who took care of choleric patients remained healthy while others who had no contact with cholera patients died of *wabâ*."[155] However, Tabrizi stated that the first way to avoid *wabâ* is to leave the infected area.[156] Leaving a pestilential area, according to Tabrizi, was not for fear of contagion, since what caused the transmission of *wabâ* was the alteration (*takayyof*) of the natural temperament (*mezâj*) of the healthy individual.[157] The oddities of transmission may explain this adhesion to contradictory theories. In such a situation, there was no possibility of "crucial experience" that could permit a definite conclusion.[158]

The interpretation of disease in Islamic medicine was therefore sandwiched between the theory of miasma and the idea of contagion. Nevertheless, such conceptual overlap before the end of the nineteenth century and "triumph of microbiology" appears natural due to the vague notion of contagion. Physicians recommended that those who could afford to leave noxious miasma and go to areas of healthier atmosphere should do so. While Shirâzi or others lambasted doctors who feared contamination, they never criticized those who fled miasma. This idea had its roots in the theory that epidemic disease was first caused by miasma and second by humoral predisposition. In practice, however, the situation was different and it seems that the fear of contagion prevailed. In the cholera outbreak of 1904 in Tehran, for instance, it was reported that "for the past three weeks all the public departments have been closed and secretaries and clerks have been advised to disperse. Ministers have shut themselves up in their private apartments, and a strong cordon of guards has been placed around the palace to prevent anyone approaching the Shah, except the ministers who may do so only on most urgent matters."[159] In other words, fear of contagion was based on practical experience even though it was not supported by miasma theory, and miasmaphobia was founded on pure theory not always verified in medical experience.

The impact of theory on medical practice was also fundamental in contemporary Western medicine. While still imbued with miasma theory, contemporary Western physicians tended to identify the contagiousness or epidemicity of cholera and plague. Their experience often supported the former or the latter, or both, but often provided evidence in support of either contagiousness or epidemicity.[160] Very rarely did physicians rely solely on their experience regardless of the prevalent theories. Those who did were not believed, like John Snow, who, instead of reflecting on theories and applying them to his experience, proceeded to gather data. He analyzed a number of contaminated people using a source of water in a specific point in Soho in London. Snow concluded that cholera was caused by the contamination of potable water by the sewage system.[161] As a result, we find International Sanitary Councils harboring contradictory ideas, and conflicting factions continuing to form around political or economic interests until the bacteriological discoveries that deprived the debate of its substance were known.

Paradoxically, although the distinctive characteristic of *wabâ* was foul air, the illness itself resulted from bodily deficiency, namely, the infected humors accumulated in the body. It was these putrid humors that absorbed the putrid air. The body became ill when its humors were of a quality similar to those of the polluted air. Thus the individuals of sanguine, hot, and humid temperaments were predisposed to epidemics. On the contrary, the foul air could not overwhelm or affect those temperaments (*amzâj*) that were opposed to the temperament of the foul air. This also held true for the region or area. If the city or region where foul air dominated was cold and dry, epidemics did not occur because the cold and dry weather rendered the body of the inhabitants firm, whereas in a hot and humid region, the bodies became soft and porous and therefore suitable for absorbing the putrid air.

This paradox held center stage given the theoretical uncertainty surrounding the concept of contagion. Although it did not occupy a major place in the Persian, Greek, Chinese, or Indian literature,[162] the concept of contagion was certainly not new in the nineteenth century. However, as with the notions of miasma and *wabâ*, further reflection came with the evolution of medical thought. Sâvaji's classification illustrates this change. Epidemics occurred from a combination of three factors: (a) the atmosphere (*bâdiya*), (b) predisposition (*sâbeqa*), and (c) infection (*wâsela*). In a choleric fever, the (*bâdiya*) causes the putrefaction of the air; a predisposing cause is found in bodies saturated with putrid humors; the *wâsela*, seems to be the infection of the humor that hosts putrid air.[163] Could we suggest that *wâsela* refers to the idea of contagion? Etymologically, *wâsela* (literally meaning "connection") implies the idea of touch and/or mutual touch, but it is by no means certain that the term contains any modern connotation of contagion. In his commentary on Dr. Polak's treatise on diseases affecting soldiers, an anonymous physician, who differentiated clearly between contagious diseases (*ʿadwâ*) and epidemic diseases (*wâfedah*), maintained that contrary to Polak's claim that dysentery can have symptoms similar to measles, smallpox, and *mothbeqa*, neither dysentery nor *mothbeqa* and *mohreqa* are contagious. He then referred to ancient and contemporary authors, quoting their different experiences of contagiousness and epidemicity, and concluded that diseases like *mohreqa* and *mothbeqa* were epidemic because they were caused by atmospheric alterations (putrefaction of air), while a contagious disease was transmitted via a (putrid or infected) matter that was transferred from one person to another.[164] This was a far cry from the general concept of miasma, suggesting that traditional physicians were beginning to differentiate between epidemic and contagious diseases.

The theory of miasma was also dominant in Europe. Even some of the contagionists were influenced by this theory. The French surgeon Jean-Baptist Rousseau believed that cholera-morbus, originating in India, was contagious or transmissible and that this was the cause of the pandemic. He specified that only patients affected by diarrhea could transmit the disease to a healthy individual. However, it was not by the contact of skin but by the respiration of miasma emanating from the sick and their clothes. Even close proximity to a choleric person did not transmit the disease unless the aeration and sanitation of the area was neglected. According to Rousseau, noxious miasma with a certain degree of concentration transported by air was the vector of the cholera pathogen.¹⁶⁵ The preventative measure he advised was to clear the environment of noxious miasma by ventilation of the living area.¹⁶⁶ We find this overlap, or switching back and forth, between the concepts of contagion and miasma among traditional physicians in nineteenth-century Iran. What all this indicates is that there was a continuity rather than a break between concepts of miasma and contagion. Often they complemented rather than opposed each other. This held true for both Western and Islamic medicine in the nineteenth century. As Lois Magner noted, "when contagion was defined loosely enough to include harmful material that was indirectly, as well as directly transmitted, it was not incompatible with equally vague definitions of miasma as diseases inducing noxious, contaminated air."¹⁶⁷ Extrapolating on the observation of Vivian Nutton on the Hippocratic or Greek concept of contagion, one might speculate that this overlap was due to the fact that in these concepts of contagion, "what passes is an emanation, an effluxion, a breath, a poison, a putrid effusion, an excrement, or a miasma."¹⁶⁸ These shared concepts played an important role in the transmission of modern medicine to Iran.

Theoretical Transition

There were thus two major factors in the nineteenth century that stimulated medical development: state sponsorship and the recurring waves of epidemics. From the second decade of the century following the introduction of vaccination by Jukes and Cormick, treatises were written on smallpox with the support of the royal heir apparent, Prince ʿAbbas Mirzâ. However, it was not until the second pandemic in 1829 that literature on epidemics began to flourish, when Mirzâ Mohammad-Taqi Shirâzi wrote an essay on plague (*Tâʿuniah*) in Rasht, followed in 1835, by his first treatise on cholera, called *Wabâiyeh kabireh*. We have not found any other essay written on epidemics by local physicians before the cholera of 1846. Contact with Western physicians, which began under Mohammad Shah (1834–1848), coupled with increasing outbreaks, stimulated research. A treatise on cholera was written in 1846, based on French sources.¹⁶⁹ From the beginning of the reign of Nâsser al-Din Shah, and the reforms implemented by his first Prime Minister Amir-Kabir, there are an increasing number of treatises on cholera, which resulted either from the pure curiosity of some physicians or from state sponsorship and commission. It is significant that between the first cholera epidemic, in 1821, and 1850, only a few tracts were written on cholera, while during the 1850s several were written on cholera and plague.

More significant than the quantity of texts on epidemics is the intellectual dynamism within humoral medicine we witness. The main purpose of Shirâzi's first treatise on cholera (1835) was to identify the causes and signs of the epidemic *asbâb va alâmât dâlleh bar hodus*, then preventive measures, then the symptoms of cholera, and then different treatments to cure it.[170] In an interval of one year (1269 Hijra—straddling between 1852 and 1853), at least three treatises were written on cholera, commissioned by the government.[171] All these authors underline that their work was motivated by the mortality caused by the epidemics, and encouragement from the government.

In the following decades, the epidemic literature significantly increased.[172] Physicians were bewildered by the different degrees of intensity in cholera epidemics, and by the similarity of its symptoms to ordinary dysentery. It was in response to this question that nearly 30 years after his first treatise on cholera (1835), Shirâzi wrote on the differences between cholera and dysentery, known traditionally under the name of *heyzeh*. Later, in 1852, the same year as Sâvaji wrote his treatise on cholera, Mohammad-Hossein-e Afshâr, who was one of Dr. Polak's students at the Dâr al-Fonun, referred, in his treatise (which seems to be a liberal translation of Polak's lectures on cholera), to the Greek equivalent of *wabâ*, namely, cholera. He said that the term (*wabâ*) had previously referred to a disease characterized by the expulsion of whitish yellow bile through diarrhea and vomiting. "This disease," he noted, "is of two kinds: when it is individual (*khâss*) it is called *heyzeh*, and when it is public (*ʿâm*), it is named *wabâ*." This indicates that for Afshâr, obviously following his master Polak, cholera was of two kinds. However, in the public form of *wabâ*, Afshâr identified three phases, the inception, intensity, and decline, but with similar symptoms. In other words, for Afshâr and Polak, cholera and *heyzeh* possessed a similar nature.[173] Nonetheless, elements of anatomical pathology are obvious in Afshar's treatise. For instance, Afshâr refers to the pain of the opening of the stomach (*famm* = cardia) and the coagulated blood in the viscera.[174] Stomachache is mentioned in all traditional tracts but not the coagulation of blood, which was observed through the dissection of choleric victims in Europe. The other symptom of cholera in traditional medicine was what they called *habs al-bowl*, literally meaning retention of urine. This was an interpretation of the symptom in choleric patients who had diminished urine due to dehydration. According to traditional medicine, urine was withheld in the blood, and the solution was bloodletting.[175] Afshâr, familiar with elements of anatomical pathology, indicates that the urine retention resulted from the fact that all fluids (*rotubât*) of the body were directed to the stomach and intestines and expulsed through diarrhea and vomiting.[176]

In Galenico-Islamic medicine, often *maraz* (disease) and its *ʿaraz* (symptom) were taken for one and the same phenomenon. With the increasing number of observations and descriptions of symptoms, this identification gradually tended to dissipate. Efforts to differentiate between cholera and cholerin furthered the idea of distinction between diseases and their symptoms, insofar as some of the symptoms of these diseases were similar, while clinical observations indicated that they were two different diseases. One significant example is the differentiation made by Mirzâ Mohammad-Taqi Shirâzi between *wabâ* (cholera) and *heyzeh* (acute enterocolitis, with symptoms similar to cholera). This distinction was significant insofar as in Islamic medicine before the nineteenth century *heyzeh* was identified with *wabâ*. In India, cholera, identified as an endemic disease since antiquity, was confused with other kinds of diseases. Visiting India in the middle of the sixteenth century, Christoval Acosta reported that he found "a dreadful and virulent disease that the Arabs called *hachaiza*, which is a corruption of the term *al-heyzeh* used by Avicenna and Râzi."[177] The merit of Shirâzi was in the fact that he found the two diseases were different in both their symptoms and their nature. This differentiation was based on humoral theory, and also on Shirâzi's experience and observation, and it makes sense despite its pathological misconceptions: (1) From the nature of the air, a physician can distinguish whether the fever in question is *wabâyee* (epidemic) or not and investigate which of the humors is affected by the unnatural heat; (2) there is fever in cholera but not in *heyzeh*; (3) in *heyzeh*, indigestion or food corruption is the cause but not in cholera; (4) in cholera thirst and dehydration are inevitable but not in heyzeh; in *heyzeh* the very low temperature of the surface of the body is excessive, while in cholera the body is warm, due to the unnatural fever (*harârat-e ghariba*) and the force of this fever is greater in the inner body than on the surface and in some cases the extremities become cold; (5) in cholera the matters expelled via diarrhea and vomiting are putrid and sticky but in *heyzeh*, they consist of corrupted food; (6) there is no cardia ache in cholera, while there is a putrid sweating, but in *heyzeh* there is no sweating and if there is it does not smell; (7) the retention of urine in *heyzeh* can be dissipated by bloodletting, but if this happens in cholera, it is very difficult to overcome.[178]

Despite these clarifications, however, Shirâzi's confusion between the symptoms of cholera and *heyzeh* is significant. He attributes the *bard al-atrâf* (the coldness of the extremities) that occurs in cholera to *heyzeh*, stating that not only in cholera is the body warm, but also the fever resulting from the unnatural heat appears, although with more strength in the inner body than on its surface and in some cases the coolness of extremities can also be observed. The other confusion relates to the retention of urine that was a common symptom of cholera according to Galenico-Islamic medicine, but Shirâzi attributed this to *heyzeh*. Both these misconceptions were rooted in Shirâzi's strict observance of humoral pathology. In the first case, since fever in cholera was caused by the unnatural heat that resulted from the putrefaction of air, it was inevitable and logical. Consequently, the low temperature of the extremities that sometimes occurred was deemed exceptional. In Galenico-Islamic medicine, the retention of urine in cholera patients meant that urine is returned to the body, while in modern medicine, urine production is reduced.[179] For Shirâzi, the fact that the symptom of urine retention occurred in both *heyzeh* and cholera was not helpful in the pathological differentiation that was the aim of his study. According to humoral physiology, Shirâzi believed that urine retention could be cured by bloodletting, but if this happened in cholera, its cure was difficult. This was probably because for Shirâzi, in *heyzeh* it was the blood that absorbed the urine, while in cholera the urine was absorbed by the putrefied humors (*akhlât*).

Nineteenth-century physicians differ from Avicenna or al-Majusi in their discussion of epidemics insofar as they point to specific and clinical observations of their own, and whenever they find that their observation corresponds to some aspects of Avicenna or Galen's descriptions, they quote them in support of it. There is therefore a selective rather than systematic reference to the classical authors. The symptoms of cholera, for instance, included diarrhea and vomiting of various substances depending on which humor, yellow bile, blood, or phlegm, was affected. Boils or inflammations appeared in various parts of the body in the case of plague, again depending on which humor was affected.

Humoral theory was ingrained in Iranian culture. Therefore, belief in treatment based on that theory informed the whole process of medical practice. The faith of the patient in humoral theory helped the efficiency of the doctor's humoral treatment. For centuries, different techniques and forms of bloodletting were used in the treatment of illnesses and their efficiency could not be questioned because they operated as part of culture and ideology.[180] Nâsser al-Din Shah, who suffered from chronic hemorrhoids, writes in his diary of September 1885 about his cold, which he described as intermittent fever (*nowba*) and *qoshᶜarireh* (trembling, horripilation). For his sore throat doctors prescribed a purgative, called *sedlis*. But the night before the application of this remedy, he had a hemorrhage of the anus. He noted that due to this (natural) bleeding, his sore throat disappeared completely.[181] In another case in June 1886, he had toothache for several days, which was complicated by dizziness. Doctors applied three leeches at his anus, and the Shah's toothache and headache were both healed.[182] Belief in humoral theory and the hot and cold distinction of all diseases and remedies was so entrenched that European physicians, like Tholozan and Schneider, who practiced in

Iran, could not avoid such methods, and despite their disbelief in the theory, they added to rational therapies a tisane or harmless herbal drug considered to be hot or cold, according to the illness.[183]

Despite their cultural attachment to humoral theories, consumers took into account the efficiency of drugs, and were not entirely against the use of new products. On the other hand, traditional physicians were opposed to chemical drugs that threatened their market. The conquest of the medical market was thus the aim of both traditional and modern medicine, and this competition furthered the medical transition, insofar as traditional medicine endeavored to indigenize new drugs or new medical treatments of Western origin to expand its market. Quinine was one of the new drugs introduced into Iran that had already undergone this experience before the nineteenth century. Diluted in water or lemon juice, it was used to treat malaria and different kinds of fevers caused by infectious or epidemic diseases, such as intermittent fevers (*nowba*), or headaches that could lead to *nowba*.[184] Since quinine was well assimilated in the local market, it became the object of competition between traditional and modern medicine. Dr. Polak recommended that army physicians should use only quinine sold in Western packages and bottles and beware of fake Iranian quinine. The anonymous author of MSS 506, which was written in response to Polak's treatise (MSS 6164) on the diseases affecting soldiers, rebuked Polak's warning and said that European quinine was impure and mixed with other substances. "Not only (European) quinine, for which 500,000 tomans was spent in Iran each year, but all drugs introduced from the West were adulterated."[185]

These economic and social factors were fundamental in the encounter with modern medicine. The market was not able to substitute herbal with chemical drugs overnight, nor could the medical literature that provided information on drugs and formed the foundation of the medical establishment be replaced by the creation of one Dâr al-Fonun. The medical texts and market underpinned and informed the cultural and ideological dimensions of traditional medicine. Although the use of chemical drugs remained marginal, their very presence in the market opened a pathway for traditional medicine to see beyond, and to enter the territory of modern medicine. Five or six of the physicians of Nâsser al-Din Shah, including Mirzâ Mohammad-Taqi Shirâzi, Mirzâ Kâzem-e Rashti, ᶜAlinaqi Hakim al-Mamâlek, Dr. Tholozan, and a physician of the army, treated the Shah (ca. 1860s) by a European purgative. The European purgative (*namak-e farangi*) used in this treatment is significant considering the hostility of Mirzâ Mohammad-Taqi Shirâzi toward modern medicine and his treatise on the refutation of chemical drugs, called *resâleh-ye jowhariyyeh*.[186]

Conclusion

Galenico-Islamic medicine in nineteenth-century Iran was not fundamentally different from its Hippocratic cousin in eighteenth- and nineteenth-century Europe, which gradually evolved into neo-Hippocratic and then anatomical pathology. The Qâjâr physicians were witnessing this evolution as they were actively engaged with Western medical literature. In his treatise, the author of MSS 506 criticized Polak, not on principles, but on the way he simplified the classification of fevers and their treatment. Through the reinterpretation of diseases based on humoral theory, and under the pressure of the epidemics, the Qâjâr physicians experienced a gradual change in their understanding of the body and its diseases albeit to a smaller extent than their Western counterparts. However, this effort of interpretation alone was not enough for substantial change. Anatomical pathology, introduced through the Dâr al-Fonun's teachers and the Western-educated physicians, played a crucial role in this transformation as it threw new light on clinical observation, and assisted physicians in the (re)interpretation of old theories by providing them with new conceptual tools. The next chapter will expand on this issue through the study of local physicians who were directly involved in the study and translation of modern Western medicine.

CHAPTER 2

The Physicians and Their Encounter with Western Medicine

With the exception of a few outstanding physicians who flourished under the Timurids and the Safavids, such as Bahâ al-Dowleh (d.1506) and ᶜEmâd al-Din Mahmud-e Shirâzi (sixteenth century), Iran had no significant medical authority in contrast to the so-called Golden Age, from the eighth to the fourteenth century.[1] This was to some extent caused by the general sociopolitical dislocation following the Mongol invasion and the decentralization of power across the Islamic world. In Persia, this situation continued until Shah ᶜAbbas I Safavid (1571–1629) created a strong central state. Under the Safavids, however, increasing number of Persian physicians emigrated to India for various reasons, including internal conflicts, religious intolerance and/or more lucrative income at the Mughal courts.[2] In India, they encountered not only Ayurveda and other Indian systems but also Western medicine.[3] Hakim ᶜAlavi Khan was one of these physicians. He first traveled from Shirâz to India in 1699. When Nâder Shah invaded India in 1738, he took ᶜAlavi Khan with him. Still later, ᶜAlavi Khan left Iran for Hejâz under the pretext of making a pilgrimage to Mecca, but in 1743 he returned to Delhi, where he died in 1747 and was buried in the mausoleum of Nezâm al-Din, one of the Sufi saints.[4]

The situation for medicine, particularly in terms of state sponsorship and contact with the outside world, worsened considerably during the political and social instability of the eighteenth century before the Qâjârs restored a durable central state, although it may have slightly improved under the short-lived reign of Karim Khân-e Zand (1759–1779), when, according to Rostam al-Hokamâ, in every city there were a good number of learned physicians.⁵ It is likely that learned medicine, whether under court patronage or not, benefited from the relative prosperity and stability of the liberal Zand dynasty.

While medicine in Iran declined in the eighteenth century, in neighboring India, Persian medical literature flourished, notably because of the princely patronage that began with the Mughal Empire, founded by Babur in 1526. One of the outstanding Persian physicians in India was Mohammad-Hossein b. ᶜAqili-ye Khorâsâni, who produced an encyclopedic work in three large volumes, which were widely studied from the second part of the nineteenth century in Iran. This work included the *Qarâbâdin-e kabir* or *Majmaᶜ al-Javâmeᶜ* on compound drugs; the *Makhzan al-advieh* on simples, which he completed in about 1771; and the *Kholâsat al-hekmat* (*Digest of Medicine*) on general medicine, which he wrote in 1782.⁶ The first two volumes introduced the best-known drugs, using the works of his predecessors, and the *Digest of Medicine* was a detailed study of all medical knowledge in the Galenico-Islamic tradition. ᶜAqili was one of the first eighteenth century physicians to be in contact with Western medicine, but he remained unshaken in his Galenico-Islamic views.

This chapter will examine the encounter between Galenico-Islamic medical literature and nascent modern medicine in the nineteenth century that took place at several theoretical levels both coinciding with and succeeding each other in time. It will describe how traditional medicine was reinterpreted in the light of its relationship with the state and with modern Western medicine.

Characteristics of "Traditional" Medicine in the Nineteenth Century

By "traditional medicine" we mean the Galenico-Islamic tradition, which was an offshoot of Galenic and Hippocratic medicine through translation into Arabic under the Abbasid Caliphate (750–1256). As in all transfer of culture and knowledge, the original features of Greek medicine during its transmission to, and its practice in, Islam were subject to change due to social, environmental, and linguistic factors. It is important to examine, in the light of these changes, the extent to which Galenico-Islamic medicine in nineteenth-century Iran did benefit from the principles of Galen and Hippocrates so as to renovate itself. The idea is to see if Galenico-Islamic medicine in the nineteenth century in Iran underwent experience similar to neo-Hippocratic medicine in the West.

Since medicine and other sciences were transferred through the establishment and development of Islam as a religion, they underwent philological or conceptual alterations. Islamic medicine also inherited the format of Greek medical literature, which greatly shaped its development. This left room for two kinds of dichotomy that eventually characterized Islamic medicine: the first between religion and medicine as science, and the second between format and content or theoretical principles. The dichotomy between religion and medicine was the natural consequence of the necessity of non-Islamic sciences to serve practical purposes. ʿAbdul-Rahmân al-Suyuti, in his introduction to the *Medicine of the Prophet*, laid particular emphasis on the aim of medicine, which is, he says, to make the body suitable for the worship of Allah.

> It is obligatory upon every Moslem that he draw as close to Almighty God as he can and that he put forth all his powers in attention to His commands and obedience to Him and that he make the best use of his means and that he succeed in drawing near to Him by conforming to what is commanded and refraining from what is forbidden and that he strive for what gives benefit to Mankind by the preservation of good health and the treatment of disease. For good health is essential for the performance of religious obligations and for the worship of God.[7]

Al-Suyuti could not be clearer. However, the idea of medicine as an instrument for achieving good religious practice could imply the priority of medicine over religion. This concept was taken up by physicians who, reiterating the Prophet's saying that science is twofold, that is, science of religion and science of body, stated that the science of the body is superior to religion because without a healthy body one cannot perform religious duties. Medicine was thus perceived as receiving the blessing of religion, while being considered nobler than it. No doubt, this perception aided the development of medicine in Islam but it was contradictory in principle and evasive in its terms. Logically, medicine was superior, but practically it was a mere instrument to reinforce faith. ʿAqili emphasized the importance of experimenting with drugs,[8] practicing animal anatomy and even vivisection if it was not too painful.[9] His emphasis on dissection (*tashrih-e ʿamali*) indicates that ʿAqili, and by extension the Unani medicine in India, distanced themselves from the "descriptive anatomy" of Islamic medical literature.[10] It is also significant that ʿAqili was impressed by how the Europeans challenged the hardships associated with traveling around the world for the sake of knowledge and discovering new drugs.[11] Yet, ʿAqili himself did not follow this perception of science. It was, in his view, peculiar to the West. According to him, a physician should study ten branches of knowledge, including religious sciences (jurisprudence and hadith, Tradition of the Prophet), logic, philosophy, natural science, and geometry. About philosophy, ʿAqili warns that doctors should not study this science for the sake of their own intellectual curiosity, but to realize the might of God and to defend their faith.[12]

There was also a tension in Islamic medical literature between theory and practice, which resulted from the translation of Greek medical texts. In these texts, ideas and concepts were framed in a literary format for pedagogical purposes, and included commentaries, questions and answers, Hippocratic aphorisms, clinical case histories, and the six non-naturals (*Setta Zaruriya*), that is, six factors that should be controlled, namely, air, food and drink, sleeping and waking, exercise and rest, retention and evacuation (including coitus), and mental states, such as anger, joy, and love.[13] The influence of these literary formats (or curriculum of medical study) that conveyed humoral theories on the development of medicine should be examined in its own terms. To what extent did original works, such as clinical observations by Razi or Bahâ al-Dowleh, follow the format of the medical texts translated into Islam in the eighth and ninth centuries? How far were such original works inspired by the vision, methodology, and perspective of those medical texts? How far was the adoption of their format followed by the assimilation of the concepts and approach of those medical texts? In the nineteenth century, traditional Persian physicians boasted about having acquired their knowledge not only through books but also through clinical experience. To what extent did these physicians effectively practice clinical medicine and to what extent did they merely followed a literary format without really practicing it? These questions are yet to be thoroughly explored, but for the purpose of this section, it will suffice to examine the shares of format and perspective, or curriculum and content, in Galenic and Hippocratic medicine in Islam.

Medical education was based on reading certain books, and once students had read and assimilated this literature, they were considered fit for practice. Furthermore, the claim of most physicians that their writings on epidemics were based on clinical experimentation does not agree with the general theory dear to all of them, that when signs of *wabâ* appear, the inhabitants of an affected area should leave their homes for a more healthy area.[14] Indeed, physicians were among the first who fled the infected area, and they often avoided patients for fear of contagion even though we cannot exclude that some of them did occasionally or systematically examine cholera patients. In the 1892 cholera epidemic, for example, we are told that the only doctor who remained in Tehran was a Jew.[15]

The Place of Surgery in Traditional Medicine

Dichotomy between theory and practice was even better pronounced in the dissociation between anatomy and surgery. Being fundamentally bookish, practitioners of Galenico-Islamic medicine found the distinctive features of modern Western medicine in practical anatomy and surgery, given the primary importance that these fields had acquired in modern Western medicine. Dissection of human cadavers was a major chapter in modern medicine. Vivisection was known through Galen's *Fi tashrih al-hayvân al-hayy* (*On the anatomy of live animal*) translated by Hunain b. Is'hâq, and some physicians refer to vivisection and its necessity.[16] But with the exception of Ibn Mâsawayh's dissection of apes, we do not know of any Islamic physician having carried out the dissection of even dead animals, let alone vivisection. As dissection was not allowed and thus not practiced in Iran, even with the introduction of modern medicine in the nineteenth century, students were sent to Europe to study it. During the period under study, despite the admittance of dissection and practical anatomy in the Dâr al-Fonun's curriculum, they remained framed in modern medical literature without being practiced in operational theaters, due to the age-old division between medicine and surgery. One can see here how traditional medicine affected the education and practice of modern medicine.

The division between surgery and medicine, and lack of practical anatomy, was founded on the distinction between *tebb-e nazari* (theoretical) and *tebb-e ᶜamali* (practical) medicine. The former referred to the study of general principles, such as temperaments, humors, fevers, symptoms, and their classification and descriptive anatomy. The latter consisted in putting those principles into practice. In the reality of medical practice, it can be translated as "pathology" and "treatment," which constitute the main chapters of medical writings: the first dealing with principles and the second with healing practices linked to the use of simple and compound drugs. Even this division was entirely theoretical with no practical purpose, as Avicenna clearly emphasized by saying: "Thus, if you learn what theoretical and practical parts of medicine are, you [can be considered to] have mastered theoretical and practical medicine even though you have never worked in these fields."[17] From such a perspective, anatomy, dissection, and surgery are considered as knowledge and not practical skills. However, sometimes practical medicine was also equated with manual medicine or surgery. According to Ibn Ridwân (988–1061), for instance, practical medicine signified "the study of restoration of fractured bones, luxation, incision, stitches, cautery, perforation, ophthalmology and all other surgical procedures."[18] Although, as Ibn Ridwân indicated, surgery dealt with superficial operations and ophthalmology, it went sometimes far beyond, and also covered invasive operations dealing with casualties or victims of punishment. One may, for example, suggest that the copy, dated 1209 Hijra (ca.1794–1795 AD), made of Hakim Mohammad's *Zakhira-ye Kâmela*, was a response to the need for surgery in the period of civil war and armed conflict that followed the collapse of the Zand Dynasty in 1779.[19] In addition to skin diseases and inflammations, the *Zakhira-ye Kâmela* also discusses a number of treatments and drugs used for injuries caused by guns.[20] With the introduction of modern weapons in the eighteenth century, the number of injuries increased and the injuries got worse.

We find thus different perceptions, or applications, of the division between theoretical and practical medicines across time and space. How can we explain the persistent Galenic and Hippocratic principles of clinical observation, experimentation, and dissection in Islamic medicine even though they were not fully put into practice? The answer lies in a dichotomy between theory, or respect for the text, and practice. The respect for Galen's books appears to be related to the importance of "text," and the "divine" representation of "text" that derived from the sanctity of the Koran and the Hadith in early Islam. This respect for "text" and "tradition" was introduced by Islam and constituted one of the major features of Islamic medicine. At least insofar as Iran is concerned, the importance of "text" and "book" increased after the arrival of Islam. After Islam, knowledge and science was set down in writing systematically, whereas before they were rarely conveyed through written texts. The *Vandidad*, the sacred texts of *Avesta*, were not widely available in written format, but were orally transmitted. It was with the introduction of the Koran that the "book" and "written texts" became important. Studying consisted of reading aloud— students reading texts aloud in the presence of their masters. The action of reading aloud, *qarâʿat*, was a tribute paid to the text, indicating its importance. The concept of reading aloud, from which the name "*qur'ân*" is derived, might also be a legacy of Judaism and Christianity, but with Islam it took a religious and ideological importance unparalleled in other religions. The method of education based on reading a text might also be a legacy of the Alexandrian school, where the students read books out loud when learning.[21] Whatever the truth of the matter, the Islamic tradition of respect for text influenced the way knowledge, including the Greek science that was integrated with the expansion of Islam, was passed down.

Both forms of dichotomies, that is, the association of format/content and science/religion, were already ingrained in Galenico-Islamic medicine when it came into contact with Western medicine in the eighteenth and nineteenth centuries.

The Encounter with Western Medicine

Hardly one can find a starting date for Persia's first encounter with Western medicine. It could have happened via Western physicians, surgeons, or merchants who traveled in Iran or were captured in wars or through the introduction of a Western drug or disease. All these could have occurred at different periods, during the Crusades, the Mongol invasions, the Safavid

Empire's wars with the Ottomans, and commercial relationships with Europe. Hakim Mohammad, a surgeon in the Persian army who dedicated his book to the Safavid Shah Safi I (1629–1642), frequently refers to diseases like *âtashak-e farang* (syphilis or smallpox) and, among other surgeons and physicians he had met or read about, "Plato the wound dresser" (*aflâtun-e zakhm band*), and Masih al-Zamân.[22] Some drugs were first mentioned by physicians at the court of India in the seventeenth century. But during the reign of Shah ʿAbbas I and his successors, commerce with India and the West introduced these "drugs" to Persia.[23] Although many products entered Iran via the Ottoman Empire and the Levant, the (re)conquest of the Persian Gulf ports by Shah ʿAbbas 1 opened a new phase in relations with the West, as before that time, the "Portuguese stayed at Hormoz and did not trade directly with the Safavid court."[24] In the seventeenth century, Qâzi ibn Kâshef al-Din Mohammad Yazdi, writing about china root, tea, and coffee, frequently referred to Western authors, although he did not mention their names,[25] but before the nineteenth century, the encounter was sporadic. Persia was not of sufficient interest to Britain, for example, to warrant a permanent diplomatic relationship.[26] The Industrial Revolution and scientific progress in Europe in the eighteenth century, which could have had an important impact, coincided with the fall of the Safavids (1722) and the rise of tribalism and civil wars.[27] To this was added the rise of Russian imperialism. With increasing Russian expansionism in the second part of the eighteenth century, Britain preferred to deny Iran any facilities that Russia might use to advance toward India.[28]

It was not before peace was partially restored by the end of the eighteenth century that Western countries resumed their relationship with Iran, and Western physicians arrived with the first diplomatic missions in the early nineteenth century. However, these physicians did not have a strategy to engage with local medicine; their presence was merely required because unaccustomed climate and local diseases threatened European missions.

The first European physician to enter nineteenth-century Iran was assistant surgeon George Briggs, who accompanied John Malcolm's first mission in 1800. He returned to India the following year.[29] Probably the second European physician to visit Iran under the Qâjârs was Dr. Salvatori, who was the first physician and surgeon to General Gardanne's mission, which arrived in Iran in November 1807 to secure Persian help against British India. The chief physician of the court of FathʿAli Shah in this period was Mirzâ Ahmad, who treated the Frenchman Amédé Jaubert, who had caught a fever shortly after his arrival to Tehran on June 5, 1806 (to prepare for Gardanne's mission).[30] Although Dr. Salvatori stayed in Iran until the departure of the mission in February 1809, he does not seem to have been active apart from opening a dispensary in Tehran, about which no detail exists. The only documents we have of Salvatori is a report of February 1808, mainly on the epidemic of plague that was raging in Anatolia, a clinical report on a Lieutenant Bernard, who finally succumbed to plague near Khoy, one on vaccination for *ophthalimia*, which was widespread in Iran, and, finally, one on measures to be taken to preserve the health of France's troops during a campaign through Persia.[31] After the departure of Gardanne, the remaining French delegation, Jouannin and Lajard, had no medical service at their disposal.

Unlike the French physician(s) who, due to the failure of the Gardanne's mission, did not settle in Iran, British physicians were able to do so. The first was James Campbell, surgeon to Sir Harford Jones's mission. Jones, the, ambassador of the British government, entered Tehran on February 14, 1809, a day after Gardanne's mission had left. On his way to Tehran via Bushehr, Harford Jones took with him Dr. Andrew Jukes, a surgeon of the Bombay Establishment, who at this time had political duties at the Residency at Bushehr.[32] Later that year, Lord Minto, the new governor general in Calcutta, while Jones was still in Tehran, sent John Malcolm as his envoy. Malcolm's mission was accompanied by the surgeon John Cormick.[33] Jukes returned to Bushehr to fulfill his diplomatic mission, but died in Isfahan in 1821.[34] Campbell stayed in Tehran serving the British mission until he died of illness in 1818, and his post was not filled until 1821, when John McNeill, a graduate of Edinburgh and a surgeon of the East India Company, replaced him as the assistant surgeon to the British Legation.[35] Besides serving their mission, these physicians also provided medical assistance to the court and occasionally to members of the public, but they were not under Qâjâr patronage or control. Mirzâ Ahmad, chief physician to Fathᶜ Ali Shah, perceived the presence of Western doctors at court as a threat to his own position and to his medical "system."[36] This was confirmed by Salvatori, who found that his efforts to approach native doctors were fruitless: "Arrogance and interest prevent them from revealing their ignorance."[37] Even so, there were Persian physicians who adhered to Western medicine, as indicated by a manuscript on *Tebb-e jadid-e kimiyâyee taᵓâ berklesus* (*Modern chemical medicine of [based on] Paracelsus*), which was translated into Persian from Arabic in 1810.[38] When, in 1807, Salvatori visited Iran, among the authors that Persian doctors read, he mentioned Avicenna, Galen, and Paracelsus.[39]

One of the Western physicians, Dr. John Cormick, served the army of ᶜAbbas Mirzâ in Azerbaijan, and also acted as ᶜAbbas Mirzâ's personal physician. In 1813, at the request of ᶜAbbas Mirzâ, he vaccinated the children of the royal family in Azerbaijan and later, "nearly 80,000 children of the province of Azerbaijan were vaccinated."[40] He wrote a treatise, entitled "Smallpox inoculation and the need for its universal use,"[41] which was translated into Persian by ᶜAbdol-Sabur-e Khoi, on the order of ᶜAbbas Mirzâ.[42]

ᶜAbdol-Sabur also composed another treatise called *Morakkabât-e jowhariyyeh* that was based on Paracelsus's iatrochemistry. The author mentions that

> thanks to the good education of the heir apparent, several kinds of modern sciences and discoveries have appeared in Iran... and several doctors have been trained who sometimes base their treatment on Western drugs and sometimes on traditional herbal drugs.... As the European drugs from several viewpoints are superior to the Iranian herbal drugs, the prince ordered that these drugs are also prepared and distributed across the country.[43]

Local attitudes toward Western doctors differed according to social position, belief, and material needs. ᶜAbbas Mirzâ preferred English doctors to his traditional *hakimbâshi*s (chief physician), although he was accompanied by both.[44] Fathᶜ Ali Shah, on the other hand, did not seek the assistance of the British doctors, and as long as he was healthy and surrounded by his own traditional *hakimbâshis*, he did not believe in Western treatment.[45] However, with failing health and repeated illnesses toward the end of his life, he was persuaded to accept the service of Dr. McNeill. ᶜAbbas Mirzâ, who suffered from cirrhosis and a chronic anal fistula, had frequent recourse to Drs. Cormick and McNeill.[46] The populace seemed to be motivated by medical needs rather than by ideology or etiquette, and flocked to the homes of Western physicians as soon as they were informed of their arrival. When Dr. Jukes first introduced the vaccine against smallpox, ladies in Bushehr came with their children to be vaccinated. Dr. Treacher Collins, a British ophthalmologist who was invited by Zell al-Soltân, governor of Esfahan, states that as soon as it was "known that an eye doctor had arrived in Julfa from Europe... my house before long was invaded by patients of all classes suffering from eye disease." This experience was repeated throughout the country whenever he stopped on his travels.[47] Dr. Odling, of the British mission in Tehran, operated a dispensary in the grounds of the British Legation and received the poor for treatment.[48] The attitude of the local physicians toward Western medicine was informed by various factors, including economic, ideological, political, and professional. Their position changed throughout the period of our study depending on which of these interests was threatened or promoted.

Diplomacy and the Introduction of Modern Medicine

The first European scientific mission to Qâjâr Iran was sent by the Directorate of Revolutionary France, for a political purpose. This mission consisting of Jean-Guillaume Bruguière, a physician, and Guillaume-Antoine Olivier, a naturalist, was intended to negotiate an alliance with Persia against both Russia to check its advance and Britain to prevent its influence.[49] Both Bruguière and Olivier had undertaken scientific research in the Levant and Ottoman Empire before going to Iran. The choice of this mission might be explained by the experience that Bruguière and Olivier acquired during their scientific work, but it also reflects the role of science in revolutionary France. The mission, however, failed to achieve its goal mainly due to the Iranian situation. On their arrival, in September 1796, Agha Mohammad Khân, the founder of the dynasty, was engaged in war in the Caucasus, and was unable to make such a strategic alliance. The only result of Bruguière-Olivier's mission was a book written by Olivier that included some observations on customs, arts, religion, diseases, medical practice, and epidemics.[50]

France's interest in Iran by the end of the eighteenth century was determined by its relation with other European powers, particularly Britain and Russia. When Napoleon realized that the superiority of the British navy prevented an attack on Britain, he decided to fight Britain in her Indian colony. In 1807 a mission of 28 members, including engineers, a physician, and army officers, under General Gardanne, was sent to Iran in order to secure a Qâjâr alliance against Britain and to train an elite army to assist the French army in its march to India. Despite the setbacks it suffered from the political consequences of the Tilsit peace treaty between Napoleon and the Tzar in the same year, and from British pressure on the Qâjâr court, Gardanne's mission sent two officers, Lamy, captain of engineering, and Verdier, captain of infantry, to create an army on the French model.[51] However, a Franco-Persian army did not materialize. At a time when the Qâjâr state was confronting Russian expansion in the Caucasus and needed powerful allies, Iran turned to Britain and signed a Preliminary Treaty of Friendship and Alliance in 1809. Nevertheless, unlike France, in order to protect its interests in India and contain further Russian advances, Britain occupied the Persian Gulf coasts.

During the colonial era, the introduction of modern sciences followed colonial lines. However, as the colonial powers used Iran as well as Afghanistan as buffer zones, neither Britain nor France pursued a colonial agenda in Iran. This situation remained the same until the first two decades of the twentieth century, despite the secret agreement between Russia and Britain, dividing Iran into zones of influence. Commenting on Samad-Khân, Mozaffar al-Din Shah's minister in Paris, Maurice Spronk noted: "To our sense, such concerns are exaggerated! Not only is the dismemberment of an old empire not an easy task, the territorial integrity of Persia is guaranteed until new conditions arise—and probably for long time—because of incontestable advantages its potential conquerors have in preserving Persian independence."[52] Iran was in a weak position and experienced foreign interventions and occupations, which resulted in territorial loss but not in colonization, or even "semi-colonization."[53] The project of the great powers was decided by their rivalries outside Iran. Iran was a means and not a goal. Napoleon's plan to invade India through Iran came in consequence of its war with Britain. All treaties of alliance between Britain and Iran, and France and Iran, were based on external interests and Iran could never create a sustained relationship with them on the basis of its own interests. The Qâjâr state negotiated with the three major powers, Britain, France and Russia, endeavoring to strike a balance in their influence in order to reduce their power of intervention by playing off one against the other. According to James Morier, FathᶜAliShah refused to enter into war or permanent peace with any one of the powers at the expense of any other, unless they were prepared to help Iran in the military field or else pay money.[54]

ᶜAbbas Mirzâ, the heir apparent, endeavored to introduce modern sciences and techniques in Azerbaijan, but the Qâjâr government as a whole in that period had no clear strategy of modernization, not least because it had neither the material means of implementing reforms nor the expertise.[55] As a result, any attempt in this direction took on an ad hoc character and had no long-lasting effect.

In October 1811, at the end of Harford Jones's mission to Iran, ᶜAbbas Mirzâ requested that Jones take to England two Persian boys, Mohammad-Kâzem and Hâji Bâbâ Afshar, for schooling. According to D. Wright, the two boys knew no English and were unable to read or write in Persian, a fact that suggests their study in England was not a thought-out project of modernization. No bursary or funding was assigned to them, so the students were left at the mercy of events.[56] Their fields of study were painting, for Mohammad-Kâzem, and medicine, for Hâji Bâbâ Afshar, with no scheduled activity. In fact, as Denis Wright states, Harford Jones "dumped the two Persians at the doorstep of the Foreign Office," and nearly two years were wasted before a proper program was drawn up for them, by which time Mohammad-Kâzem had died of consumption (March 25, 1813).[57] At the time that ᶜAbbas Mirzâ proposed to send another group of students to England in 1815, in the aftermath of Waterloo, the British government was no longer interested in an alliance with Iran. Britain abruptly withdrew its military mission (led by Lieutenant Colonel D'Arcy, who had drilled Persian troops). ᶜAbbas Mirzâ requested that D'Arcy take ten students with him. D'Arcy agreed upon five, but he secured ᶜAbbas Mirzâ's commitment to pay £1,200 annually toward their studies.[58] However, this proved to be inadequate, and more than a year into their stay in London, the students, having learned next to nothing, threatened to return home. Arrangements were finally made for them to continue their studies according to a program, but again due to diplomatic circumstances, in May 1819, Mirzâ Abol-Hasan Khân, Ambassador of the Shah, brought to London instructions that all five students and Hâji Bâbâ return to Iran. Accordingly, all but Mirzâ Ja'far, who was allowed to stay to complete his medical studies, returned to Persia.[59] It is said that Mirzâ Ja'far tried to delay his return by making every possible excuse, until he "received a sharp note from the Foreign Office telling him leave before his allowance was cut off."[60] Whether or not Mirzâ Ja'far was a tiresome person, his departure echoed the experience of the four students a year earlier, and indicates a lack of British official interest in the modernization of Iran or treaty arrangements.

After the death of 'Abbas Mirzâ, his son Mohammad Mirzâ, the governor of Tabriz, entered Tehran with the military and pecuniary help of the British mission to counter opposition from his rivals. This assistance awakened the attention of Russia to match British influence at the court of the new Shah. Mohammad Shah (r. 1834–1848) suffered from chronic gout and often fell ill. English physicians were frequently requested, not only by the Shah, but also by members of his government and his harem. This provided the British mission with an opportunity to penetrate the inner court, and secure British influence over the Qâjâr state. That is why British physicians often fulfilled political duties and reached high diplomatic rank as with Dr. McNeill and later Dr. Riach, a practice that also responded to Persian custom since pre-Islamic times.[61] Political influence favored British medical provisions, and medical advice was used to leverage political influence. Was it not for his numerous medical services to the court of Fath'Ali Shah that Dr. McNeill, surgeon to the British mission, was appointed British Ambassador in Iran?[62] Dr. Riach, the mission's surgeon in Tehran, also assisted Mohammad Shah in the capital, and Dr. Bell remained at the court of Tabriz. Despite his position, Dr. McNeill continued to offer medical services to the Shah and his court. However, the political clash between the two countries following the military campaign to Herat in 1836 undermined the favorable position of British doctors. Dr. Riach, who accompanied Mohammad Shah on his march toward Herat, was dismissed, alongside other military officers, and from 1838, with this rupture in diplomatic relations between Britain and Iran, British doctors "lost the position they had enjoyed as the royal family's closest medical advisers."[63]

The situation had changed, however, by the end of Mohammad Shah's reign. Britain was no longer willing to provide his successor with financial or military help. On the other hand, the Qâjârs were dissatisfied that the British Legation, finally established in Tehran in 1809, was in fact a representative of the East India Company and not the British government. This was the beginning of a process of diplomatic deterioration that reached the point of no return with Britain's opposition to the Herat campaign.

In such a situation, the heir apparent was bound to appeal to Russia. Colonel Shiel, the chargé d'affaires of the British government in London, thought that this would advance Russian influence. Realizing that Russia and Britain used medical assistance as leverage for political influence, Hâji Mirzâ Âghâssi, the new prime minister, persuaded the Shah not to use British doctors, including Charles Bell, the British Legation's physician in Tehran from 1835 to 1845.[64] Soon after he came to power in 1836, Âghâssi requested for a physician from the French government and at the end of 1837 Dr. Barrachin received a three-year mission to establish a school of medicine in Iran. On his way to Persia, he stopped in Istanbul with gifts for the Sultan and to do business with the Ottoman government. But Mustafâ Rashid Pasha, the minister of Foreign Affairs, whom Barrachin had met in Paris, advised him against going to Iran and invited him to stay in Istanbul instead. Having obtained the consent of the French ambassador, Barrachin began there the work that he had planned to undertake in Iran, namely, the creation of a medical service for the military, and a school of medicine in Istanbul.[65]

Despite his failure to engage a French doctor, Âghâssi did not have recourse to British or Russian doctors, and even dismissed Mirzâ Bâbâ, the chief physician (hakimbâshi) to the Shah, who had been educated in England and who was reportedly an Anglophile. After placing the ailing Shah under a Jewish doctor, whose traditional remedies, according to Sheil resulted in "a severe paroxysm of the Shah's disorder," Âghâssi called on a Frenchman, Dr. Labat, who happened to be in Tehran at the time and also requested a qualified physician from the French government. But despite high remuneration and decorations, Dr. Labat left Tehran before the Shah's illness became critical.[66] Although British and Russian physicians were consulted and succeeded in restoring the Shah's health, they never regained the presence and influence they had had before Âghâssi. In fact, despite the poor impression left by Labat's departure, François Guizot, the French foreign minister, was asked to send another physician, "a physician that should be deeply knowledgeable about all aspects of medical science and principally knows how to treat the arteritis and the gout and all requisite medicaments."[67] Dr. Ernest Cloquet, a permanent doctor for the Shah, arrived in Tehran in the spring of 1846.[68]

Âghâssi's reluctance to have recourse to Britain can be explained by the failed attempts in this direction under ʿAbbas Mirzâ, as explained earlier. His recourse to France, on the other hand, was an attempt to revive a relationship prematurely interrupted under Napoleon. In December 1808, three months before leaving Iran, General Gardanne wrote to Comte de Champagny (Napoleon's minister of Foreign Affairs) that "the state of affairs in this empire is such that it will always be under the influence of and dependent on the nearest neighbour who has a preponderant force at its disposal. I even believe that France cannot hope to consolidate its powers here while its armies are too far away."[69]

There were, however, two reasons that explain the Qâjâr predilection for France in the field of medical modernization: The first was that the Qâjâr elite believed a French cooperation would not aim at imperial or colonial influence. The second was that, while France was deemed without colonial goals in Iran, it represented a potential Western ally against Russia and Britain when needed and that a medical or scientific cooperation could strengthen this alliance. In his farewell conversation with Gardanne in January 1809, Fath‑Ali Shah confided that France was the only "true friend and protector of Iran surrounded by perfidious enemies."[70] France had never attempted to invade Iran, other than making a tactical or strategic alliance that permitted her to cross the country to fight Britain or Russia. The Qâjârs believed that had Napoleon not lost, he would have come to assist Persia against Russia and Britain. This assumption remained alive in the mind of the Qâjâr elite. A combination of these two motives helps to explain the pro-French diplomacy of Haji Mirzâ Âghâssi.

Hâji Mirzâ Âghâssi and the New Era of Modernization

While Prince ‑Abbas Mirzâ, under the immediate effect of war with Russia, was more concerned with military reforms, Hâji Mirzâ Âghâssi, Mohammad Shah's prime minister, conceived of the introduction of modern science as fundamental for the self-strengthening of the country. This required a new direction in the relationship with the West. Initially, Âghâssi was of the belief that a strategic alliance with France similar to that under Napoleon could be renewed and that such an alliance could help his project of modernization. The relatively good relationship with France under Âghâssi's administration indeed encouraged some French citizens to come to Iran, although with no official link to the French government. At the end of 1844, five Iranian students went to France, including "Hussen Qouli Aga [Hossein Khân-e Qâjâr] for the study of military engineering, Mirzâ Zaki for civil engineering and Mirzâ Rezâ for painting and fabrication of crystals and porcelain and each one was given a 400 toman bursary."[71] In a letter of June 13, 1845, Âghâssi requested experts in the construction of deep wells in order to plan the irrigation of uncultivated land.[72] But it was not before 1847, in the last year of his government, that Âghâssi sent Mirzâ Mohammad‑Ali Khân-e Shirâzi, the deputy minister of foreign affairs, as extraordinary ambassador to France in an attempt to secure a French alliance and to invite a number of instructors, technicians, and scientists to Iran. Mirzâ Mohammad ͨAli Khân-e Shirâzi was instructed to sign a commercial treaty with France, according to which French and Iranian merchants could be allowed to import and export with equal custom duties. He was also instructed to ask François Guizot, the French Minister of Foreign Affairs, to dispatch specialists in all branches of science for a period of seven years with the aim of training Iranian specialists. He insisted that for each science and art five or six specialists were required to educate experts. Books were to be purchased and brought to Iran.[73] But, this project never took off as the return of Mirzâ Mohammad ͨAli Shirâzi coincided with the death of Mohammad Shah and the fall of Âghâssi's government (1848). However, the main reason for this failure was that France, under Louis Philip, unlike in the Napoleonic era, held no interest in Iran. As a result, in their answer to the embassy of Mirzâ Mohammad ͨAli Shirâzi, the new French government under Cavaignac endorsed the presence of Comte de Sartiges as French ambassador in Tehran, but did not address Âghâssi's request.

Âghâssi's policy not to recruit physicians from Britain or Russia continued under his successors, beginning with Mirzâ Taqi Khân-e Amir-Kabir, the first prime minister of Nâsser al-Din Shah, who succeeded Mohammad Shah in 1848. Amir-Kabir took up the project, initiated by Âghâssi, but did not have recourse to France, either because he considered France's lack of interest particularly evident after Comte de Sartiges left Iran and ended the regular diplomatic relationship in June 1849 or because he wanted to pre-empt opposition from Britain. Consequently, he sent Mirzâ Dâvood Khân, the government translator, as an envoy to Austria and Prussia, with instructions to bring six instructors for a period of six years.[74] This was obviously a project far less ambitious than that of Âghâssi. Amir-Kabir sought only seven instructors: one for infantry, one for artillery, one for geometry, one for mines, one for medicine, one for surgery and dissection, and one for the cavalry.[75] It is true that Amir-Kabir, in establishing the Dâr al-Fonun, was also inspired by the modern system of higher education in the Ottoman Empire after the *Tanzimât* reforms of 1839 and more particularly by the *Mekteb-i Tibbiye-i Adliye-i Shâhâne,* or the Imperial Military Medical Academy, established in 1839.[76] However, Mirzâ Mohammad ͨAli Shirâzi, who Âghâssi had sent to France to recruit instructors of modern science and who was appointed by Amir-Kabir as the Minister of Foreign Affairs, might have encouraged Amir-Kabir in the introduction of European science. Through Mirzâ Mohammad ͨAli Shirâzi, who was also appointed as head of the Dâr al-Fonun, we can see Âghâssi's footprint in the creation of this school.[77] Furthermore, a number of instructors who joined those employed from Austria and Prussia were already active in Iran under Âghâssi's government, including Dr. Cloquet from France, Borowski from Poland, who taught geography and French, Bohler, instructor of mathematics from France, Richard from France, and Schlimmer from Holland, who taught pharmacology.[78]

State Medical Institution

There was another major difference between the influence of British and Russian physicians and the physicians of other European countries. While the former were principally acting under the authority of their respective legations, and only assisted members of the court or the public, the latter were mostly employed or sponsored by the Qâjâr state itself. Throughout the nineteenth century, British doctors were attached to their diplomatic and military missions and later to the telegraph stations. British physicians like Dr. Dickson and Dr. Cormick junior fulfilled duties as first and second physicians of the British Legation in Tehran and were not doctors to the Shah or paid by him. For a short time, Dr. Cormick, the son of the personal physician of ʿAbbas Mirzâ, attended the court of Nâsser al-Din-Mirzâ, governor of Tabriz, but at the death of Mohammad Shah and the accession of Nâsser al-Din Shah to the throne, he was relieved of his post while Dr. Cloquet was confirmed as senior physician to the court.[79] Another British physician, Dr. Dolmage, who had arrived in Persia in 1859, held a teaching post at the Dâr al-Fonun for a short time, owing to his proficiency in Persian, but soon he was dismissed. Other members of the Western medical corps in Tehran were from Austria or the Netherlands, countries that had no precedence in the Qâjâr state. There were, however, exceptions, and we also find British instructors who were employed at the Dâr al-Fonun, like the Englishman who replaced Captain Charnota in mineralogy, who died in 1854.[80] But this also depended on who was at the head of the Dâr al-Fonun or the Ministry of Education. Mokhber al-Dowleh, who was educated in Germany, employed a mine engineer from England, as well as a physician, Dr. Isidor Albo from Berlin, who was also hired to teach medicine.[81]

The political factor had an impact on the rank and responsibility of European doctors and the role they played in Persian medical modernization. Although British physicians attended the court and collaborated with French doctors in the treatment of the Shah, the latter were given first priority at the court. When Mohammad Shah suffered from erysipelas in 1848, Dr. Cloquet called Dr. Dickson and the physician of the Russian Legation.[82] After the accidental death of Dr. Cloquet in 1855, other French physicians succeeded him: first, Dr. Barthelemy and, then, Dr. Tholozan, who was nominated as the physician to Nâsser al-Din Shah in September 1858.[83] Initially, the priority given to French rather than British doctors at court was entirely political, as no distinction was made between French and British medical schools. During the first tour of Europe by Nâsser al-Din Shah, Dr. Tholozan accompanied the Shah, the principal ministers, and his harem, and Dr. Dickson was with the rest of his party.[84]

C. Elgood attributes the decline of British medicine in Qâjâr Iran to the fact that, with the establishment of the telegraph line, the Legation doctor was no longer a candidate for the highest post in the mission, insofar as the telegraph made it possible to consult Whitehall on any question and to receive a decision within 24 hours.[85] Nevertheless, the British government did not stop employing the medical service as leverage for attaining its political and commercial goals. It was through Dr. Dickson, the physician of its Legation in Tehran, that the contract for the Indo-European Telegraph line was signed in 1867. Even so, the Qâjârs did not accept that the line should be built by Britain, but wanted Iran to construct the line and Britain to maintain it.[86] The telegraph company created a new opportunity for the medical presence of Britain in Iran, but British physicians never regained the prominence they previously enjoyed.

From the outset, the Qâjâr state was directly involved in the introduction of modern medicine to Iran. Physicians who came without a state connection, like Dr. Dolmage, did not succeed. The Dâr al-Fonun's instructors, including Dr. Polak, Kazollani, and Schlimmer, were invited to Iran from Austria, Italy, and the Netherlands and worked under state or court patronage.[87] Sending students abroad required the cooperation of the host country. The enduring distrust between the Qâjârs and Britain made such cooperation difficult, as it was the host country that provided the medical literature and medical education. The policy inaugurated by Hâji Mirzâ Âghâssi in the 1840s continued for more than a century and left a legacy of French medical influence in the country.[88] The "transition from the medicine of Galen and Avicenna to the medicine of Harvey and Pasteur," to use Elgood's words,[89] was largely assumed by French medicine. It is true that, as Ackerknecht indicates, in the nineteenth century Paris was the Mecca for medical students of all nations.[90] It is, however, obvious that this was not the sole criterion that decided the sending of Iranian students to Paris.

However, if political reasons were behind Âghâssi's agenda of modernization through French assistance, the continuation of this trend was rather informed by developments of an intellectual and organizational nature. Along with direct commercial and diplomatic relationships with the West and a certain degree of economic improvement, a new class of intelligentsia emerged and assumed responsibility for the transfer of modern sciences. The matrix of this development was the court medical institution that acted as a crucible for social and professional networking. Court medicine provided physicians with an institutional home. Traditional physicians held a strong position at court, even though the state encouraged the development of modern medicine and sponsored the activities of Europeans and European-trained Iranian doctors. Prominent physicians who flourished within the context of state sponsorship and relations with the West illuminate the special influence of French medicine in medical modernization.

The privileged place that France enjoyed in Qâjâr Iran in the cultural domain was not so much the result of the French government as it was the work of the Iranians themselves. The seeds of "francophily" were sown in Iran by Napoleon, as the princes admired his conquests, and his history was translated into Persian, and read for the Shah. The introduction of French culture began under Hâji Mirzâ Âghâssi, with the return of the Lazarits and the construction of their schools in Iran from 1837 onward.[91] France endeavored to establish its cultural influence in Iran to make up for its failure in politics and economics, and to use its cultural influence for political and economic purposes. But whatever French intentions might have been, it was the Iranian intelligentsia that decided to build on French ideas.

However, it is not certain that the medical presence of France in Qâjâr Iran paved the way for French commercial or political interests or that France had a definite project of using this presence to gain an economic foothold in the country. When in 1876, Dr. Tholozan attempted to obtain the Kârun River navigation concession from the Shah, following the example of the Reuter concession, the French government opposed such an enterprise.[92] The primacy of the French language only favored and furthered this cultural influence. In the realm of political ideas, J.-J. Rousseau was a favorite. In medicine, most books translated into Persian were from French authors. This gained momentum under the Constitutional Revolution (1905–1909), when much French literature, including novels and the history of the French Revolution, were translated into Persian.[93] There were opportunities in England, Germany, and elsewhere in the West for the Iranian intelligentsia to study. But, as Homa Nategh indicates, "to avoid Russia and Britain, there was no other solution for Iran than having recourse to France."[94] From the mid-nineteenth century on, the cultural relationship with France had its own dynamics. It seems that the rationale was the concern to form a social group that would intellectually be in harmony, enabling them to develop modern science and education in the country.

Local Physicians and Their Encounter with Modern Medicine

From the early nineteenth century, local physicians adopted different attitudes toward modern medicine when it was first introduced through the European missions. Mirzâ Ahmad Hakimbâshi, chief physician to the court of FathᶜAli Shah, was against Western medicine, but some of his contemporary court doctors helped Dr. Cormick propagate vaccination and translated his treatise into Persian, and others translated the works of Paracelsus, which contradicted humoral theories.[95] The medical profession included a wide spectrum of physicians with different persuasions. This section will examine the three major groups of physicians in their encounter with modern medicine.

Traditional Doctors

All learned Persian physicians, including the traditionalists, gradually became acquainted with modern medicine but their reaction varied according to their vested interests, their beliefs, or their perceptions. Mirzâ Mohammad-Taqi Shirâzi, one of the traditional doctors under the Qâjârs, was known for his staunch hostility toward Western medicine not so much because he was the most orthodox, but rather because of his critical reaction to the increasing use of *jawhariyât* (chemical drugs) introduced from the West.[96] Shirâzi, who was also a mathematician, was probably the most prolific and dynamic traditional court doctor in the nineteenth century, writing on several aspects of medicine, including the epidemics of plague and cholera. Several of his treatises were written in Arabic, including the *Jowhariya, Tâᶜuniya, Kowthariya,* and *Resâleh-ye Bohrâniya*.[97] Mirzâ Mohammad-Taqi Shirâzi, who was initially known as Hâji Âqâ Bâbâ-ye Hakimbâshi-ye Rashti (as he was living in Rasht for some time), served under FathᶜAli Shah, Mohammad Shah, and Nâsser al-Din Shah. He received the title Malek al-Atebbâ in January 1863, as well as the military rank of Second General (*sartip-e dovvom*),[98] from which time he was called Hâji Aqâ Bâbâ-ye Malek al-Atebbâ (prince of physicians). Shirâzi adopted a polemic style in his writings. Just as in his *Tâᶜuniya* (treatise on plague), where he lambasted physicians who avoided plague patients, in his *Jowhariya*, he criticized chemical drugs. In February 1831, he wrote *Resâleh-ye Bohrâniyeh* in response to Mirzâ Ahmad-e Tabib-e Tonekâboni, one of FathᶜAli Shah's physicians, who had questioned Shirâzi's method of prescribing laxatives in the fourth and sixth days of a fever but not on the third and fifth days.[99] Shirâzi pioneered a pathological and symptomatic comparison between cholera and *heyzeh*, explained in Chapter 1. In fact, his work echoed the debate fought in Europe over the diagnosis of cholera (Figure 2.1).

Figure 2.1 Mirzâ Mohammad-Taqi Shirâzi (Malek al-Atebbâ) (Aqâ-Bâbâ Rashti)

Despite his opposition to Western medicine and drugs, Shirâzi advised, among other herbal medicines, "a purgative salt that is introduced from England," which was magnesium sulfate, known in Iran as "*namak-e farangi*" or *namak-e englisi* (English salt) as it was produced in Epsom, Surrey.[100] It is interesting that Shirâzi preferred this product introduced from Europe to the

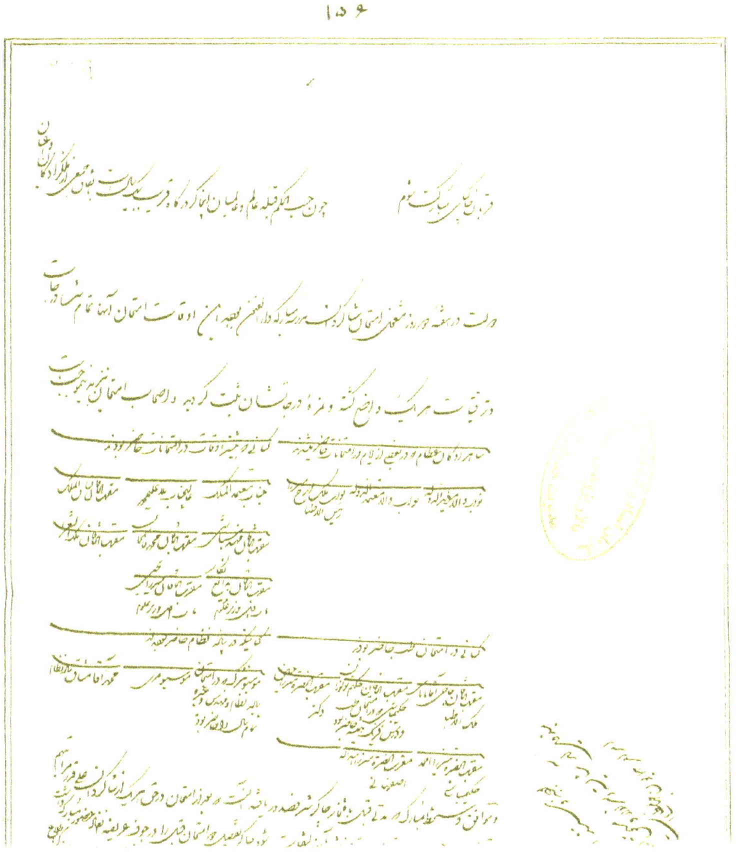

Figure 2.2 Malek al-Atebbâ and Tholozan, examiners at the Dâr al-Fonun

sodium sulfate that was produced locally, called "the artificial European salt" (*namak-e farangi-ye masnuʿi*). According to Polak and Schlimmer, it was not possible to extract a chemical compound similar to Epsom salt from minerals in Iran. The local "sodium chloride" contained elements of magnesium sulfate, and could not be consumed because of its extreme bitterness.[101] Shirâzi subscribed to this drug introduced from Europe because of its unmatched efficiency. Shirâzi, as other traditional physicians, worked with Western physicians at the court and at the Dâr al-Fonun. In an undated document, probably in the early 1870s, he is one of the examiners of students, alongside Dr. Tholozan, Mirzâ Hossein-e Doktor, and Mirzâ Ahmad-e Hakimbâshi (Figure 2.2).[102]

A similar attitude toward the West was experienced in the use of mercury (*zanbiq*). Diagnoses were based on the principle of hot and cold and therapeutics on administering the drug of a temperament opposite to that of the disease. Consequently, mercury, which was considered cold, was not used for "cold" diseases. According to an anecdote recounted by James Morier, when Fath‹Ali Shah's chief physician, Mirzâ Ahmad (Tonekâboni), realized that the British Legation's physician had cured the indigestion of the prime minister, he endeavored to procure the tablet prescribed in order to find out its components to make use of it in his own practice. Through Hâji Bâbâ he found that this remedy, namely, calomel, was none other than mercury and concluded that mercury was cold and so could not cure an illness caused by the cucumber and lettuce that the prime minister had eaten, which were also cold. While the chief physician tried to discredit the European doctor and his treatment, he also attempted to appropriate the new technique. Both calomel and magnesium sulfate were introduced to Iran by British physicians, Drs. Campbell and Cormick, during their missions in Iran under Fath‹Ali Shah.[103] By his actions, Mirzâ Ahmad acknowledged the effectiveness of mercury, a fact that was not without effect on his ideas. Indeed, in his books, Mirzâ Ahmad "occasionally deviated from traditional practice."[104]

The other traditional court physician of importance under the Qâjârs was Mirzâ Mohammad-Kâzem-e Rashti (ca. 1820–1905). Rashti, after the death of Shirâzi in 1872, inherited the title of Malek al-Atebbâ. Later, in 1888, when his book, *Hefz al-Sehheh-ye nâsseri* was lithographed, he also received the title of Filsuf al-Dowleh.[105] Rashti may have written other books, but the only volume of his that remains is the *Hefz al-Sehheh*, written between November 1859 and January 1860, while he resided in Anzali.[106] From 1861 onward, Rashti "in recognition of having written his book of *setta zaruriya* [*Hefz al-Sehheh*] and writing (medical) journals" joined the personal physicians of the Shah and was allowed to attend at the Shah's lunch table.[107] When Rashti was the physician to the Governor of Guilân Amir-Aslân Khân Majd al-Dawla, the maternal uncle of Nâsser al-Din Shah, he also acted as the physician of the army stationed there. It was during this period that he wrote his book. Rashti was in contact with other European doctors through his participation in the Sanitary Council (*Majles-e Hefz al-Sehheh*), of which he was a member.[108]

Unlike Shirâzi, Rashti did not so much oppose modern Western medicine as consider it imperfect.[109] In his mission to the Anzali Port, he referred to a European doctor of the army, most likely Dr. Schlimmer of the Netherlands, who had been sent to Guilân by Mirzâ Taqi Khân Amir-Kabir in December 1849.[110] Rashti's criticism was that in Europe the treatment of disease was often limited to drugs like antimon (antimony, a kind of vomitive), calomel (or mercurous chloride), and quinine. He did not explicitly refute the value of these drugs, but he stated that they should not be prescribed for certain diseases. He claimed that the European doctor prescribed antimony to soldiers affected by *zât al-janb* (pleurisy), who as a result died.[111]

Both Shirâzi and Rashti were close to the Shiite ulama. Toward the end of his life, Shirâzi left Iran for Karbala in Iraq, where according to Astarâbâdi, he was practicing during a cholera outbreak.[112] It is significant that Rashti classified the medical profession according to the Shiite hierocracy in two groups: the *mojtahed,* who reached the *ejtehâd* by mastering medical knowledge to perfection, and the *moqalled* (or followers), who had not reached, or were not able to attain, perfection and therefore had to seek advice from their *mojtahed*.[113] It was in the newly established *Ruznâmeh-ye ᶜElmi* (*Journal of Science*, 1877) that Rashti first published his ideas, before his book was lithographed about ten years later. As we will see later, his collaboration with the *Ruznâmeh-ye ᶜElmi*, hereafter *RE*, seemed to have mitigated his orthodox opinion on the question of surgery. Orthodox humoral medicine considered that all fevers (*hommâ*) could be treated without manual (surgical) operation. However, Rashti, warned that while in treating tonsillitis bloodletting is necessary, the ulcers in tonsillitis require surgery, as humoral treatment (*eslâh-e mazâji*) does not suffice.[114] In this single step, Rashti bridged the gap between humoral and anatomical pathological medicine.

The other traditional doctor who worked with modern institutions and who deserves mention is Mirzâ Mohammad-e Râzi-ye Kani Fakhr al-Atebbâ, a court physician and a pupil of Mirzâ Ahmad-e Tonekâboni, chief physician to FathᶜAli Shah.[115] Kani wrote a treatise on cholera and plague, in which he claimed that these epidemics had existed for centuries, and there was nothing new to discover as previous physicians had thorough knowledge, and if they had only touched upon them briefly, it was because the rarity of these diseases meant that it was not necessary to go further.[116] With the exception of Epsom salt (*namak-e farangi*), which he advised as a treatment for cholera, Kani does not refer to any European authors or drugs.[117] His references included Avicenna, Hippocrates, Râzi, Galen, and Elyâs, and his analysis was based entirely on humoral and miasma theory. Shirâzi, Rashti, and Kani were all members of the sanitary council (*majles-e hefz al-sehha*) who sat together with Dr. Tholozan and other Western-trained doctors.[118] However, Shirâzi and Kani retained their official view on epidemics without referring to infected water as a potential means of transmission, an idea that was quite widely acknowledged at that time.

Hybrids: Traditional Physicians Studying Modern Medicine

A few years before the inauguration of the Dâr al-Fonun on December 29, 1851, Dr. Cloquet taught modern medicine to physicians at his home. Eʿtemâd al-Saltaneh, reporting the death of Amir-Kabir (January 11, 1852), who had established the Dâr al-Fonun, adds that Dr. Cloquet, hakimbâshi (chief physician), "was teaching anatomy in his house and this was the beginning of the development (*takmil*) of this science."[119] These anatomy lectures seem to have been descriptive and not practical or based on dissection, which remained prohibited throughout the nineteenth century in Iran.[120] Nevertheless, the term "manual medicine" (*tebb-e yadi*) was increasingly used in medical discourse. By *tebb-e yadi*, it was meant surgery, which became necessary in military medicine. The practice of surgery brought the importance of anatomy to the fore, inspired by the work of anatomists like Drs. Polak and Cloquet who operated on gunshot wounds, or performed common surgical procedures such as extracting gallstones.[121]

Far from being a new subject, anatomy (*tashrih*) constituted one of the major chapters of each medical compendium in traditional medicine. According to ʿAbd al-Razzâq, who followed Avicenna's and Jorjâni's writings on this question,[122] anatomy was a prerequisite for the study of medicine. "If a physician does not know anatomy, in other words, the exact location of each organ in the body, often he makes a mistake in the treatment of diseases."[123] However, even though theoretical medicine acknowledged the necessity of practical anatomy as part of medical knowledge (*tebb-e nazari*) and even though it was recommended by all learned physicians, this did not necessarily mean that they put it into practice. This paradox echoes the dual meaning of the term *tashrih*, which semantically combines both explanation (literal description) and dissection (cutting an object into slices).[124] But in traditional medical literature, *tashrih* means merely the description of an organ based on knowledge provided by earlier sources. Such "descriptive" *tashrih* was founded on classical works like the *Tashrih-e Mansuri* of Ebn Elyâs and its figurative illustrations.

As a result, the relationship between medicine and anatomy also was theoretical, and did not benefit surgery, which was in any case considered a filthy job by learned physicians. Some changes, however, began to appear among local doctors' view on surgery. In 1877, Mohammad-Kâzem-e Rashti Malek al-Atebbâ acknowledged the role of surgeons in the treatment of some inflammations for which restoring humoral balance was not sufficient. He warned colleagues that throat diseases could not be cured by humoral treatment (*eslâh-e mazâji*) alone, but also through surgical procedures, even though he emphasized that for such a disease, he prescribed humoral treatment and referred the ulcer to a surgeon.[125] Rashti's advice indicates a significant change because according to traditional medicine "all ulcers and inflammations occurred in peace time require humoral treatment."[126] In other words, surgical operations were limited to injuries caused in war.

The significance of Drs. Cloquet and Polak's anatomy lectures reside in the fact that they were going to introduce the intrinsic relationship between medicine and surgery from the prism of modern medicine.[127] Two years after he began teaching at the Dâr al-Fonun, Dr. Polak authored a textbook on anatomy for his students. When challenged to explain how practical anatomy was possible given the fact that dissection was prohibited he distinguished his "practical" anatomy from the traditional *tashrih* ("descriptive" anatomy) by answering that his anatomy lectures were based on mummified human bodies, including bones, veins, nerves, and muscles, which he had brought with him from Europe, adding that some of the dissections he carried out for his students were on animals (most probably on sheep and goat).[128] Thanks to Polak's lectures, medical students became increasingly interested in dissection of the human body to the extent that they volunteered to find skulls for their work.[129] The strength of tradition and religion, however, was such that the principle of "learning by doing" that Cloquet and Polak wanted to introduce remained elusive.

In these conditions, the Dâr al-Fonun anatomy did not go far beyond the texts, although the content of the texts did change by the end of the nineteenth century. What was now understood by *tashrih*, however, was not descriptive anatomy, but "dissection," even if this was not fully practiced. In a treatise translated from an anonymous French book on general pathology (*pathologie générale*), the translator observes that "for pathology and physiology we need *tashrih* (dissection). However, since in Iran dissection of the body (*tashrih-e abdân*) is forbidden, we explain here all severe illnesses that have been analysed based on dissection undertook in Europe."[130]

One Persian physician, who seems to have studied Western medicine before the creation of the Dâr al-Fonun, and very probably via Drs. Labat and Cloquet, was the anonymous author of a commentary[131] on Polak's treatise on diseases affecting soldiers.[132] He states that he has been practicing medicine based on the Canon since 1833;[133] in other words, toward the end of Fathᶜ Ali Shah's rule. The significance of this commentary (MSS 506) is that while it is fundamentally based on Galenico-Islamic medicine, it refers to the anatomical pathology of contemporary Europe, using French authors and French medical terms. Considering that he was a senior doctor by 1845, when the first group of Persian students was sent to France, he could not have acquired his knowledge of French medicine from studying in France.

The first group of students who were sent in 1845 included Mirzâ Rezâ, and Mirzâ Yahyâ, who aimed to study medicine but returned in 1848, following the revolution in France and diplomatic mishaps between the two countries.[134] He also mentions that he worked with "a certain European doctor," who could be Cloquet or Foccetti, the instructor of pharmacology who accompanied Polak.[135] Considering that he was a senior *hakimbâshi*, was his interest in modern medicine driven by mere intellectual curiosity or professional concerns? Or was he in contact with modern medicine because he attended Dr. Polak's lectures?

After the death of Dr. Cloquet in 1855, Dr. Polak became the first *hakimbâshi* of Nâsser al-Din Shah. The nomination of a European doctor as the first chief physician since the administration of Hâji Mirzâ Âghâssi angered the traditional court doctors. The author of commentary on Polak begins his book in these terms:

> I have heard that Dr. Polak the Jew who is now the first *hakimbâshi* at the court and superior in rank to all European and Iranian physicians, has claimed that the illnesses that the soldiers stationed in the barracks of Tehran are often subject to are of three kinds only and none of these diseases are now beyond the power of modern medicine introduced by the Europeans. Therefore His Majesty asked the Prime Minister to see into this issue by receiving Dr. Polak. In a meeting with the Prime Minister, Polak requested that the author of the book (MSS 506), together with the army physicians, should attend his lectures at the Dâr al-Fonun and learn about the new treatment and commit themselves that they should not detract from that treatment and that if they do otherwise they should be subject to punishment.[136]

This echoed the fears of Mirzâ Ahmaq [Mirzâ Ahmad, *hakimbâshi* of Fathᶜ Ali Shah] the fictional character in Morier's Hâji Bâbâ, that the practice of British physicians Dr. Cormick and Dr. Campbell earlier in the century would "take the bread out of [their] mouth."[137]

Polak spent five months writing up his lectures in a book on four major diseases that affected the soldiers, namely, intermittent fevers (*nowbeh*), dysentery (*eshâl al-dam*), typhus (*mothbeqa*), and influenza (*sarmâkhordegi*).[138] In his response, our anonymous author (MSS 506) went into great detail on these four major diseases by adopting one of the classical forms of medical literature, commentary,[139] quoting Polak section by section and then refuting his ideas and/or elaborating on the principles of humoral medicine. For instance, in section 18 of the chapter on intermittent fevers (*nowbeh*), Polak recommends that if a regiment enters an area where this fever is endemic, the preventive measures to be taken include camping in a dry area, avoiding humid and cold weather, and using quinine that eliminates or reduces humidity. In his commentary, the author refers to the six non-naturals, the most important of which, he states, is air. "For this reason Polak recommends dry air and land for living."[140]

Our author's point in commenting Polak's treatise is that Polak's work is an oversimplification, as diseases found among soldiers are countless and not of four kinds only. In support of this opinion he refers to the French physician August François Chomel (1788–1858), whom he praises.[141] Throughout a convoluted style and repeated digressions by lapsing into redundant descriptions accompanied by numerous Persian aphorisms, the author taps into both Galenico-Islamic sources and the work of contemporary European physicians, some of whom, due to poor transcription, can hardly be identified, including Rostan, Zemirma, Merik, Mesdikhâ, among others.[142] References to all these Western authors who represented neo-Hippocratic medicine alongside Galenico-Islamic sources helps the author to support his ideas based on humoral theory.

Nevertheless, our author was not the only one who believed in the infinite number of fevers; a lack of adequate identification of fevers caused many Europeans to have similar ideas as well. During his diplomatic mission to Iran from 1855 to 1858, Arthur de Gobineau observed that "fevers were rife in Asia and they ravaged Iran as elsewhere. Like cholera, fevers can be treated in high and elevated places... the variety of fevers are countless: from the fever of Guilân that takes away the patient in the third access to intermittent fevers that can last several years, there is an infinite nuance of fevers, all abominable."[143]

In his *Persien (Safarnâmeh),* Polak indicates that "after being appointed personal physician to the Shah, due to his frequent trips to accompany him, the hospital's management fell into the hands of irresponsible and conscienceless hakims." He wrote the book in question, *Resâleh dar moᶜâlejât...,* which was the object of criticism in MSS 506, for the attention of the commanders of the army after he discovered the mistreatment of soldiers by army physicians both at the hospital and in the field. Polak claimed that his recommendations decreased troops' mortality.[144] This statement confirms conflict between Polak and local physicians. The author of MSS 506 voices clearly his concern about the issue of rank and institutional position. He indirectly criticizes the court and the government's decision in choosing a foreigner as first hakimbâshi, when, he believed, Iranian doctors were more knowledgeable than their European counterparts. Rivalry between physicians in the same institution is rather a matter of course, but the historical significance of this conflict rests in the fact that it triggered an intellectual dynamic that caused traditional medicine to evolve.

There is another document (MSS 505) that seems to be also written by the author of MSS 506. In this document, the anonymous author advocates the creation of state and military hospitals in the major cities of the country, a project in line with the views of Polak. These hospitals would serve for medicine as an institutional home. Attempts to bring medical practice under control had been frequent since the Middle Ages. In 1851, the state stipulated that in order to prevent unskilled people from entering the army as physicians, Dr. Kazollani, the chief physician of the army (*hakimbâshi-ye kolle nezam*), should examine and assign physicians to each regiment (*fowj*). Our author proposed a different approach, an establishment through which the state could control both medical practice and training. Unlike examination, which aimed to regulate the activity of most physicians, whether skilled or not, this proposal warned against the inadequacies of most physicians, and [instead of assessing them in order to allow them to practice] it advised that it was not possible to train enough doctors in a short time, in the traditional way, for the army and society at large. However, a few knowledgeable doctors could serve the army if they worked at hospitals, and through hospitals, more physicians, pharmacists, and surgeons could be trained.[145] In other words, hospitals could concentrate resources and provide more efficient service. Beyond the hospital, pharmacists and the quality of their products could not be controlled and contaminated drugs could cause diseases or even death. According to the author, physicians should spend one or two years working in hospitals before they qualify.[146] Although the author prides himself on the history of Iran, where, he believes, hospitals originated, he refers to contemporary European hospitals, where skilled physicians were trained, as role models.[147]

In the past, although physicians debated the opinions of others, their debates were based on the same theory and principle.[148] By the nineteenth century, however, the introduction of modern medicine gave a new dimension or format to critical literature, as it opened up new perspectives, widening the field of comparison and commentary. Mirzâ ᶜAbdol-Karim-e Tabib-e Tehrâni was one of the physicians who studied modern medicine before the establishment of the Dâr al-Fonun. He boasted having studied both traditional and modern medicine, and wrote a book in 1851 on the symptoms of disease, in which he referred to both. If the author of MSS 506 drew on European sources to defend his argument against Polak, Mirzâ ᶜAbdol-Karim was for modern concepts. He also begins with the two traditional branches of medicine, the treatment of disease, and the preservation of health (*hefz al-sehheh*). But, his first step away from traditional medicine was to support the idea that, more important than physiology, which was concerned with preservation of health, was pathology. This reading of the terms physiology and pathology according to humoral concept is peculiar to the traditional physicians of Qâjâr Iran. They understood and translated modern terms through the prism of old concepts. It provides an example of semantic transition in medical knowledge in nineteenth-century Iran. Such a perception of physiology and pathology is a far cry from the definition proposed by Claude Bernard, who classifies physiology and pathology as science of observation and science of experimentation respectively. Nevertheless, priority given to pathology over physiology in Mirzâ ʿAbdol-Karim-e Tehrâni's work echoes C. Bernard's objection to those who "subordinate pathology, a more complex science, to physiology, a simpler science."[149] Unlike Rashti, who, following Hippocratic tradition, believed that the main duty of a physician was to preserve health rather than treat disease, Mirzâ ʿAbdol-Karim maintained that pathology was the main goal of medicine, and physiology, which dealt with *hefz al-sehhah,* was secondary or complementary.[150] On each question Mirzâ ʿAbdol-Karim refers to various authors, before providing his own opinion. On the definition of "illness" (*maraz*), he refutes Galen's idea that "illness is a state where the body does not function naturally," by arguing that the "science of anatomy (ʿ*elm-e tashrih*) has shown that diseases originate in the lesion (*âfat*) of matters [tissues], which are not felt but produce change in the function of the body [symptom] that is perceived."[151] In other words, Mirzâ ʿAbdol-Karim clearly subscribes to the anatomical pathology of Broussais.

Modern Trained Physicians

The establishment of the school of Dâr al-Fonun in 1851 was initially aimed at training engineers, physicians, technicians, and translators for the state and the army. The most important result of this school was the development of Western languages, particularly French. Earlier, a few treatises were translated into Persian from Arabic and English, particularly works of Paracelsus on chemical medicine and of Dr. Cormick on smallpox.[152] From the 1850s, thanks to the students of the Dâr al-Fonun, Iran witnessed a surge in the translation of scientific texts. Most of the modern medical literature in the Qâjâr period was translated from French. The first translations were made from textbooks written by the instructors of the Dâr al-Fonun. Mirzâ Mohammad Hossein-e Afshâr translated several treatises by Polak on surgery, ophthalmology, and cholera. But there were also books written by French physicians, like the *Traité élémentaire de pathologie interne* of Grisolle, translated by Mirzâ Reza, one of the first students sent to France under Mohammad Shah, and books compiled from French sources on specific subjects, such as surgery and anatomy.

The major lesson these translations in the second half of the nineteenth century conveyed was the centrality of anatomical pathology. While Mirzâ Mohammad-Kâzem-e Rashti Malek al-Atebbâ, in the *ruznâmeh-ye ʿelmi*, had distinguished physicians from surgeons, and medicine from surgery, Mirzâ Mohammad-Hossein-e Afshâr went on to say that:

> Surgery in Greek means manual operation and consists of a science that helps to understand the source of diseases that can be felt or seen through the five senses on the surface of the body and often it is by manual operations (*aʿmâl-e* yad) that some illnesses can be treated and healed. It is therefore obvious that there is no strict border line between the two sciences of medicine and surgery... and it is necessary that surgeons master medicine.[153]

While traditional medicine did not see any need for dissection to gain anatomical knowledge, Mirzâ Hossein-e Afshâr insisted that "surgeons must master anatomy and in order to attain this they must dissect cadavers so that they do not need vivisection."[154] However, anatomical pathology, as introduced by Polak, was imbued with humoral concepts. Polak's division of fevers into inflammatory (*hommâ-ye warami*), hectic (*deq*), and humoral (*khelti*) was quite similar to the classification of traditional medicine.[155] His opinion on humoral (infectious) fever (*hommâ-ye khelti*) conformed entirely to the view of a fever "caused by the infection and putrefaction of blood, that, in turn, is caused by the putrefaction of the air as in epidemic fevers (*hommiyât-e wafeda*) or by putrid food."[156] Even so, Polak and other Western doctors at the Dâr al-Fonun introduced elements of anatomical pathology into humoral medicine. This synthesis was taken up by others. Mohammad Hossein-e Afshâr observed that "fevers belong to internal [humoral] diseases (*amrâz-e bâteneh*), but since most ulcers intrinsically or symptomatically relate to fevers, or sometime fever arises after surgical interventions, we examine them in this book [on surgery]."[157] Khalil ebn-e Mirzâ ʿAbdol-Bâqi Eʿtezâd al-Atebbâ, who translated a large volume by Dr. Albo based on Albo's lectures at the Dâr al-Fonun, elaborates on various techniques of bloodletting through the eyes of modern physiology:

> Bloodletting, *fasd*, is of two kinds, the first is applied to a place on the body (*mowzeʿi*) and is practiced by scarification or cupping (*hejâmat*), or by applying of leach. The general bloodletting (*fasd-e ʿomumi*) consists in opening a vein and letting blood out.[158] In the past, the *fasd* was called *enserâf* (diversion, inflection) as surgeons extracted blood of a place next to the infected place in order to divert the matters that, mixt with blood, was directing there. However, modern surgery and physiology [theory of blood circulation] have discovered that liquid in the body [veins] is composed of blood and aliments. Thus if blood is extracted, it does not mean that the blood of that specific place of the body is removed but from other parts of the body as well.[159]

Dr. Albo acknowledges that there are benefits to be gained from bloodletting and explains them in the following way: it reduces the blood of the whole body; thus the ill part will receive less blood than usual, and thus bloodletting is a technique that should be applied when there is (an urgent) need for preventing a large amount of blood in one of the organs, such as the brain, the lungs, and so on.[160]

It is certain that the revision of Galenic medicine originated with Harvey's theory of blood circulation but, in nineteenth-century Iran, surgery played both a crucial role in calling into question the efficiency of bloodletting and an important one in enabling physicians to come into contact with the body. Mirzâ Nosrat-e Quchâni, one of the pupils of Polak and Albo, referred to an article by Mirzâ Rezâ-Qoli Jarrâhbâshi (chief surgeon), in which Mirzâ Rezâ-Qoli claimed that he had operated on a young boy who had torn his abdominal wall and whose bowels had eviscerated. By replacing them in the abdomen and sewing up the tear, in five days the boy was completely recovered. In another case, the bowels of a boy had eviscerated when a petrol container made of glass was broken on his belly. As the bowels could not be reduced readily, the surgeon extended the lesion before reintroducing the herniated internal organs into the abdominal cavity. He then stitched it, and the boy recovered. Mirzâ Nosrat rebuked this claim by saying that in these incidences, if the peritoneum is not cut, bowels cannot come out and if it is, it is impossible for it to heal (so quickly); how, he asks, can the injury be healed in five days when the skin, the peritoneum, and the five longitudinal and traversal muscles were damaged?[161] Instead of humors, pulse, and urine the objects of study and examination were tissues and internal organs.

From the point of view of anatomy, local physicians and particularly the students of the Dâr al-Fonun began to see all branches of medicine in a new light. In pharmacology, they wanted to know the effect of drugs on tissues, and the mechanism of this effect. The humoral perception of cold and hot, humid and dry, in which the temperaments of organs and drugs were examined, was no longer sufficient. With the help of one of his pupils, J. Schlimmer, instructor at the Dâr al-Fonun, wrote a book on the quality of drugs and their effects on the body, and entitled it *Meftâh al-Khawâss-e nâsseri* (*Key to the Qualities [of Drugs] of Nâsser [al-Din Shah]*).[162] According to humoralism, drugs affected the body under the weight of the inner or core temperature (*harârat-e gharizi*), but remained unchanged and were not assimilated by the body. According to the new theory, the *dawâ* (drug) affects the tissue of the body and it functions by assimilation into the tissue or organ, such as phosphate de choux, which enters the bones and strengthens them.[163] Drugs were now classified according to their apparent effects, such as caustic, inflaming, constipating, or narcotic. The caustics, called *akal*, are those drugs that, through a chemical operation, eat away the tissue that they come in contact with.[164] Unlike traditional medicine, in which the qualities and effects of drugs were all known, new questions were generated. Thus, alongside the alteration of body tissues in a process of healing, "it also becomes clear that the drug itself is altered, but we do not know how the alteration occurs."[165]

The translation movement in the nineteenth century was far from random, and it targeted the major branches of Galenico-Islamic medicine dear to traditional physicians. Just as descriptive anatomy and surgery changed ideas in pathology and therapeutics, so the principles of experimental sciences, illustrated in these translations, changed accepted views about the preservation of health. A comparison of two treatises illustrates this development: the *Hefz al sehheh-ye nâsseri* by Mirzâ Mohammad-Kâzem-e Rashti (on traditional medicine) and the *Hâfez al-sehheh-ye nâsseri* by Mirzâ Nosrat (based on modern medicine).

Rashti's work spoke to one of the major branches of humoral medicine; the preservation of health based on the six nonnaturals (surrounding air, food and drink, exercise and rest, sleeping and waking, retention and evacuation [including bathing and coitus], and mental states (anger, sadness, joy).[166] The "six nonnaturals" refer to "circumstances external to the body that a person could control" as opposed to the term "natural," which referred to the inner body and its components, such as humors, elements, and forces.[167] Mirzâ Nosrat, who compiled his book from French sources, also divides *hefz-e sehhat* into two categories: externals, including air, water, light, food and drink, and internals, including age, sex, generation, and habit. Just as Polak and Albo began their surgery by discussing humors and bloodletting using the physiology of Claude Bernard, so Mirzâ Nosrat explained the external and internal factors involved in the preservation of health using the chemistry of Lavoisier and the physics of Newton. By referring to Lavoisier, Mirzâ Nosrat explains that what was known, in traditional medicine, as inner or animal heat (*harârat-e haywâni*) "is produced by decomposition and composition of the air inside the lungs."[168] In traditional medicine, fever was the unnatural heat (*hommâ-ye ghariba*) ignited in the heart that, via the veins that carried blood and spirits, spread throughout the body. Or, fever could be generated in another organ, and carried, by the blood and spirits *to* the heart and from there to the rest of the body, thus perturbing the natural, animal, and spiritual functions. Fever or heat caused by anger, anxiousness, and fatigue was not fever, but if it exceeded a certain level and affected humors and the heart, it transformed into fever. Other than this, the heat belonging to the body was *gharizi* (inner or metabolic) and not unnatural.[169] In anatomical pathological medicine, fevers were also identified by an increase in the temperature of the body. But, it was the anatomical signs that accompanied fevers that were of interest here. Both traditional and modern physicians reexamined humoral physiology. Not only was traditional medicine revisited in the light of anatomical pathology, but also the latter was taught with reference to humoralism. This required that some of the teachers of traditional medicine at the Dâr al-Fonun were modern trained physicians like Mirzâ Hossein-e Doktor, who taught both modern and traditional medicines,[170] and established a link between the two. Some physicians responded passively to the new ideas and the drug market, while others eagerly called into question humoral principles through detailed analysis. This process could not have happened if traditional medicine had been simply rejected and replaced with modern medicine by force. Change required adherence to both conceptual frameworks. It is not a coincidence that the first major medical treatise translated into Persian was the work of Augustin Grisolle (1811–1869). Referring to the classification of diseases, Mirzâ Reza Doktor—as he was called—who was sent to Paris under Mohammad Shah, subscribed to Grisolle's rejection of key humoral precepts. On the question of whether fever is the disease itself or a symptom of disease, he rejected the theory that fever was a mere symptom of an inflammation, because at times it caused inflammation. In some fevers, like typhoid, no physical change in the bowels and mesentery (*mâsâriqâ*) was observed after death.[171] Referring to the clinical signs of typhoid, Mirzâ Reza observed that earlier observers "ignored anatomy [of the body] and they were unable to distinguish its hidden signs and contented themselves with the apparent symptoms and, as a result, different diseases with similar symptoms were identified or they mistook one disease for another one."[172] The *mothbeqa* fever whose identity was quite hard to establish through the precepts of Avicennian literature was now described as continuous fever (*dâyem wa mothbeqa*), as distinct from intermittent fevers (*dâyer wa monfasel*), and was subdivided into seven different fevers, including typhoid.[173] Grisolle's work shed light on Mirzâ Reza's studies of humoral medicine and, in particular, helped to quantify what in Galenico-Islamic medicine was called *harârat-e gharizi* (see Chapter 1). *Harârat-e gharizi* (innate heat) was now "the natural temperature of the body in the state of health and it is 33 °C that could increase up to 40 °C." Mirzâ Taqi Kâshâni explored the concept, noting that since Hippocrates and Galen the source of *harârat-e gharizi* was considered to be the heart, but that it was not clear how innate heat was produced until 1777, when Lavoisier discovered that it is ignited when the air in the lungs is decomposed. In this operation, oxygen taken in through breathing is composed with the hydrogen and carbon of the blood.[174]

Changes in the description of diseases and their classification soon followed. Unlike Mirzâ Reza who, following Grisolle, divided fevers into five major groups and subdivided each group into several others, Mirzâ Abol-Hassan Khân-e Tafreshi, writing in 1882, noted that "nowadays fevers are of three kinds, continuous, inflammatory (*bothuri*), and intermittent (*nâ'eba*)." As the symptoms of fevers were quite similar and lead to confusion, quantification helped to distinguish them from each other. For example, the *hommâ-ye yawm* (literally a fever that, according to Hippocratic medicine, lasted one day), was now fine-tuned into different degrees. According to Schlimmer, it could last 6, 12, 24, or 36 hours and according to Tafreshi it lasted two and three days.[175] As *hommâ-ye dâyer* (intermittent or recurrent fever), presenting short-term accesses, could be confused with *hommâ-ye yawm*, if it lasted more than 12 hours, it was not *dayer* but *yawm*.[176] Likewise, Mirzâ Abol-Hassan Khân, who was encouraged by Tholozan to write his *Matla'al-tebb-e nâsseri*, introduced a variation by combining Andral's theory of the chemical change of blood in fever with the inflammation theory of Broussais. Mirzâ Abol-Hassan Khân maintained that the anatomical signs of fever were not limited to inflammation, but could also be seen from changes in the composition of the blood, characterized by the reduction of the fibrin.[177] Mirzâ Reza, who translated the first edition of Grisolle's book, does not mention Andral, even though both Andral and Jules Gavarret are referred to by Grisolle in the fifth edition of his book.[178] Anatomy furthered diagnosis. Anatomical changes occurring in consumption (*sel*), for instance, revealed that it was wrong to give a definite verdict. There was not one kind of tuberculosis, but several, and of different stages and degrees. It must therefore be possible to cure consumption at certain stages of its development.[179]

The large number of medical treatises translated from, or based on, French sources may have been a response to the idea of Dr. Tholozan, who proposed a sweeping replacement of Persian traditional medicine by modern medical literature.[180] From Mirzâ Reza, who translated Grisolle, to Mirzâ ʿAli and Mirzâ Abol-Hassan ebn-e ʿAbdol-Vahhâb-e Tafreshi, the inspiration of modern medicine in Qâjâr Iran remained both neo-Hippocratic (Pinel, Chomel, Sauvage) and in line with anatomical pathological sources, such as Grisolle, and Andral. Tholozan's project was to some extent realized. However, what Polak, Tholozan, Albo, and others considered replacement was regarded as adaptation and revision by the local physician. The translations were by no means literal, but adapted and attuned versions, and did not replace medical knowledge but instilled new concepts and questions into the medical discourse, revisiting humoral medicine, and comparing it with anatomical pathology.

In the same vein, modern European medicine did not embody a complete set of modern treatments or theories entirely distinct from humoral medicine. In his treatise on cholera, for example, the British-educated Dr. Basil in 1892 provided contradictory ideas as to the nature of disease and its treatment. Based on the experience of John Snow, cholera was thought to be transmitted by infected water, but he did not exclude the role of air. Therefore, Basil advised confining the sick inside, if they were unable to move to healthier areas with pure and uninfected air, but this idea contradicted Basil's opposition to quarantine. He stated that "infectious diseases are of two kinds: those like measles and smallpox that are transmitted from one person to the other via air and those like cholera that are transmitted via food and water. Although one cannot completely deny the role of air in cholera epidemics, recent researches indicate that it can be transmitted via water."[181] When discussing preventive measures, Basil seemed unsure about the distinction between diarrhea and cholera, even though he pointed to distinctive symptoms, such as a low and weak pulse in diarrhea and a high pulse and fever in cholera.[182]

There was some gap between anatomy-pathology as taught in Iran by Polak and Tholozan and the anatomy-pathology in Europe. Although elements of humoral theory and miasma were present in European medicine, the gap between humoral pathology and John Snow in Britain, for example, was greater than the gap between humoral pathology and representatives of modern medicine in Iran. Instead of the inflammation of the stomach, in the pathology of cholera explained by Polak, Snow referred to the reproduction of the morbid matter and to its structure similar to a cell,[183] and while Snow pointed to physical conditions, such as community houses where several families crowded in one room, or the mixture of cholera evacuations with water used for drinking and culinary purposes, as the cause of the fast spread of cholera,[184] Dr. Basil, although agreeing that water was a vector, still wondered whether air, or miasma, did not represent another factor.

In fact, this gray zone of knowledge on cholera that characterized medicine in both Iran and Europe facilitated the dialogue between humoral medicine and anatomy-pathology. Dialogue was also facilitated by the conscious sharing of ideas between traditional and modern physicians. In 1890, a health board (*majless-e hefz al-sehheh*) that included, on the one hand, Drs. Babayev, Castaldi, Cormick Junior, and Mirzâ ʿAli (physicians of modern medicine), and, on the other (Mirzâ Abdolʿali) Seif al-Atebbâ, Mirzâ Hakimbâshi, and Mirzâ Jaʿfar-e tabib (practitioners of traditional medicine), signed a guideline for the prevention of cholera, in which emphasis was laid on the cleanliness of water and the avoidance of burying the dead in shallow graves. However, we find several elements in the report that suggest both groups shared humoral concepts, such as miasma, choleric season (spring or summer), and putrid air.[185] The borderlines between Western and Iranian medicine remained blurred. This provided a favorable conceptual framework for negotiation and dialogue. Polak, for instance, stated that "as some Iranian doctors praise churned sour milk, *dough-e âhantâb* (literally boiled in iron vessel) a great deal for treating dysentery, I advise it for those who wish to use it, but if this does not work, that which I prescribe [calomel] should be used."[186]

Conclusion

Contact with modern medicine was made possible through the agency of the state. The court and later state-funded modern institutions provided spaces where European and Iranian doctors met and worked together. State patronage gave an impetus to the development of medical literature. Most medical tracts, whether modern or traditional, were written in the name of the Qâjâr princes, and particularly in the name of Nâsser al-Din Shah (r. 1848–1896). In fact, the state in Iran fulfilled the same task, in introducing modern science, that the colonial administration did in India, with the difference that, in Qâjâr Iran, the defense of traditional knowledge did not become a point of contention. Western medicine was adopted by all non-Western countries but its localization or indigenization varied according to political situation and the scale of Western influence. Transfers of knowledge have always raised questions of cultural identity,[187] particularly acute in the age of colonialism. The major aim of the trips of Nâsser al-Din Shah to Europe was to become acquainted with Western technology and science, culture, army, schooling and education, and political organization. In a letter, Nâsser al-Din Shah further explains:

> In each age or century, sciences and techniques are developed in one region or country in accordance with God's will; as 1500 years ago education and science flourished in the East, and in other times in Greece, Rome, China or India. For example, the gunpowder that today decides all wars and peace treaties in the world was originally invented in China and from there it was transferred to Europe. Also one of the presents of the Caliph Hârun al-Rashid was a clock to Charlemagne. Thus, the industry of clock making, which is now so advanced in France, was first developed in Islamic countries. Therefore it is the duty of awakened and intelligent people to adopt every science and technique that they find wherever in the world and make use of them.[188]

It is in the light of this that we can explain the extent of the translation movement under Nâsser al-Din Shah. In other words, the Shah tried to justify the adoption of Western science. Sanctioned by the Shah, modern science was no longer considered foreign. In the prefaces of almost all the medical and scientific tracts written in the second half of the nineteenth century in Iran, Nâsser al-Din Shah is celebrated for having created modern schools, and propagated modern as well as traditional sciences. Often, books that were translated into Persian were given a new title after the name of the Shah, even when the original source was mentioned. One of these books was *Jawâher al-ḥekmat-e nâsseri,* by Dr. Mirzâ ʿAli, who studied medicine in Paris. This title is highly significant, as this book, which was a translation of one or several French textbooks on general pathology, was entitled *The Quintessence of the Knowledge of Nâsser [al-Din Shah]*, associating modern medicine with the Shah (presumed to be) a symbol of Persian identity. *Meftâh al-khavâss-e Nâsseri*, written by Dr. Johan Schlimmer, on modern pharmacology and the effects of chemical drugs on tissues and organs, affords another example; this title means *the Key to the (drug) properties (of Nâsser) al-Din Shah,* in the guise of dedicating the book to the Shah.[189]

Most authors specified that they were fostered (*parvardeh*) by the state. The idea this statement conveys goes beyond what we might perceive as state patronage. In the absence of other social structures, the state provided the social and organizational framework for an emerging professional intelligentsia that would not otherwise have been feasible. This framework allowed for regular contact with Europe and European ideas. Within this context, Western doctors at the Qâjâr court played an important role. Their lectures and writings stimulated local doctors, while their presence at the court or in state-run institutions or mission hospitals prompted a sense of professionalization among local physicians. Alongside conceptual and epistemological changes, physicians endeavored to cultivate a new professional ethos. The overall transition from traditional to modern medicine was made possible through a process of social construction that will be discussed in the next chapter.

CHAPTER 3

The Reform Movement and Medical Institutionalization

Modernization and Medicine

Modernization projects in nineteenth-century Iran experienced long spans of inertia and failure, not least because they were sporadic, inconsistent, and limited to the military. The reform pattern that took shape during the consolidation of the Qâjâr regime, gripped by nearly three decades of sporadic conflict with Russia, continued throughout the Qâjâr period. The limited scope of reform was also rooted in the social and political structure of a country that, unlike other Middle Eastern or Eastern countries, experienced little direct European trade. For instance, when French merchants, since the early eighteenth century, began to promote commerce with Persia, they failed to secure the support of their governments, and Napoleon's relationship with Iran, at the beginning of the nineteenth century, was entirely for military purposes.[1] Despite the expansion of Western imperialism, contact with the West was difficult. The rivalry between Britain, Russia, and France turned Iran into a buffer zone, which rendered any substantial Western trade scarce. In the first two decades of the nineteenth century, British and French missions were accompanied by military, engineering, and medical experts. However, their activities were patchy, and owing to political changes in Europe, there was even less incentive for French and British involvement during the reign of Mohammad Shah (r. 1834–1848).

The two wars with Russia, the subsequent loss of territory in the Transcaucasia, and increasing British interference awakened the need for modern knowledge and self-strengthening, as illustrated in literary texts in the nineteenth century.² Defeat by Russia shook Iranian society, and stimulated movements against the existing system in the guise of religious reform, illustrated in the Babi movement that revised the Shiite eschatology (mahdism).³ Many Europeans visiting Iran under Mohammad Shah wrote of "the progress of Sufism, whose followers formerly dared not avow their sentiments, now openly profess their doctrines, the main object of which is to keep more to the spirit than to the letter of the law; although many have gone beyond the prescribed limits, and have become Freethinkers."⁴ Iran under Mohammad Shah also witnessed the emergence of nationalism in the form of the revival of pre-Islamic history, as one can infer from several chroniclers discussing the history of pre-Islamic Iran.⁵ Reforms attempted by Hâji Mirzâ Âghâssi responded to sociopolitical development and the emergence of a modern intelligentsia. The Iranian economy suffered from a scarcity of labor and the inability of the state to collect taxes due to a system that divided the country into fiefs, "which paid only a feeble rental charge to the state and employed instead a considerable number of people with nothing to do and a large domestics." Hâji Mirzâ Âghâssi was the first to try to reduce the power of these magnates, but with no success, he "abandoned his reform."⁶ Slow changes in medical vision and landscape appeared with the waves of reform in state organization, the military establishment, and education in the second half of the nineteenth century.

The purpose of this chapter is to examine this medical change within the framework of the reform projects, with particular emphasis on the construction of new institutions, such as medical schools, hospitals, and sanitary councils. It is from this perspective that the chapter challenges the received idea that medicine taught at the Dâr al-Fonun was representative of Western medicine. In Chapter 2, we have discussed how the persisting neo-Hippocratic elements in anatomical pathological medicine left the doors of the Dâr al-Fonun open to traditional medicine. This chapter will examine the social composition and features of modern medical institutions set up in the second half of the nineteenth century. The chapter argues that, while introducing modern science, medical institutions were both permeated by local ideas and staffed by traditional physicians. It will indicate how the latter, far from impeding reform, played an important role in the transition.

The Dâr al-Fonun's Medical School

In its expanding relationship with the West, the Qâjâr government was in need of skilled people, knowledgeable about Western politics, culture, and languages. The Qâjârs had recourse to their subjects among religious minorities, such as the Armenians or Jews, who were in contact with the West through trade or religion.⁷ The state of relative isolation from the West that the Qâjârs inherited intensified this problem. The caricature of the first envoy to the West under Fath‹Ali Shah, pictured by James Morier, is telling.

The Shah ordered Mirzâ Firuz (pseudonym of Askar Khân-e Afshâr, the first ambassador to France) to first visit Istanbul in order to "discover what was the extent of France, how many tribes it had, who was the infidel Bonaparte, what was England that was known in Persia by means of their broadcloths, watches and penknives; who was the company, and so forth." The Shah also ordered him to write a general history of the Franks. The ambassador commissioned Haji Baba to make inquiries, and Haji Baba gained information from a certain Reis Effendi, a *kâtib*, or scribe, in Istanbul. What Reis Effendi explained to Haji Baba was collected for the Shah in a handsome volume, entitled *History of Europe*.[8] Although Morier's account is a fiction based on his experience at the Qâjâr court and his engagement with the Qâjâr elites, it can provide us with an idea of the state of relationship between Iran and Europe and the state of Qâjâr knowledge on European countries in the early nineteenth century. Nearly a century later in March 1900, instigated by his Minister of Foreign Affairs Mirzâ Nasrollâh Khân-e Moshir al-Dowleh, Mozaffar al-Din Shah (r. 1896–1906) ordered the creation of a school of political science (*madrassa-ye siyâsi*) to train individuals, "so that in our relationship with other countries we know what to do and what to say."[9] The creation of the Dâr al-Fonun, or polytechnic school, was initially aimed at training officers for the army and translators and bureaucrats for the government.[10] However, far from providing public education, its aim was limited to train children of the Qâjâr nobility, and in the first place the Qâjâr princes.[11]

As the Dâr al-Fonun was established on the Western model, it faced the problem of appropriate background education. Primary and secondary education within the *maktabs* and *madrasas* were under the ulama. No establishment of higher education propagating modern Western science could be funded privately or by *waqf* without state protection, not least because it would not have survived the opposition of the ulama, who controlled the *waqf* endowments, the major resource for education. The ulama did not systematically oppose the Dâr al-Fonun because it was a completely new institution with no equivalent in the traditional education, and in any case remained limited to one or two schools up until the end of the nineteenth century.[12] Therefore, in the absence of a social and economic class that could support nontraditional or non-Islamic education, the state (or court patronage) was the only avenue for the introduction of modern science. Nevertheless, as J. Schlimmer observed, students were barely accustomed to serious studies. In order to gain admission, medical students were required only to be able to read and write the Persian language and to know some Arabic in order to understand grammatical rules and derivations.[13]

Aware of the need for a pre–Dâr al-Fonun education so that candidates could learn preliminary sciences, the Qâjâr elites tried to create their own primary education system. These attempts were first made through the education of religious minorities. In 1852, the government established a primary school within an Armenian church in Tabriz and appointed instructors and teachers.[14] In April 1858, Prince Ardeshir-Mirzâ Rokn al-Dowleh, the governor of Azerbaijan, visited this school (*Maktab-khâneh*) in Tabriz and tested selected children in the Arabic, Persian, French, and German languages, observing that the Shah and his ministers wished to educate all children in the same way, regardless of their religion. After the visit, the prince expressed his hope that "soon, capable students who would be able to serve the government would be trained in that school."[15]

In January 1859, the state journal announced that the Shah had ordered the governors of all "protected" cities and provinces "to first of all provide the names and number of the madrasas, their teachers and students and discipline of their studies that comprise religion, mathematics and literature, and secondly, to order all teachers that from now on they closely supervise the progress of their students through monthly examination, punish those who neglect their study and reward those who make progress."[16] This project aimed to bring primary education under state control while preparing children for study at Dâr al-Fonun. It did not, however, succeed because these primary schools were not funded by the state. More than 20 years later, ʿAliqoli Khân-e Mokhber al-Dowleh, the minister of education and science and head of the Dâr al-Fonun, was ordered to create a new primary school alongside the Dâr al-Fonun to which everyone could be admitted and from which, after completion, they could enter the Dâr al-Fonun. The construction of this new (primary or foundation) school began in July 1882,[17] and was operational by 1885.[18]

The school aimed to teach "useful" sciences that were taught in Europe but "neglected" in Iran, including "practical engineering, mathematics, medicine, and surgery."[19] In parallel with this, the expansion of the Dâr al-Fonun was on the agenda. In addition to the establishment in Tehran and Tabriz, it was a stipulated aim "to create a polytechnic school wherever at least ten candidates can be recruited."[20] However, owing to the lack of financial and human resources, the project did not extend beyond Tehran and Tabriz and Esfahan.[21] Initially the expenses of the Tehran Dâr al-Fonun came from the state treasury (*Divân-e aʿlâ*). When the Ministry of Education, the *vezârat-e ʿolum*, was created in 1859, Eʿtezâd al-Saltaneh, the minister of education as well as the managing director of the Dâr al-Fonun, was also appointed as governor of Malâyer and Toyserkân and used the revenue from those places to fund the school.[22] In a ministerial reorganization in 1866, each minister was entrusted with several provinces or cities, which were overseen by their deputies. The minister of war, for instance, was also the governor of 14 *welâyat*s, including Khorâssân, Semnân, and Dâmghân. The finance minister had control of five places, including Azerbaijan.[23] It was from the revenue of these provinces and cities that ministers drew their budgets. The Ministry of Education had the fewest provinces to govern, and therefore the least revenue. Even so, only part of this revenue was allocated to the Dâr al-Fonun. The reason given by the state journal for allocating the Malâyer and Toyserkân revenues to fund the Dâr al-Fonun is telling: "[S]ince the revenues of these places are not high in any case, the Shah decided to give the governance of these two places to Eᶜtezâd al-Saltaneh so that after paying the expenses of the school, he could return the remaining money to the state treasury."[24]

The Tabriz school (*madrasa-ye nezâm*), established on the Dâr al-Fonun model, was active in May 1859 under the supervision of Mohammad-Sâdeq Khân, colonel of artillery.[25] However, due to lack of adequate finance, this school was closed and reopened several times.[26] Apparently it reopened under the name of *Madrasa-ye Mozaffariyya* or *Dâr al-Fonun-e Tabriz* in October 1893, but three years later, in May 1896, after Mozaffar al-Din Mirzâ went to Tehran to occupy the throne, the school's administration fell apart.[27]

Tehran's Dâr al-Fonun recruited only small numbers of students throughout the second half of the nineteenth century despite the fact that it operated regularly. The first intake, of about 100 students, in 1851 does not seem to have increased a decade later. The list of the students, who took their exams in October 1861, amounted to 107 and in January 1866 to 123 students.[28] By the end of the century, the number did not exceed 250 students.[29] In 1882, the journal *Dânesh* reported that over the course of ten examination periods since 1269/1853, only 237 students had graduated from the Dâr al-Fonun and been employed by the government, while current student enrolment was 185, with an additional 50 students who were not registered and had no duties, but attended language classes.[30] An undated document provides the names of 76 alumni, including staff and graduates of the Tehran Dâr al-Fonun and the posts or missions to which they were assigned. This document also includes lists of 11 current staff and 94 students. The total figure is 181, which includes all previous and current students and staff who were employed by the government.[31] All of those listed in this document held political, military, or medical posts in the service of the state, including Nazar Aqâ, a diplomat and translator at the Iranian embassy in Paris; Mirzâ Rezâ, who was commissioned to purchase weapons in Paris; ᶜAbdol-Hossein Khân, son of Eᶜtemâd al-Saltaneh, who was the colonel of the Regiment of Khalkhâl; and ᶜAliqoli Khân, who was the general director (*ra'is-e koll*) of the *Telegraph*. ᶜAli Khân, another alumnus of the Dâr al-Fonun, who was appointed to build a military fortress in Mâzandarân, had previously (ca.1878) been instructor of artillery at the Dâr al-Fonun after completing engineering studies in Paris.[32]

The conceptual overlap between the content of "modern" medical textbooks and traditional medicine, discussed in the previous chapter, mirrored the social composition of the Dâr al-Fonun. Despite its exclusivity, the Dâr al-Fonun could not function in isolation from the existing system. It is true that the Dâr al-Fonun introduced modern science and technology, but it also taught traditional science from its inception. Mirzâ Seyyed ᶜAli (an ex-pupil of Dr. Cloquet, who continued studying modern medicine with Dr. Polak) taught, in addition to modern medicine, the *Sharh-e Nafisi* (by Ibn Nafis) and *Qânuncheh* (by Mohammad b. Mahmud Chaghmini), considering that "he was an expert in Iranian medicine and knowledgeable about Western medicine."[33]

The teaching system within medicine was similar to that of the military, where adjutants in the army, as well as students, had to attend the classes. In medicine, too, the traditional doctors and surgeons of the army had to attend Dr. Polak's classes (in medicine and surgery) and Dr. Focceti's classes (in physics and pharmacy).[34] Furthermore, instructors were not only the Europeans or modern educated Iranians. The youngest students were taught Persian, Arabic, and religion, following the traditional curriculum, by Sheikh Mohammad-Sâleh, who was also the Imam of collective prayer (*namâz-e jamâᶜat*) at the school.[35] After graduation, the students of Sheikh Sâleh became clerics, knowledgeable about the *shariᶜa* and religious science.[36] This pattern continued to the end of the century. In 1859, the *RVE* reports that because some Iranians did not believe in Western medicine, it was necessary that for their benefit, traditional medicine also be taught, by Mirzâ Ahmad-e Hakimbâshi.[37] Mirzâ Hossein-e Doktor was another who taught traditional as well as modern medicine.[38] Mirzâ Seyyed ᶜAli, who trained in modern medicine, also taught traditional medicine. In 1871, Mirzâ ᶜAbdol-Wahhâb taught Iranian medicine and Mirzâ Rezâ Doktor taught Western medicine.[39] In 1880, Mirzâ Abol-Qâsem taught traditional medicine while Mirzâ ᶜAli Doktor taught modern medicine.[40] All this is indicative of the curriculum of the Dâr al-Fonun, as well as its social composition.

Whether they were pro-traditional or pro-modern, all medical students studied both traditional and modern medicine. Mirzâ ᶜAbdollâh studied traditional medicine with Mirzâ ᶜAbdol-Wahhâb and modern medicine with Mirzâ Rezâ Doktor and physics and chemistry with Mirzâ Kâzem.[41] Instruction in traditional *materia medica* was provided, and the pharmacy of the Dâr al-Fonun contained both Iranian and Western drugs.[42] In 1863, despite the fact that the pharmacy had equipment imported from Paris, which produced drugs swiftly and in large amounts,[43] the school was unable to dispense with local drugs. During Nâsser al-Din Shah's visit to the Dâr al-Fonun in January 1877, Hâji Mohammad Ja‘far, the head of the Corporation of Drug Sellers of Tehran, was present alongside Mirzâ Kâzem-e Mahallâti, instructor of chemistry.[44]

The curricular mix and social composition of the medical school responded to social, cultural, and educational requirements. Many patients either did not believe in modern medicine or had eclectic tastes. The state journal reported:

> As the medical students of the Dâr al-Fonun are knowledgeable about Iranian as well as European medicine, many people consult them and often their treatments have been effective. However, since some of them have not perfected their skills enough, E‘tezâd al-Saltaneh strictly forbids them to treat people unless they are examined and their skills approved by Dr. Tholozan in Western medicine and by Mirzâ Ahmad-e Hakimbâshi-ye Kâshâni in Iranian medicine. Once their capacity to treat patients is approved, Dr. Tholozan and Mirzâ Ahmad will provide them with a licence to practice; otherwise they must not receive any patients. After exams, the names of those who have received such a licence will be published in the journal.[45]

The persisting need for traditional medicine in the curriculum of the Dâr al-Fonun became more evident with the increasing number of medical students who wished to study modern medicine. Schneider observed: "Attracted by the successes of modern surgery [many] medical students have deserted their old masters for practicing modern medicine based on available textbooks but without sufficient preparation."[46]

Modern medicine had no institutional body of its own, and depended entirely on the same state that had Galenico-Islamic physicians as its court doctors. As a result, the traditional medical establishment was willy-nilly involved in medical reform. However, while traditional physicians were increasingly aware of the importance of modern medicine for practical purposes, their expertise in traditional medicine was equally appreciated by their patients. This is evident from the diploma that was being issued as late as the second decade of the twentieth century. Diplomas awarded to Sheikh Mohammad-e Tehrâni and Mirzâ ‘Abbâs‘Ali Khân in 1899 and to Mirzâ Seyyed Hossein Khân-e Nezâm al-Hokamâ in 1917 state that after mastering traditional medicine, the candidates had studied modern medicine and surgery.[47]

This medical configuration was more pronounced in the army, where each division was provided with surgeons and/or physicians. Regardless of the medicine they practiced, many were unskilled. In 1869, a royal decree was published concerning army physicians. It said:

Since some army physicians and surgeons are without knowledge and requisite practical skills and create disorder in treatments and surgical operations, they are now all under the supervision of (the minister of education) Eʿtezâd al-Saltaneh so that the unskilled shall no longer work in the army and only those who are knowledgeable and skilled in medicine and surgery shall continue. From now on all physicians and surgeons have to attend the medical classes of Mirzâ Rezâ Doktor and Mirzâ Seyyed Razi-ye Hakimbâshi (chief physician) of the army and learn medicine and surgery according to European and Iranian (traditional) systems respectively.[48]

The purpose was clearly "skill," irrespective of the kind. In the same announcement, the Shah honored Mirzâ Seyyed Razi-ye Hakimbâshi and Mirzâ Rezâ Jarrâhbâshi (the latter, surgeon in chief, and both of traditional education) with a robe of honor.[49] Six months later, in November 1869, after a warm summer, which interrupted the activity of the Dâr al-Fonun, Jaʿfar Qoli Khân, the deputy head of the school, ordered that all army physicians and surgeons had to attend their theoretical and practical courses.[50] This demonstrates the essential role of the Dâr al-Fonun in tightening state control on medical education. It is in this light that we may explain how Eʿtezâd al-Saltaneh, the modernist head of the Dâr al-Fonun, who was opposed to traditional knowledge, accepted the activity of the traditionalists under his supervision.[51]

Nâsser al-Din Shah was praised for having "revived both traditional and modern sciences" and for being "the propagator of modern science while he perfected traditional sciences."[52] Likewise, most authors boasted of having studied both traditional and modern medicine.[53] Mirzâ ʿAbdollâh Shirâzi (1856–1925), one of the renowned physicians of Fars, was a traditional doctor who also studied modern medicine.[54] This emphasis indicates that modern educated physicians could not bypass traditional medicine, deep rooted in society. It also suggests that with a few exceptions, most physicians had a practical, rather than orthodox, attitude toward medical knowledge. A significant number of medical students were the sons of traditional physicians. Mirzâ Mohammad-Hossein, who was one of the first students of the Dâr al-Fonun to study modern medicine and who was particularly skilled in making chemical drugs (*jowharoyyât-e farangi*), was the son of Mirzâ Ahmad-e Tonekâboni, one of FathʿAli Shah's hakimbâshis, while Mirzâ Ahmad, probably Mirzâ Mohammad-Hossein's son, studied modern medicine at the Dâr al-Fonun in 1882.[55]

The significance of traditional medicine in the Dâr al-Fonun's curriculum resides in its educational aspect. Far from representing rivalry between local and Western medicines, it indicated that modern medicine needed the assistance of traditional knowledge, not only because of concepts and practices shared by both but also because the only way of introducing modern medicine was through existing medical language. The significance of Johannes Schlimmer's medical terminology lay in the way it introduced the instructors of modern medicine to local specialist terms and classifications and in bridging the gap between Western and Iranian medicines.

Administrative Reforms and Medical Modernization

The Dâr al-Fonun was part of a general project launched by Amir-Kabir that included sanitary councils, hospitals, and the modern army, and operated in tandem with these institutions. The outbreak of epidemics stimulated medical thought, as discussed in Chapter 1, but it also brought about institutional reforms and gave the state new opportunities to expand its patronage toward medicine in its different dimensions: medical knowledge and education, the medical market, and medical practice.

Since the establishment of the Qâjâr state, the number of court doctors had risen, in response both to an expanding bureaucracy and to the military modernization that became pressing with each epidemic. The creation of sanitary councils was a response to the increasing waves of epidemics and also supported the centralization of power that was dear to proponents of reform.

The project of smallpox vaccination, made compulsory for children in 1851, can be considered the foundation of public health in Qâjâr Iran.[56] Amir-Kabir employed health officers (or inoculators of smallpox [*âbeleh-kub*]) on a good salary and commissioned them to go to different provinces to provide vaccinations.[57] In Mâzandarân, Karbalâyee Khur-Mohammad-e Abeleh-kub vaccinated about 400 people, and in Guilân, Mirzâ Abol-Qâsem vaccinated 60 children.[58] In Yazd, as in Rasht, the population first resisted the scheme and feared the new technique, but once the governor of Yazd allowed Mirzâ Hassan-e Abeleh-kub to vaccinate his own children, other families took their children to be vaccinated, too.[59] The project aimed to reduce mortality rates due to smallpox (considered an endemic disease), which killed or handicapped large numbers of people every year.[60]

Having learned about quarantine during his earlier mission to settle boundary disputes in the Ottoman Empire, Amir-Kabir created quarantine stations, particularly on the Western borders, which were crossed by pilgrims to Najaf and Karbala.[61] After the death of Amir-Kabir, other state officials, including Mirzâ Mahmud-e Kalântar, the mayor of Tehran, Amir al-Omarâ (the head of the army), and Amir Assadollâh Khân commissioned physicians to write and distribute treatises or manuals on the preservation of health. In early 1852, 60 copies of a guideline on the treatment of cholera (*qawâ'ed-e mo'âlejeh-ye wabâ*) were distributed in Tehran among physicians, clerics, and others of note, to inform them about this epidemic.[62] Aqâ Mirzâ Mohammad-e Tehrâni and Mirzâ Musâ Sâvaji published treatises on *hefz-e sehhat* (preservation of health) in May 1853.[63]

State intervention in public health was under way by the early 1850s. However, after Amir-Kabir, and despite the recurrence of epidemics almost every year, no health reforms were made under the premiership of Mirzâ Aqâ Khân-e Nuri (1852–1858). A period of reform began when, following Nuri's dismissal on August 30, 1858, the Shah divided the duties of the prime minister between several ministries, including finance, justice, the army, foreign affairs, and pensions and endowments (*wazâyef wa owqâf*).[64] There was no Ministry of Education in the new government, but in June 1859, 'Ali-Qoli Mirzâ E'tezâd al-Saltaneh, who was head of the Dâr al-Fonun and the Telegraph Company, was appointed minister of education (*vazir-e 'olum* [sciences]) as well.[65] In October 1859, the Shah ordered the creation of a Government Council (*Majles-e-Showrâ-ye Dowlati*) or *Dâr al-Showrâ-ye Kobrâ* (Greater House), also called the Council of the Ministers (*mashwerat-khâneh-ye khâsseh-ye wozarâ*). Mirzâ Ja'far Khân-e Moshir al-Dowleh was appointed *ra'is* (head) of this council, which was to meet twice a week to discuss matters that Nâsser al-Din Shah assigned to them. In November 1859, an Advisory Civil Council (*Mashweratkhâneh-ye 'Ammeh-ye Dowlati*) or Expediency Council (*Majles-e Maslehat-Khâneh*) was established, in which all those who were knowledgeable about finance, agriculture, industry, education, and public health (*Hefz al-Sehheh*) met and advised the State Council, but were not permitted to discuss political matters. The Advisory Civil Council in Tehran was composed of 25 members and included Mirzâ-Musâ Vazir (administrator) of Tehran, Mollah Bâshi (court cleric), Mirzâ Mohammad Taqi Sepehr (state chronicler and author of *Nâssekh al-Tavârikh*), and Rezâ Qoli Khân (manager of the Dâr al-Fonun). Curiously there was no physician to advise "on issues of demographic growth and the preservation of the health of the population".[66]

These administrative reforms were accompanied by others also destined to improve the efficiency of the state and the military. Merit and education, rather than inheritance, were to be the criteria for obtaining a post. The state journal frequently reminded governors and officials who traveled to abstain from forcefully obtaining *suyursât* (provisions) from local inhabitants, or if they did, to pay the appropriate price.[67]

Nâsser al-Din Shah also set up a Court of Justice (*Divân-e Mazâlem*) that he himself supervised. Court sittings took place every Sunday, so that anyone who wished to lodge a grievance or complaint could report directly to the Shah.⁶⁸ Similar courts of justice were also set up in other major cities. These projects were not always implemented, because the transfer of positions, rank, and salaries from father to son continued. But it is significant that the government questioned this age-old practice and planned to change it.⁶⁹

The reorganization of administration after the dismissal of Mirzâ Aqâ Khân-e Nuri witnessed a new era in medical modernization. The idea of a Consultation Assembly (*majles-e mashwerat*) was also applied to medical organization. In 1860, following one of the epidemics, a Health or Medical Consultation Assembly (*Mashverat-Khâneh-ye Tebbi*) was created in the office of the chronicler, Rezâ Qoli Khân-e Hedâyat (who was also director of the Dâr al-Fonun at the time), to deal with public health and take preventive and curative measures against epidemic disease. The council was formed under the supervision of Prince Iraj-Mirzâ Ra'is al-Atebbâ (chief physician).⁷⁰ Its members included Dr. Tholozan and Hâji Âqâ Bâbâ Hakimbâshi (Mirzâ Mohammad-Taqi-ye Shirâzi), Mirzâ Ahmad-e Kâshâni Hakimbâshi, and Mollâ Mohammad-e Qoboli. The medical students and notable physicians belonging to traditional education could also take part.

Politics and Medical Modernization

Another era of reform began toward the end of the ten-year contracts awarded to the Austrian and Italian instructors at the Dâr al-Fonun. A link between this phase of reform and a new stage in the diplomatic relationship with France cannot be dismissed. The fact that the first group of students sent to France, who accompanied Farrokh Khân-e Amin al-Dowleh in his mission in 1856, signed a treaty of peace and alliance, suggests that the political relationship was decisive in choosing France as the destination for the students to study modern science. During this period, Iran endeavored to improve its relationship with France at a time when its relationship with Britain was at boiling point over Afghanistan, a circumstance that led to the British occupation of southern Iran (December 1856–April 1857). However, an experience similar to the one that occurred under Napoleon earlier in the century was repeated here. Iran hoped to gain the support of France in its conflict with Britain over Herat. But during an Iranian visit to Napoleon III and his government, the French Ministry of Foreign Affairs informed Farrokh Khân that Britain opposed the pressure but did not mind Russian interference.⁷¹ In other words, France did not want to support Iran if it meant conflict with Britain. Iran did not represent an important strategic interest for France.

Three of Dr. Polak's medical students who had learned French accompanied Farrokh Khân-e Amin al-Dowleh's mission: Mirzâ Rezâ, son of Mirzâ Moqim-e Mostowfi; Mirzâ ᶜAlinaqi (later Hakim al-Mamâlek); and Mirzâ

Hossein-e Afshâr. Leaving Tehran on July 14, 1856, the Legation arrived in Paris on January 17, 1857.[72] All three students completed their studies and wrote their thesis in 1860 in Paris. With the impending departure of Dr. Polak, the Shah, via his envoy in Paris, requested that the French government send a physician to Iran. Dr. Tholozan, an *agrégé* of the military school at Vale de Grace, left for Iran in September 1858. Five months later, in February 1859, a group of 42 students from notable families were selected to go to Paris, with Hassan͑Ali Khân's mission. They finally arrived in September that year and began their studies, in various schools, in October.[73]

Although France was not very helpful to Iran in its conflict with Britain, the Qâjâr government found her a potentially powerful ally and a valuable "card" to play in its relationship with Britian and Russia.[74] This, together with the French sociopolitical developments there throughout the nineteenth century, made France attractive to Iranian elites. Significantly the arrival of the Persian mission was marked by the reception of the ambassador, Mirzâ Hassan͑Ali Khân, at the *Société d'Ethnographie,* on October 17, 1859, at which the president, the Baron de Bourgogne, laid emphasis on Persia's cultural and literary wealth and on the Persian language, which occupied "the first rank amongst the Oriental languages." He reminded his audience that "the language of Hâfez and Sa͑adi belongs, just as French does, to the old and grand Aryan family. This secular kinship, although in the remote past, should be seen as a sign of our literary fraternity and as the contribution of your luminous and distinguished spirit." Mirzâ Sa͑id Khân took part in meetings of the Ethnographic Society and presented papers in which he discussed the progress of science as a result of collaborative work.[75]

The admittance of Iranians into the French Masonic Lodge, the Grand Orient de France, was particularly important in the formation of the Iranian intelligentsia. From the early nineteenth century, most students, envoys, and ambassadors were put in contact with Freemasonry societies. Both Mirzâ Askar Khân-e Afshâr in Paris in 1808 and Mirzâ Abolhasan Khân-e Ilchi (ambassador) in London in 1810 joined the local lodge. This trend continued until the second half of the nineteenth century. On his return to Iran from his studies in Paris in 1858, the reformist and intellectual Mirzâ Malkam Khân created a pseudo-masonic association, called *Farâmushkhâneh.*[76] Other prominent reformers and statesmen, such as Mirzâ Ja͑far Khân-e Moshir al-Dowleh, Mirzâ Hossein Khân-e Sepahsâlâr, and ͑Alinaqi Khân-e Mokhber al-Dowleh, were reputed to be masons. Chroniclers, such as Rezâ Qoli Khân-e Hedâyat and Mohammad-Taqi Khân-e Lesân al-Molk-e Sepehr, were members of the Farâmushkhâneh of Malkam Khân.[77]

Membership of these societies, which represented modern ideas, created a social and institutional network that helped the transfer of modern science, while it was also used as a channel for a closer relationship with Western politics and intellectuals. The role of the members of these societies in initiating a number of reform projects is undeniable. It is significant that Freemasons were involved in the first modern institutions and scientific journals. Mirzâ Ja‛far Khân-e Moshir al-Dowleh, who was one of the students sent to London under ‛Abbas Mirzâ and knew the British parliamentary system well, became head of the first State Council, or "*showrâ-ye kobrâ-ye dowlati*," in 1859. Likewise, ‛Aliqoli Khân-e Mokhber al-Dowledh, son of Rezâ Qoli Khân-e Hedâyat and the head of the Dâr al-Fonun, created the journal *Dânesh* (*Science*) in 1882. Significantly, the head of the next reformist government, Mirzâ Hossein Khân-e Sepahsâlâr, studied with Mirzâ Ja‛far Khân-e Moshir al-Dowleh before being sent to Paris for further education.[78]

The intellectual interest in French culture and science can be understood in the context of the political rapprochement with France dating from the government of Mohammad Shah. However, over time, this intellectual interest acquired its own dynamics independent of diplomatic events. The role of the French diplomats or intellectuals who were employed in Iran was crucial in this process.[79] While acting as French envoy, Comte Arthur de Gobineau made inquiries about religion, culture, and society; met with local intellectuals; and wrote his observations in several volumes. At the request of Iranian philosophers, and with the help of local scholars, including Molla Lâlazâr-e Hamadani and Prince E‛tezâd al-Saltaneh, he translated Descartes's *Discours sur la method* into Persian.[80] In the same spirit, Dr. Tholozan, in his role as personal physician to the Shah, endeavored to modernize medical literature in Iran by introducing new ideas. Dr. Tholozan's long career illustrates this cultural relationship, which continued despite political and diplomatic incidences.

Tholozan replaced Dr. Polak as chief physician to Nâsser al-Din Shah and professor of the Dâr al-Fonun, and recommended his best students for study in Paris.[81] In 1861, he commissioned a translation into Persian of a French treatise on auscultation and percussion, "in order to transmit and propagate amongst Persian physicians the two greatest and most useful discoveries that have enriched the medicine of our time."[82] During nearly 40 years of service in Iran, he undertook research on cholera, plague, and other local diseases, including leishmaniasis (locally called *sâlak* or disease of Bushir). However, not long before he settled in Iran, the winds of counterreform began blowing.

The state's new organizations were initiated by an intelligentsia with its umbilical cord attached to the court. While this relationship could sometimes accelerate reform, it could also have an adverse effect, as modern institutions, like the state councils, were infested with counterreform elements, who were easily alarmed at the prospect of losing their privileges.[83] The dismissal of reformist Prime Minister Amir-Kabir in 1851, and the abandonment of the subsequent waves of reform during the 1860s and 1870s, illustrates the influence of these counterreform elements. Due to their intrigues, Mirzâ Ja'far Khân-e Moshir al-Dowleh, head of the State Council, was sent to Mashhad as administrator of Imam-Rezâ's *owqâf*, and died shortly thereafter. In 1861, the Freemason society of *Farâmushkhâneh*, which had introduced ideas of law and progress, was closed by the order of the Shah and printing became closely controlled. In December 1863, the government published an announcement forbidding the publication of ideas or treatises that went against "*shari'a* and the interests of the state and nation."[84] In 1865, the government prohibited travel to Europe without the permission of the state.[85]

Mirzâ Mohammad Khân-e Sepahsâlâr, who was minister of war, was motivated more by improving the efficiency of the system than by an ideology of progress. This view was also held by the Shah, but whenever a new organization seemed to threaten his absolute authority, modernization was discouraged. Time and again, Nâsser al-Din Shah supported reform but then reversed his position and eliminated the reformers, hence the lack of consistency and continuity in the reform projects.[86] The first Health Council (*Dâr al-Showrâ-ye Tebbi*), which was created in 1860/1861, seems to have been discontinued after only a few months.[87]

The state journal, the *RVE*, played a pioneering role in the promotion of public health by publishing articles on preventive measures and hygiene and snippets of medical information about the nature of epidemics and the treatment of patients.[88] However, it also reflected popular and folk views, stating, for example, that cholera was the population's punishment for their sin and debauchery and so, to remedy this, people went to mosques and prayed to God to be pardoned.[89] In other words, it welded modern and traditional views: "Physicians", it said, "have found that cholera spreads in filthy environments and amongst dense populations." It then concluded that the inhabitants of affected areas should leave their homes, a solution recommended by traditional physicians.[90]

Not long after, other journals, such as *Mellati* (belonging to the nation, *Mellat*) (in 1868) and '*Elmi* (in 1877), began to publish articles on science and public health on a more regular basis. These coincided with the activities of the reformists during the era of Mirzâ Hossein Khân-e Sepahsâlâr (1871–1881).[91] In May 1872, Sepahsâlâr drafted a report on "modern customs" (*rosum-e jadideh*), and had it underwritten by the Shah, and later created new ministries, including a ministry of commerce and industry, with a minister at the head of each, responsible to the prime minister (*sedârat-e 'ozmâ*).[92] The new rules were very probably designed with the assistance of Mirzâ Malkam Khân, who was granted the title of *Nâzem al-Molk* (registrar or regulator of the kingdom) in the same year, and appointed as adviser to the prime minister.[93]

The new journals, including *Mellati*, *ᶜElmi* (scientific), and *Nezâmi* (military), which was later called *Merrikh* (Mars), were independent of the state and focused on science and education.[94] In 1868, seven years after the creation of the Health Council (*Dâr la-Showrâ-ye Tebbiyyeh*), another assembly, called *Majmaᶜ al-Sehheh* or *Majles-e Hâfez al-Sehheh* (Sanitary or Medical Council), was established and began its activity by the publication of guidelines written by Tholozan on preventive measures against cholera in two consecutive issues of *Mellati*.[95] During the 1870s, other journals with a particular interest in science, medicine, and public health appeared. From April 1871, the *RVE* and the journals *ᶜElmiyyeh* and *Mellati* merged into one journal, *Iran,* which regularly published scientific articles alongside political and official news.[96] The journal *ᶜElmi* published more on medicine; its first article, after an outline explaining that the aim and approach of the journal was based on the dissemination of knowledge and science, talked about the Dâr al-Fonun, the state hospital, and the Sanitary Council.[97]

During the 1870s, the functions of the Sanitary Council expanded. In 1876, coinciding with a meeting of the council, its duties were extended to the establishment and implementation of medical rules among the profession, including physicians, surgeons, pharmacists, and even bloodletters (*fassâd*). The head of the Sanitary Council at this time was Eᶜtezâd al-Saltaneh, minister of education and director of the Dâr al-Fonun. Members met every Friday to examine physicians in order to assess their ability to practice. Prince Iraj-Mirzâ Raʾis al-Atebbâ; Mirzâ Kâzem-e Rashti Malek al-Atebbâ; Mirzâ Seyyed Razi, Hakimbâshi to the army; Mirzâ Abol-Qâsem, teacher of traditional medicine at the Dâr al-Fonun; and Hâji Mirzâ Abolfazl-e Sâvaji examined the candidates in traditional medicine, and Mirza Rezâ Doktor, Mirzâ Zeyn al-ᶜAbedin-e Kâshâni, and Mirzâ ᶜAli-Akbar-e Kermâni, chief physician at the state hospital, examined students in modern medicine and surgery. Successful candidates were given licenses to practice; those who displayed only limited knowledge were allowed to practice independently, dealing with minor illnesses, but were not allowed to treat serious diseases without the supervision of skilled physicians. Those who failed were prohibited from practicing but were allowed to continue their study at the Dâr al-Fonun or with a well-known physician. Such a system of control was also extended to the provinces.[98]

In fact, the existence and activities of the sanitary councils became more widely known through the journals. The first council, formed in 1861, became inactive shortly afterward, and there was no news about other councils until 1868, when the journal *Mellati* published Tholozan's guidelines for cholera epidemics, and 1871, with the publication of new guidelines in the journal *Iran*. However, the sanitary councils published medical articles that were for the benefit of both physicians and laymen. Following the formation of the Sanitary Council in December 1876, the journal *ʿElmi* published, in two successive issues, an article on tonsillitis or inflammation of the tonsils (*waram-e lawzatayn*) by Malek al-Atebbâ, one of the members of the Majles-e Hefzal-Sehheh.[99] In an article in *Iran*, the Sanitary Council warned about the devastating role of bloodletting in causing anemia.[100]

The first practical use of the *sehhiyeh* (public health offices or the Sanitary Council) was initially linked to the creation and control of quarantines. In 1867, when a plague epidemic spread among Iraqi cities bordering Iran, the Ottoman Pasha of Baghdad established quarantine to prevent the movement of pilgrims. It was made clear that the transport of people or even their clothes carried the epidemic. Mirzâ Hedâyat, the physician of the Dâr al-Fonun, was dispatched to oversee the implementation of quarantine on the Persian side of the border.[101] Furthermore, the *sehhiyeh* endeavored to apply order to and supervise the practice and sale of drugs. In 1877, the Sanitary Council issued pharmacists, in each quarter of the capital, licenses to sell specific drugs, and as a result, many were forbidden to sell chemical drugs.[102] The *sehhiyeh* also monitored the state hospital. On March 11, 1877, members of the Sanitary Council visited the hospital and after reading the reports of physicians from other provinces checked all patients in each ward and were informed of the state and condition of their medication and food.[103] Later, sanitary councils trained medical officers (*hâfez al-sehheh*), who were sent across the country. At the beginning, their duties consisted of informing the central Sanitary Council in Tehran, but gradually they formed the nucleus of health centers in each province.

The Majles-e Hefz al-Sehheh (Sanitary Councils) and the Expansion of Medical Institutionalization

The Sanitary Council was the operational arm of the Dâr al-Fonun, and recruited a wide range of physicians and medical officers. Although a modern institution, it aimed to take all kinds of physicians and medical practices under its wing because the social and professional fabric of medicine at that period did not allow otherwise. When Schneider arrived in Iran in 1893, he found that

> [a] good number of Persian doctors dressed like mollahs, wearing a shawl around their waist and turbans on their head; and on their feet, heelless and pointed sandals. They were as religious and fanatic as the mollahs themselves and to some extent sharing their sacred characters. When they went to visit

their patients they proceeded with great pomp, riding a mule of high size and surrounded by their numerous disciples. They disdained European medicine, whether religious or scientific and expressed their astonishment at not seeing diseases divided into hot and cold and being treated by drugs of opposite temperament.[104]

Although this description of traditional physicians in Iran is true to a point, it is not in keeping with the participation of these physicians in modern medical institutions. Most traditional court doctors sat on the sanitary councils alongside Western physicians or modern trained Iranian doctors. For example, Mirzâ Âqâ Bâba (Mirzâ Mohammad-Taqi Shirâzi Malek al-Atebbâ), famous for his orthodox attachment to Galenico-Islamic medicine, participated in the Sanitary Council (*mashverat-khâneh-ye tebbi*) from 1861. Yet, when at the second meeting of the Sanitary Council, Hâji Âqâ Bâbâ asked if anyone could suggest a treatment for the paralysis of a certain Khâjeh, the clock maker, both traditional and modern physicians exchanged views and debated medical issues. However, this was not without tension and disagreement. As reported in the state journal, Mahd-e ʿOlyâ's (the Queen Mother's) illness was first treated by Shirâzi, but her condition worsened. It fell to Dr. Tholozan to restore her to health.[105]

Anatomical pathology marked the borderline between modern and traditional medicine. In dealing with issues of public health, there was no clear borderline between Western and Iranian medicine. Concept of public intervention was present in the works of traditional physicians such as Sâvaji and Tehrâni. Without doubt, the new institution of *Majles-e Hefz* (or *Hâfez*) *al-Sehheh* was formed on a European model. Health officers (or *hâfez-e sehhat*, literally health keepers) were copies of the French *officier de santé*. But the concept itself was rooted in Galenico-Islamic medical literature. Mirzâ Mohammad Kâzem-e Rashti had written his book, *Hefz al-sehheh-ye nâsseri*, 20 years before he became a member of the *Majles-e Hâfez al-Sehheh* (MHS), while Sâvaji in 1852 indicated that the Head of the Army ordered him to write his treatise on *Setta Zaruriya*, the six nonnaturals, which was the Galenico-Islamic version of public health.[106] Significantly, Sâvaji used the term *hâfez-e sehhat* for healthy individuals. The concept of *hefz al-sehheh* in traditional medicine thus related to individual health. Each healthy individual must take measures to remain healthy and observe the rules of medicine to preserve their health during an epidemic.[107] There were differences in measures and treatments suggested by Sâvaji and Tholozan, but both stressed the importance of the preservation of health. In the guidelines written by Tholozan, in 1869 reference was also made to the *Setta Zaruriya*.[108] From Sâvaji in 1852 to Tholozan in 1869, however, there was a shift in focus from individual predisposition as the main source of epidemic disease to vectors of disease residing outside the body, in other words, environmental factors. The strategy of preserving health shifted from the individual to the public sphere.

As a new forum for physicians, journals like ʿ*Elmi* and *Dânesh* created new opportunities for them and promoted and increased their social standing. Meanwhile, the central Health Office in Tehran increased its recruitment of physicians or health officers, who were appointed to create local Sanitary Council branches across the country. Traditional doctors continued to participate in the Sanitary Council, but the numbers of modern physicians increased. In 1881, the Sanitary Council in Tehran had 18 members, including Dr. Tholozan, Dr. Dickson (physician of the British embassy), Dr. Arno (representative of the Sanitary Council of Istanbul), the Austrian Dr. Begmez (an army physician), Mirzâ Kâzem-e (Rashti) Malek al-Atebbâ, Mirzâ Seyyed Razi (chief physician of the army), Mirzâ Abol-Qâsem (instructor of traditional medicine at the Dâr al-Fonun), Mirzâ Seyyed ʿAli; Mirzâ ʿAli Doktor, Mirzâ Mohammad (director of the state hospital), and Mirzâ Mohammad Kâzem (instructor of chemistry at the Dâr al-Fonun).[109] Dr. Tholozan was instrumental in the organization of public health, and shortly after his arrival began the formation of the Sanitary Council in 1861. However, in this capacity, he worked in the shadows for nearly three decades, even while being recognized as chief physician to the Shah, until he officially became head of the *Majles-e Hefz (*or *Hâfez) al-Sehheh* in 1888.[110] Meanwhile, Tholozan published articles and guidelines on smallpox and cholera. In 1862, he published a short treatise on vaccination, in which he established the link between popular inoculation (variolization) and Jenner's model of vaccination, clearly explaining how it could medically and physiologically prevent smallpox. At the same time he underlined the role of the state in smallpox vaccination across the country.[111]

The creation of the *Majles-e Hefz al-Sehheh* in 1861 was aimed at tackling epidemics.[112] Located at the school of Dâr al-Fonun, it observed and registered epidemics and other diseases in Tehran and in other parts of the country. In 1882, for instance, it reported that "the country was this year exempt from infectious or contagious diseases and the illnesses that were mostly seen were different kinds of intermittent fever and in some places dysentery also dominated, but they were not fatal and mostly were treated."[113]

The creation of Sanitary Council responded to medical institutionalization triggered by the introduction of modern medicine. For centuries, rank and file doctors were practicing across the country, but it was with the Dâr al-Fonun and the introduction of modern medicine that the shortcomings of "unskilled" doctors became increasingly evident as new tools of measurement and assessment were introduced. Examination and control of traditional physicians that was not inexistent previously intensified with the introduction of modern institutions and one of the functions of the Sanitary Council was to assess medical licenses (*ejâza*), articles, and essays, as mentioned in a report of the council held in July 1889.[114]

Despite the mixed composition of the Sanitary Council, judgments were made through the eyes of modern medicine. The Sanitary Council became the point of reference in the establishment and reinforcement of order. In 1882, an article critical of two physicians, Hassan Khân and Mirzâ ʿAbdollâh, was published in issue 35 of the journal *Ettelâʿ*. Mokhber al-Dowleh, minister of science and education (*vazir-e ʿolum*), reacted to the *Ettelâʿ*'s article in his own journal *Dânesh*: "instead of writing a journal article against physicians the author should have referred to the Sanitary Council, which is equipped with knowledgeable and skilled doctors. Scrutiny reveals indeed that the author of this article (in the *Ettelâʿ*.) is devoid of the requisite knowledge and therefore his article is worthless." This was followed by a detailed explanation, written by Mirzâ Kâzem-e Mahallâti, instructor of physics and chemistry at the Dâr al-Fonun, which refuted *Ettelâʿ*'s article from a technical viewpoint.[115]

Once the *Majles-e Hefz al-Sehheh* was established, and physicians were recruited as *Hâfez al-Sehheh* (health officers), some provinces were provided with permanent health officers and/or physicians, such as Mirzâ Hessâm al-Din in Shirâz. In other cities or provinces, however, the officers were dispatched only when a disease or epidemic appeared. For example, Amânollâh Khân-e Tabib was sent from the Sanitary Council to Firuzkuh to deal with a disease called *suzeh-dâghi* (an outbreak of pustules and pimples, most likely erysipelas or akin to it), which spread among the tribes of that region. Mirzâ ʿAli-Akbar Khân-e Shirâzi was sent to Qom, where there was no Dâr al-Fonun-educated physician.[116] Later, physicians were commissioned to create sanitary councils (*sehhiyeh*) in the major cities. In 1883, health officers were sent to Tabriz, Hamadan, Brujerd, Sabzevar, Yazd, Kermânshâh, and Kurdistan.[117] Their numbers were growing; however, it was clear that most of the health officers were not graduates of the Dâr al-Fonun, but rather experienced practitioners or those who had received permission (*ejâzeh*) from a famous physician.[118]

Development of Hospitals

The often celebrated association between medicine and hospitals in medieval Islam reflects a modern historiographical bias, based on teleological analysis. Although under a few Islamic dynasties, such as the Ayubids in Egypt and the Ottomans in Turkey, hospitals flourished, and some survive today, this was far from widespread across the Islamic world and in any case was not accompanied by, or resulted in, significant development of medicine itself. In the late eighteenth century when Pierre-Jean-Georges Cabanis (1757–1808), one of the founders of clinical medicine, advocated the creation of clinical schools and the use of hospitals for their development, he referred to the hospitals built by the "Emperors of the East and by the Arabians as an example for clinical experience." According to Cabanis, these hospitals "were erected not only for the relief of the diseased people but as a laboratory for observations and experimentations by the practitioners."[119] Cabanis, however, did not provide historical evidence to support his assertion about hospital medicine in medieval Islam. Although bedside medicine was always recommended by Islamic medical literature, this did not necessarily represent hospital practice. Cabanis's reference to Islamic hospitals was more an image, used for rhetorical comfort, to illustrate his own clinical medicine, which was in full development at that time in Europe. It also seems that Cabanis's praise of Islamic hospitals mirrored the positive and sympathetic opinion of the Enlightenment philosophers, such as Voltaire and Montesquieu, about Islam, which they considered a monotheist religion with a simple morality, very little superstition, and no clergy.

Hospitals in premodern Iran were greatly lacking in resources. According to Raphaël Dumans, during the Safavid period "every large town had one or two hospitals, [usually] consisting of a large square building with a cloister [court-yard] into which all wards, large and windowless rooms, opened." Nevertheless, such buildings were not used very much, for when he visited a hospital in Espahan (in the mid-seventeenth century), he found only one moribund Indian lying on the floor and in another room a solitary madman chained to the wall. And the provincial hospitals must have been even worse.[120] Likewise, when Jean Chardin visited Tabriz (ca.1670s) he saw three hospitals "very neat and well in repair," but they "contained no inmates and were only used as a centre for the distribution of food for the necessitous."[121]

The fall of the Safavids in 1722 was followed by a long period of political instability. After nearly a century of civil war (with the exception of the short-lived Karim Khân-e Zand rule [1759–1779]), the country was devoid of an economic and administrative infrastructure that could support the construction of hospitals. Charitable foundations, which constituted the potential resources for the development of hospitals or institutions like the *dâr al-shafâs*, were also affected by the internecine feud between aspirants to the throne. As a result, the only *dâr al-shafâ* that survived to the nineteenth century was in Mashhad, and this was probably due to the religious immunity of the *owqâf* of Imam Rezâ.[122] In addition to socioeconomic factors, the fact that there was no historical, organic, or consistent relationship between medicine and hospitals can explain the dearth of hospitals in the first half of the nineteenth century in Iran.

The first hospital under the Qâjârs was established in 1851 by Amir-Kabir for the use of the army. In past centuries, there had been medical corps in the guise of camp hospitals that accompanied the military, a practice that continued into the nineteenth century.[123] It seems, however, that only major military campaigns or the army that accompanied the Shah were provided with these. In August 1851, it was not before disease spread among the troops sent from Mâzandarân to Tehran, and the death of their commander, Colonel Mahmud Khân, that a physician and tents were sent to assist sick soldiers.[124] Dr. Polak, who witnessed the situation in the 1850s, observed that there were only a few army physicians, called *hakim-e fowj*, some of whom were employed permanently, but most of whom were attached to a regiment only during military missions. Polak claimed that there was no need for surgeons insofar as soldiers injured during military campaigns were left to die; the army was concerned only with treating soldiers who fell ill.[125] Polak's testimony should be taken with a pinch of salt, but it gives an idea of the state of medical service in the army during campaigns.[126] Even in the second half of the century, supporting physicians were sent only in emergencies. In August 1864, an army led by Mirzâ Mohammad Khân-e Sepahsâlâr against the Yamut Turkmen was accompanied by a camp medical service that included physicians and surgeons under Mirzâ Razi, Hakimbâshi to the army.[127] However, in June of the same year, it was only when 80 soldiers of the Shaqâqi division were affected by typhus and typhoid (*mothbeqa* and *mohreqa*), that Mirzâ Hedâyat, a Dâr al-Fonun physician, was sent to treat them.[128]

Shortly after ascending to power as prime minister in 1848, Amir-Kabir, as a temporary solution, allocated each regiment or division stationed in Tehran a place where sick soldiers could be taken care of or treated.[129] He planned to build more military hospitals, about which contemporary sources provide confusing information, but what is certain is that a hospital was built in 1851. According to the *RVE*, from December 25, 1851, to January 11, 1853, this hospital received 1,238 patients.[130] On December 17, 1852, the same journal reported that Dr. Polak intended to take his students to the hospital the following week.[131] About two years later, another hospital was ordered to be constructed, probably the same hospital that, according to Hedâyat, the Shah ordered Mirzâ ᶜAli Khân-e Hâjeb al-Dowleh (the notorious killer of Amir-Kabir), to build for the army outside the Gate Dowlat.[132] It is not clear when, or whether, this hospital was completed, but this may be the one to which Dr. Polak refers as being built under his supervision about 1854. The data we have about hospitals constructed in the second half of the century are inconsistent. With the exception of the Dâr al-Fonun, on which the state journals reported fairly regularly, we find no information about the military (or state) hospital three years after its creation. Information given in the following decades may refer to other hospitals built within Tehran outside the gates.[133]

The formation of the State Council (*Dâr al-Showrâ-ye Kobrâ*), which took over the duties of the prime minister in 1858, gave a new impetus to medical reform. The main focus of this reorganization was to fight against corruption and the lack of regulation that was ingrained in the state and military machine. More centralization was on the agenda. A chancellery (*Divânkhâneh*) was formed, within the Ministry of War, with branches, for each division in the provinces, which were responsible for the payment of the soldiers. Each provincial or local division was also to be supervised by a representative of the Minister of War.[134] In 1863, the Shah delegated full powers for the implementation of reform in the army[135] to Mirzâ Mohammad Khân-e Sepahsâlâr, minister of war and a member of the State Council. Sepahsâlâr's responsibilities increased and he finally became de facto chief minister (*sadr-e aʿzam*) in 1865, after the death of Mirzâ Jaʿfar Khân-e Moshir al-Dowleh, who was the head of the State Council (Figure 3.1).[136]

Mirzâ Mohammad Khân-e Sepahsâlâr was barely educated, but was an energetic general with modernist leanings, who adopted modern military organization and devoted his career to improving both the condition of his troops and the welfare and public health of the population.[137] He was the first to abandon traditional military dress for modern, Western-style uniforms, and paid particular attention to keeping the army drilled and disciplined. In October 1865, a booklet on the regulation of the army was prepared by his order and direction.[138] However, in 1866, he was dismissed from the Ministry of War, replaced by ʿAziz Khân-e Sardâr-e Koll (commander-in-chief), and sent to Mashhad as governor of Khorâssân, before his death on June 20, 1867, under suspicious circumstances.[139] In line with military modernization and in order to improve the conditions of the soldiers, Mirzâ Mohammad Khân-e Sepahsâlâr accorded particular importance to the military hospital, which had been neglected. In the same way that the creation of the councils aimed to render the state system more efficient in the face of economic and social crisis aggravated by famine and epidemic diseases, hospitals for the military were intended to improve the efficiency of the army. To this end, Sepahsâlâr ordered one of the chief physicians of the army to write guidelines for the creation and organization of state/military hospitals across the country.[140] In fact, he revived the project of Amir-Kabir. Although this project was never implemented as described in the guidelines, it seems that two more hospitals were built in Tehran, one in the northwest of the city,

Figure 3.1 Mirzâ Mohammad Khân-e Sepahsâlâr

after the expansion of the capital in 1867.[141] During the famine of 1871, a relief council (*majles-e eʿânât-e foqarâ*) was created to organize distribution of money and provisions to the thousands of poor that flocked to Tehran. During a meeting on December 3, 1871, the Head of this Council, Mokhber al-Dowleh, announced that a hospital would open its doors with a capacity of 60 patients. The hospital also had a kitchen and a pharmacy, and Mirzâ ʿAli-Akbar and Mirzâ Mohammad-e Tabib were designated to run it.[142] Another hospital was built, on the European model, in 1873, when Nâsser al-Din Shah returned from his first tour in Europe.[143] Under Mirzâ Hossein Khân-e Sepahsâlâr-e Aʿzam, who also paid particular attention to military modernization, in each military barrack, some part was organized as a hospital. In 1879, in Tehran, the three barracks at the gates of ʿAbdol-ʿAzim, Dulâb, and Qazvin each had a *mariz-khâneh*, or hospital and a pharmacy, under the supervision of the Chief Physician of the Army Dr. Begmez, who visited and treated patients in these barracks but also appointed his students, as assistants, to each hospital to take care of the sick soldiers.[144] A similar project was implemented 14 years later, when a new barrack called Nâsseri (*sarbâzkhâneh-ye nâsseri*), for the regiments that protected the *Arg* (citadel) and the state buildings (*afwâj-e qarâwol-e makhsus*), was built in 1893. This one included a hospital with wards for sick soldiers and a pharmacy with Iranian and Western drugs and surgical instruments, and it was staffed by several physicians, surgeons, pharmacists, and nurses.[145]

Military Reform and Medicine

In modern history, war and military modernization have played a crucial role in the development of medicine.[146] In Iran, the introduction of modern medicine occurred when the country endured sustained military conflicts, during the first part of the nineteenth century, with Russia in the Caucasus, and with separatist tribal chiefs across the country. This resulted in the modernization of the army by ʿAbbas Mirzâ, Hâji Mirzâ Âghâssi, Amir-Kabir, Mohammad Khân-e Sepahsâlâr, and Hossein-Khân-e Sepahsâlâr-e Aʿzam, and in the development of what we may call "military medicine."

The impact of the military on medicine was not only through its need for surgery but also in the change of outlook and approach in science as a result of military modernization. In one treatise, on the advantages of cartridge ammunition compared to cannon, and its wide usage on land and at sea, the argument was based on the premise that the aim of combat is to encircle your enemies and kill them. To this end, according to the treatise, one needs a weapon that, while it can be used or carried easily, has the

Map 3.1 Tehran's map in 1859

most destructive effect on the enemy. Some "measuring" then expounds the argument further: a cannon of 6 pounds is easier to carry than a cannon of 12 pounds but its destructive effect is far less. The advantage of cartridge ammunition is that it is light to carry, and its range is far longer than that of cannon while it is far more lethal. It is therefore effective in encircling enemies and destroying them efficiently. It was also possible to increase the power of the gun without increasing its weight or its size by changing the quality of the powder of the cartridge.[147]

In other treatises on the military, the description and classification of elements involved in military action was used. Such a method of writing differed from classical literature. Just as the study of military science (ʿolum-e nezâmi) required the study of elementary principles of order and discipline, and the assimilation of mathematics, physics, chemistry, and engineering,[148] so the development of anatomical pathology and professionalization constituted the key elements of what we call "military medicine" in the Qâjâr period.

The descriptive and observational approach in anatomical pathology mirrored the military landscape description and observation that helped to dominate the enemy in a military operation. The "military medicine" born of the modernization of the army went well beyond mere manual medicine or medicine based on the need for surgical operations. It went hand in hand with anatomical pathology, discipline, order, and regulation. The growing importance of surgery in the nineteenth century should be understood in the light of anatomical pathology. This was illustrated in various diplomas in which emphasis was laid on the skill of candidates in manual medicine. In the recommendation letters (tasdiq) signed by the several physicians with whom Seyyed ʿAli Boqrât al-Molk had studied medicine, they emphasized that he had studied modern medicine and practiced manual operation (amaliyât-e yadi). These letters implied that manual medicine was an integral part of modern medicine, and that a license for practice should be awarded to him because of his skill in modern/manual medicine.[149]

In medieval education, all physicians had to master the five branches of medicine (medicine, pharmacology, anatomy, surgery, and ophthalmology) from sources including Avicenna's Canon and Majusi's Kâmel al-senâʿa. In most Islamic medical compendia all these branches are discussed, but, in reality, physicians were distinguished from surgeons, who were considered of inferior rank. There were a few instances where surgery was considered important. Even though the above-mentioned Zakhira-ye Kâmela by Hakim Mohammad ranked surgery second after medicine—"in the science of the body (ʿelm-e badan) [as opposed to the science of religion] after medicine (ʿelm-e tebb), surgery (ʿelm-e jarrâhî) is a requisite"—it associated surgery with medicine.[150]

In the nineteenth century, however, this trend gained momentum with the military hospitals. The chief physician of the army (anonymous author of Document 505) placed surgery below medicine, insofar as it did not require as much theoretical knowledge as medicine, but not only did he stress that surgery was a branch of medical art, but he also criticized the then current practice of despising surgery. In order to have skilled and knowledgeable surgeons, he reasoned, the dignity of the surgeon should be restored, and to this end, before qualifying in the practice of surgery, surgeons should have read the ancients and have observed or experienced surgery (*aᶜmâl-e yad*). According to this author, a surgeon was particularly needed during war. However, although his presence (in hospitals) during peacetime was not really necessary "because ordinarily lesions have a humoral origin and should be treated by restoring the humoral balance, since sometimes manual operations, such as excising cysts and extracting stones occur, his presence [in hospitals] in peacetime is required."[151] Significantly, the document 505 on hospital emphasized the importance of surgery with the aim of adding to it the character and identity of a specialty that was neglected or despised by learned physicians.

Military Reform and Hospital

In 1854, a treatise on military regulations, written by an Italian general who taught at the Dâr al-Fonun, was translated by ᶜAli Qarageuzlu, an army general.[152] The first hospitals that were created were for the use of the army, and military discipline was consequently applied to their organization and administration. About 3 or 4 years later (ca.1857-58), an anonymous document (MSS 505) followed a similar approach when discussing the creation, organization and administration of new hospitals. On the subject of medicine, this document recommended that students study the major classical sources, such as Majusi's *Kâmel al-Senâᶜa*, Avicenna's *Qânun*, and *Fosul-e Boqrât* (Hippocrates' *Aphorisms*). As regards the creation and organization of hospitals, on the other hand, instead of referring to medieval sources, such as ᶜOnsor al-Maᶜâli, in line with traditional medicine, it used Western hospitals and their organization as model.[153] According to the regulation set in Document 505, each medical profession was divided into three grades: the first physician, the second, and the third, the first surgeon, the second, and the third, and so forth. There were also logistics administrators (*moshref*) and orderlies (*parastâr*), who were also of three grades. Finally, there were guards, who, during the spread of epidemics and contagious diseases, were to guard the gates, preventing foreigners from entering or patients from leaving without permission, and supervising the cleanliness of the environment.[154]

In summary, all the features of a modern hospital were present in this document. As explained above, there was a traceable shift in medical vision, but for many physicians what made a difference at that time was more the expansion and/or reorganization of the existing medical system than the transformation of medicine itself. Institutional rather than theoretical reform was deemed more crucial, or more attainable, not least because institutional modernization, such as the establishment of modern schools and hospitals, could be realized faster than changes in concepts and ideas. The construction of a hospital might take a year, but the reeducation of a generation of students in modern medicine was a matter of decades. No doubt, modern institutions provided a more comfortable home for the transmission of new ideas.

Medical Professionalization

New Spaces for Medicine

The medical modernization project originated in the state and benefited the centralization of power. But the key players in this process were physicians. If traditional physicians took an active part in the running of modern institutions such as medical schools, hospitals, and sanitary councils, it was because these institutions gave them new opportunities, along with a professional identity. Physicians had previously practiced at their home or in chambers, but now they were also present in the sanitary councils, hospitals, and dispensaries. We have seen that Mirzâ Kâzem-e Rashti and Hâji Âqâ Bâbâ were sent to Rasht, before working as court doctors in Tehran. We have also seen that Mirzâ Habibollâh was sent to Hamadan to work as its Hakimbâshi.[155] Likewise, in the summer of 1861, when intermittent fever (*nowbeh*) and cholera invaded Kâshân and killed many homeless people, immigrants, and beggars, we find Mirzâ Abolfazl Tabib-e Kâshâni, the Hakimbâshi of Kâshân, and Mirzâ Soleimân working to save the lives of the affected people (see Figure 3.2).[156] In February 1864, Mirzâ Abolfazl treated a British telegraph engineer Mân Sâheb (Officer), who fell ill with *hasbeh* (typhus) and *sarsâm*, by bloodletting to expel the putrid blood. At the same time, on the advice of the British Legation, Dr. Schlimmer was also sent to Kâshân to assist Mirzâ Abolfazl.[157]

The newspapers offered further opportunities for physicians. We frequently find physicians publishing their articles in scientific journals like *ʿElmi, Dânesh*, or in the state journals. Members of the Medical Council wrote their diaries, which consisted of their observations or ideas on diseases in various parts of the country, and published them in these periodicals. Mirzâ Kâzem-e Rashti received a robe of honor in recognition of writing such reports in 1861.[158] With the expansion of lithograph printing, those who had written books decades earlier were able to publish them. Hâji Âqâ Bâbâ

Figure 3.2 Mirzâ Abolfazl-e Tabib-e Kâshâni

Malek al-Atebbâ's works on cholera and plague, written from 1830 onward, were published in a form of variorum (*Majmuʿat al-rasâyel dar ʿelm-e tebb*) in 1870.[159] Mohammad Kâzem-e Rashti's *Hefz al-sehheh*, written in 1861, was published around 1888.[160] The emerging printing industry considerably increased the readership of traditional physicians' writings.

Quarantines were often opposed and provoked public revolt not only because they disrupted the normal rhythm of commerce and ordinary life, but also because they did not conform to the ideas that dominated the views of traditional physicians.[161] Although quarantines were a medieval European and Christian institution, they were introduced and perceived in Iran as a modern institution.[162] Considering that the theory of miasma did not conform to the quarantine practice, we can suggest that the traditional physicians' involvement in quarantines was for professional purposes and that they did not necessarily believe in their efficiency. Nonetheless, although one cannot exclude the influence of personal interests in taking a position for or against quarantine, there were physicians who were more concerned with the medical effectiveness of public health measures and who rejected the usefulness of the quarantines.[163]

The State and Medical Professionalization

In Qâjâr Iran, as H. Katuzian observes, "there was no contractual security of titles to ownership," and "no social class independent of the state," and positions were based on "privileges granted by the state."[164] And as A.-R. Sheikholeslâmi notes, "The state officials did not claim their office on the basis of any independent source of power other than the Shah."[165] The state or official positions were indeed precarious and dependent on the whim of the Shah. But this was not an absolute rule for everyone in Qâjâr society. In education and science, for instance, the situation was to some extent different.[166] First, these fields were presumed politically neutral, and there was, at least formally, no restriction on their growth, as underlined in the journal *Iran*, in its inaugural issue, which divided its contents into two: "on the one hand official internal or external political news that was under the strict control of the minister and subject to censure, and on the other, non-official news that could be reported by anyone without any censorship, as was the case in Europe."[167] Articles and news on science and medicine fell within the latter category. The intellectual mix of the journals that published on traditional and modern science as well as their success was partly due to this liberal view of nonpolitical matters. Second, the promotion of modern science (as an alternative to traditional education) provided the Qâjâr state with a means to gain control over education and dent the supremacy of the ulama in education.

The growth of a modern intelligentsia in Iran was founded on the introduction and development of modern science and education and not on the grant of titles or positions. Due to the dependency of officials and bureaucrats on the state, Qâjâr society often witnessed the rapid rise and fall of individuals, but in the medical profession we see few such movements, even though some court physicians benefited from the protection of the Shah. Court physicians did not experience the same instability and uncertain situation in the exercise of their profession as the statesmen did. The patrimonial and arbitrary Qâjâr system was not necessarily detrimental to the formation or reinforcement of the medical profession, just as it did not prevent the formation of a dynasty of physicians, but rather secured their continuity.[168] Of the 22 medical students in 1861, eight were the sons of traditional hakimbâshis such as Mirzâ Ahmad-e Hakimbâshi-ye Tonekâboni, the chief doctor of FathᶜAli Shah, Mirzâ Abolqâssem and Mirzâ Massih, sons of Mirzâ ᶜAli Tabib-e Tehrâni, and Mirzâ Ahmad-e Hakimbâshi (not to be confused with Mirzâ Ahmad-e Tonekâboni).[169] Many pioneering doctors in the twentieth century were the descendants of physicians. One of these medical dynasties was that of the Nafisis. Saᶜid-e Nafisi, son of Mirzâ ᶜAli-Akbar Khân-e Nâzem al-Atebbâ, a graduate of the Dâr al-Fonun, and the head of the state hospital under Nâsser al-Din Shah, traces back the medical activity of his family to the fifteenth century, when Borhân al-Din Nafis b. ᶜAvaz b. Tabib-e Kermâni was invited by the Timurid Ologh-Beg (fifteenth century) to Samarqand. The great grandson of Borhân al-Din, Mirzâ Saᶜid-e Sharif-e Kermâni was one of the physicians of Shah ᶜAbbas the Great (r. 1587–1629).[170] Nâzem al-Atebbâ's son, Mirzâ ᶜAli-Asghar Khân-e Mo'addab al-Dowleh, became the personal doctor of Rezâ Shah and tutor to the heir apparent Mohammad-Rezâ during his studies in Switzerland. And his sons, ᶜAbbâs-e Nafisi and Abolqâsem-e Nafisi, were also physicians and graduates of the Faculty of Medicine in Paris.[171]

Court patronage offered another opportunity for physicians to establish themselves. Following the completion of their studies in Paris in 1860, Mirzâ Rezâ, Mirzâ Hossein (Afshâr), and Mirzâ ᶜAlinaqi were offered court or state positions and also opened their private practices.[172] Mirzâ Rezâ Doktor taught modern medicine at Dâr al-Fonun and also received private patients, practicing, in addition to internal medicine, surgery and ophthalmology.[173] Mirzâ (Mohammad-) Hossein-e Afshâr, son of Mirzâ Ahmad-e Tabib, also opened a medical practice in Tehran and practiced medicine, surgery, and ophthalmology while working at the Dâr al-Fonun.[174] Other graduates such as Dr. Mirzâ ᶜAli Moᶜtamed al-Atebbâ followed suit and used their position in a state institution or the court to establish their reputations.

All physicians who graduated from the Dâr al-Fonun or from medical schools in France were employed. Some, in recognition of their progress in their studies, were appointed as hakimbâshi for a province, city, or a prince before they had graduated. As a result of their success in the examination of 1861, Mirzâ Habibollâh and Eskandar Mirzâ were named the

Hakimbâshis of Guilân and Hamadan, respectively.[175] Others were appointed after graduation, like Mirzâ Mostafâ, who became the Hakimbâshi of Prince Jalâl al-Dowleh, governor of Esfahan in 1862.[176] The personal physician to a prince became the head physician of the city or province under the governorship of that prince. Mirzâ Rezâ (not to be confused with Mirzâ Rezâ Doktor), the Hakimbâshi to Prince Yamin al-Dowleh, was reported to have performed "excellent treatments" in Mâzandarân and demonstrated his skills to the prince.[177] In 1866, Mirzâ Mostafâ was invited to Mashhad to serve its population, where many were suffering from arthritis, *mohreqa*, and *mothbeqa*.[178]

The line between professional doctors and the rank and file was no longer restricted to "hereditary practice," nor was it in line with the distinction between modern and traditional science.[179] In addition to the state patronage that brought together traditional and modern physicians, the journals, sanitary councils, hospitals, and the Dâr al-Fonun were institutions where both categories of physicians could work. A new era in medical professionalization began with the new approach in education. "In the past," the state journal reported, "the students did not know anything of anatomy (*tashrih*) that cannot be acquired without seeing and dissecting, except terms and expressions from the old books. But now they have seen dissections in the West and are skilled in manual medicine."[180] The *RE*, stressed that "since the aim of the journal is to disseminate science and encourage the society at large to acquire science and knowledge, it is free from ancient rules, or from observing hierarchy and social rank in the publication of authors' articles: "We publish articles on the basis of first come first served. If we receive articles from junior students or even from common people but with educational interests, we publish them in order to encourage their authors."[181] Eventually, universal education and the liberalization of knowledge that characterized modern science and were advocated by the new journals broke the close circle of hereditary medicine and opened the field to the wider society.

Conclusion

In the process of centralization, the state introduced modern sciences and, due to their presumed political neutrality, allowed the formation of a class of intelligentsia that, in the context of Qâjâr Iran, was based on education instead of privilege or power.[182] Despite its inherent or structural link to politics, military modernization played a key and pioneering role in the formation of the intelligentsia. The class of intelligentsia remained dependent on the court and/or the state, but physicians made use of this attachment to achieve a social and professional identity and status that in turn helped bring about social transformation.

In the absence of an established and independent institution for traditional medicine, such as those in India, traditional physicians used the state-run institutions to assert or strengthen their professional identity. In this process, for further professional status and integrity, they increasingly adopted modern medicine. Proficiency in practical medicine (*tebb-e yadi*, i.e., surgery and anatomy), emphasized in most medical licenses (*ejâzeh-nâmeh*), was a major characteristic of a professional and skilled physician. It was not only modern institutions, but also modern medicine that played a crucial role in the formation of the profession, a fact that can explain why a whole host of traditionally trained physicians tried to acquire skills in modern medicine. Most of the *hâfez al-sehhehs* (health officers) were not graduates of the Dâr al-Fonun but their institutional link to the *Majles-e hefz al-sehheh* (Sanitary Council) provided them with concepts of modern medicine and public health. Reciprocally, the acquisition of modern sciences through these institutions furthered their professional integration. In other words, modern medicine, through institutional outlets, including journals, provided physicians with a new professional identity. One can argue that the transition from traditional medicine to anatomical pathology that was taking root by the end of the nineteenth century was, at least to some extent, the result of this process of professionalization.

CHAPTER 4

Medical Transition under the Constitution

With knowledge [science], everything is working perfectly;
Without knowledge [science], the world's problems become long and tedious.[1]

The military modernization that ushered in the introduction of modern science in Qâjâr Iran played a key role in the emergence of a modern intelligentsia. In the second half of the nineteenth century, as the relationship with the West grew, this modern intelligentsia, recruited from the nobility and the emerging middle class, became the driving force behind the idea of progress (*andisheh-ye taraqqi*) that, according to Fereydoun Adamiyat, constituted the intellectual foundation of the constitutional movement.[2] According to this intelligentsia, the introduction and adoption of Western ideas, science, technology, and institutions would prevent Iran from stagnating. This development attained its concrete expression in the Constitutional Revolution of 1906 with the catchphrase "science and progress" (*ᶜelm va taraqqi*).[3] It is within the context of the constitutional movement and its "ideology" of modernity that a new step was taken forward in the medical transition. This chapter will examine the new institutional and organizational devices and strategies that laid the foundations of biomedicine in Iran, in a context where humoral medicine was still present, if not prevailing, in practice.

In the middle of the nineteenth century, modernization consisted of the creation of new institutional bodies in medical education and practice. Medical, literary, and political sources inform us of a complex and problematic transformation, which occurred within a framework of medical plurality at work in the newly created institutions: While not always welcoming Western medicine, there were attempts to assimilate Western theories and practices. The creation of modern hospitals was considered as a continuation of the medieval hospitals and their forgotten prosperity. Alongside their association and collaboration, divergence and conflict between physicians of different persuasions in the army, the court, and medical schools continued in both theory and practice. There was both harmony and tension between these factors. Sponsorship of medicine by the state became part of the process of state centralization, and the intelligentsia worked simultaneously both for and against the central state just as it was wary of Western influence, while simultaneously promoting Western values. This situation characterized the ongoing medical transformation that was at work within newly established hospitals, sanitary councils, and medical schools. It was through this process that, from the early twentieth century onward, biomedicine gained prominence over humoral medicine and brought about a resolution to the conceptual dissonance that characterized medicine. As the medical modernization in Qâjâr Iran had an inherent and direct link with Western influence, it is necessary to clarify what is called Western influence by describing its mechanisms and functions.

Modern Western medicine forged a path in Iran through two major developments: smallpox vaccination and surgery based on modern anatomy. Although vaccination was not immediately accepted or universally implemented, by the end of the century it became an acknowledged and established procedure due to its efficiency. The success of surgery in treating conditions that had previously killed or disabled patients, such as bladder stones or cataracts, was undeniable. Western medicine was thus associated, to some extent at least, with an efficiency that was unrivalled in Galenico-Islamic or traditional medicine. So, it comes as no surprise to see that Western medicine increasingly gained audience among the practitioners of humoral medicine, which for cultural, educational, and practical reasons continued to be practiced for the decades more.

Western progress, when compared with the backwardness of Iran, had a shocking effect on the Qâjâr elites. Faced with an overwhelming gap between Iran and the West, militarily, technologically, and commercially, the Iranian elite surrendered to Western science and values. However, Western supremacy cannot alone explain the acceptance of foreign influence, insofar as there must be a receptive agent that takes part in this process for the influence to take effect. The change of vision of Muslim intellectuals as a result of Western influence, which is seen by some Muslims as "insidious," has been known as the most common form of influence in modern history.[4] In the relationship between West and East, both sides are therefore responsible for any "influence" or transfer of ideas. In a colonial context, passive acceptance of a culture through intrusive colonial administration might be perceived as a one-directional influence. In fact, both sides of the relationship experienced transformation.[5] In many cases, the Orientalists, who demonstrate some knowledge and understanding of the Orient, have also been influenced by Oriental culture. Their study of Eastern culture and history generates in them sympathy for the Orient, while they also experience a change in their point of view.[6]

Medical Modernization between Ideology and Politics

In the same way that referring to all Eastern countries under the heading "Orient" is inaccurate, having an indiscriminate view of Western influence is also inaccurate. The race among Western powers for influence at the Qâjâr court through the establishment of medical services generated different forms and "economies" of diplomatic relationship. It intensified medical modernization, insofar as it created, refined, and diversified methods of influence. Despite, or because of, the direct association between the medical and political presence of the British, the Qâjârs opted for politically impartial medical assistance. This was seen in the invitation of physicians from countries deemed to be without political interest in Iran, notably, Austria and Italy. On the other hand, the Qâjârs also made use of medicine to strike a balance among the different countries' influences.

In the same way that diplomacy affected medical policy, medical influence also secured leverage for political influence, a game that was played by both Western countries and the Qâjâr government. On this, there was in fact no fundamental difference between France, Germany, United States, or Britain in providing medical assistance to Iran. From the second half of the nineteenth century onward, Western governments increasingly had recourse to medical services to gain political or economic advantage. This was the general context within which the introduction of modern medicine took place. However, medical activity, in terms of research and provision of care and cure (through mission hospitals) or education (through modern schools), had its own dynamic, and it is this dynamic that should be taken into account in the study of Western influence on medicine in Iran. The succession to the post of Tholozan illustrates this dimension of the history of medical modernization in Qâjâr Iran.

Both Dr. Cherebrin of the Russian embassy and Dr. Dickson of the British embassy provided medical assistance to the Qâjâr court even though Dr. Tholozan was the chief physician to the Shah. When Tholozan accompanied Nâsser al-Din Shah on his third visit to Europe in August 1890, he contracted an illness while in France and, unable to return to Iran, he recommended Dr. Feuvrier, who had already been selected by the French Ministry of Foreign Affairs to succeed him.[7] However, Dr. Feuvrier was unhappy, as his pay in Iran was much lower than Tholozan's had been. Meanwhile, at the Ministry of War in Paris, he was put on the inactive list (hors cadre) and his salary was cut. Dissatisfied with this arrangement, he decided to abandon his post as chief physician to the Shah and return to France.[8] Marie-René-Davy de Chavigné de Balloy, the French Minister in Tehran, advised his government to encourage Feuvrier to stay, as otherwise this position would be occupied by rival powers in Tehran, which amounted to "neglecting French interests" and to "diminishing the French situation in Persia."[9] While Tholozan remained in France, the British ambassador endeavored to secure the position for Dr. Joseph Dickson, physician to the British Legation, but without success, in spite of the support of Prime Minister Amin al-Soltan, who favored the appointment to the detriment of Feuvrier.[10]

It seems that Feuvrier intended to take the place of Tholozan permanently, with the support of de Balloy, who was at odds with Tholozan.[11] However, when Tholozan decided to return to Iran, Feuvrier requested leave from Nâsser al-Din Shah, who agreed, and left for France at the same time as Tholozan returned to Iran.[12] As the Qâjâr government and the French minister were anxious to have a French doctor as Tholozan's successor, they asked Feuvrier to return to Iran. Feuvrier said he would only return on the condition that Tholozan was dismissed by the Shah. However, according to Tholozan himself, the Shah did not accept this proposition and said that he did not need Dr. Feuvrier.[13] In fact, the personal relationship between the Shah and Tholozan worked to the detriment of Feuvrier. In a letter of April 17, 1893, Feuvrier requested that the Ministry of War end his mission in Iran and reappoint him as a military physician so that he could rejoin the army and receive his salary.[14] Feuvrier's rivalry with Tholozan plagued his career in Iran, as did the issue of his rank and salary, which were lower than Tholozan's. Why the French government disagreed with Feuvrier's request and sent another physician to Iran to replace Tholozan is unclear.

After the death of Mirzâ Hossein Khân-e Sepahsâlâr, Minister of War and Commander of the Army, Nâsser al-Din-Shah entrusted the Ministry of War to the prince regent Kâmrân-Mirzâ Nâyeb al-Saltaneh.[15] Nâyeb al-Saltaneh inherited not only the ministry, but also the system that Sepahsâlâr established. He continued Sepahsâlâr's reforms by creating a military school in Tehran on the French model of the Ecole Militaire, motivated partly by rivalry with his brother, Zell al-Soltân, the powerful Governor of Esfahan, who had also established a military school.[16] In 1893, Kâmrân-Mirzâ, upon de Balloy's departure for Paris, asked him to send a physician as his personal doctor. This resulted in Dr. Jean-Etienne Justin Schneider, *médecin majeur de première classe*, being selected by the Ministry of War to be attached to the French Legation in Tehran. But the prince regent wanted Dr. Schneider as his personal physician, for a period of one year from the date of his departure from Paris, and wanted him to teach medicine at the Dâr al-Fonun, on a yearly salary of 2,000 toman with return travel expenses of 400 tomans.[17] Although de Balloy was not satisfied with these conditions and wanted Dr. Schneider attached to the French Legation rather than to the prince regent, he agreed with the conditions, but his correspondence with the government always referred to Schneider as attached to the French Legation.[18] Dr. Schneider arrived in Tehran on November 10, 1893. He reportedly sat alongside Dr. Tholozan at the *Majles-e Hâfez al-Sehheh* in March 1894.[19] At the end of May 1894, Dr. Schneider passed from Nâyeb al Saltaneh's service to that of Nâsser al-Din Shah.[20] In August that year, Nâsser al-Din Shah renewed Schneider's contract for another year, and the French government was delighted because by this decision the "British and German manoeuvres" to secure Dr. Dickson or Dr. Albo as the successor to Tholozan as the Shah's chief physician were thwarted.[21]

Mohammad-Hassan Khân-e Eʿtemâd al-Saltaneh, the Minister for Press and Publications, claimed that Schneider's promotion as Physician to the Shah was achieved due to his mediation, in recognition of which, he claimed that "Schneider visited him on 29 May 1894, accompanied by the French Chargé d'Affaires."[22] The year before, when Eʿtemâd al-Saltaneh had first seen Schneider, on November 14, he reportedly said: "We will see how many of us will be killed by this Hakimbâshi."[23] Despite such a prejudgment, his claim that Schneider's promotion was thanks to his good office points to his sympathy for the French physicians. Eʿtemâd al-Saltaneh was obviously a Francophile and opposed the British, as he himself avowed.[24] But, such sympathy was not the only reason the Qâjârs had recourse to French physicians. The recruitment of French physicians was also consistent with their general policy of deferring to France for military modernization and education. An important number of the academic staff at the Dâr al-Fonun were French: Monsieur Lemaire and (Monsieur) Dual, instructors of music, Monsieur Douvillier, teacher of mineralogy, and Joseph Richard, instructor of French.[25] As mentioned above, in his capacity as Minister of War, the prince regent created the *Madresseh-ye Nâsseri* in 1885 on the model of a French military school, very similar to the structure and organization of the Dâr al-Fonun, which had 154 students in its first year. This number increased rapidly to more than 250 by 1893.[26] The school had a department of military medicine (*Hâfez al-Sehheh-ye Nezâmi*, a copy of the French *Corps de*

Santé Militaire), and the instructor of medicine was a Frenchman, Monsieur Chalmet.[27] Schneider requested that his government send Nâyeb al-Saltaneh copies of French military regulations and army medical regulations as used by the French army so that he could use them as a model for the military school (*Madresseh-ye Nâsseri*).[28] Nâsser al-Din Shah's European tours, as the *British Journal* put it, were not only "for the purpose of appeasing a restless curiosity, but as the indispensable means of effecting a great and serious purpose [introducing modern science and technology]."[29] During these visits, all European countries endeavored to impress the Shah and his ministers. However, the Qâjârs deferred mostly to France for their military modernization.

It is, nevertheless, wrong or simplistic to argue that the only motivation for Europe to provide medical services to Iran was for their own geopolitical and commercial interests. The new waves of scientific research in non-Western countries, such as archaeological exploration and medical experimentation, went a long way toward generating scientific and comparative cultural interest among the scholars and scientific communities that played a key role in the introduction of modern science. Furthermore, although there was often empirically no distinction between a diplomatic mission or embassy and its members, we need to distinguish between the two, because members of an embassy or diplomatic mission could have different opinions, perspectives, ideologies, or cultural and intellectual interests, which had an impact on their relationships with the host country and its events.[30] By the same token, a distinction should be made between the geopolitical, commercial, and economic benefits sought by Western countries on the one hand and their cultural and purely humanitarian aims and scientific interests on the other. The role of individuals in promoting Western science and culture in Iran was fundamental. We have underlined the historical and cultural interests that French minister Gobineau had during his mission in Iran. The outstanding literary, historical, and scientific works of Henry Rawlinson, who deciphered the cuneiform inscriptions of Behistun (Bistun), or Nicolas de Khanikof, the ethnographer, as well as J.-D. Tholozan, E. G. Browne, and Cyril Elgood, in medical and literary studies, bear witness to this process.[31] In 1888, de Balloy, at the time plenipotentiary Minister of France in Tehran, thanked Nâsser al-Din-Shah on behalf of his government for allowing French archaeologists to carry out their digging in Shush, in the southwest of Iran.[32] In fact, it was Tholozan, who, through his good contacts with the Shah, acquired permission for the archaeological research of Dieulafoy in Shush.[33] It would, however, be wrong to overlook material or political aspects of the activities of these "men of culture and science." The purpose here is to highlight that their cultural and scientific activities were not a means for those material and mundane purposes. The elements, materials, and instruments for the creation of a factory of cartridge in Tehran in 1886 were indeed purchased from France via Tholozan.[34] But it would be wrong to subordinate the scientific research and medical service of Tholozan in Qâjâr Iran to such activities.

Although underpinned by the growing economic and industrial development of Western imperialism, the Western intellectual movement had its own dynamics in non-Western countries and generated scientific and cultural interests that did not always follow imperialist goals. The diverging views of the French government and Dr. Schneider regarding the Iranian medical students sent to Paris is revealing in this respect. Mirzâ Mohammad-Khân, a graduate of the Dâr al-Fonun, arrived in Paris to continue his medical studies in August 1888. After only six months of study (October 1888–April 1889), he asked to take the exams, as he was unable to prolong his studies for more than two years. General Nazar-Aqâ, the Persian minister in Paris, wrote a letter to the *Ministre de l'Instruction Publique et des Beaux Arts* indicating his approval. Moʿin al-Atebbâ was another beneficiary of such special arrangements to curtail the duration of his medical studies, which he began in October 1889. In May 1891, General Nazar-Aqâ wrote a strong recommendation to the minister, requesting special favors for Moʿin al-Atebbâ, so that he could obtain a diploma from the Paris Faculty earlier than other students because "the time of his sojourn in Paris is limited and he has to return to Iran on the order of the heir apparent to enter his service." In both cases, the minister submitted these requests to the *autorités scolaires compétentes* and in both cases the university authorities replied favorably, authorizing the candidates to take only the major and final exams and bypassing the other ones.[35]

The period of their study in France, if different from the standard length, could have been officially negotiated between the Dâr al-Fonun and the French medical faculty. But, there was no special agreement and the students were supposed to follow the full course of the medical faculty. However, their study was systematically curtailed for personal or political reasons. This had a direct bearing on the quality of modern medical knowledge in Iran. The medical texts written by French-educated physicians and used in Iran consisted of translations and compilations of French medical sources. Abol Hassan Tafreshi, a graduate of the Dâr al-Fonun in 1874, was sent to Paris in 1885 to study surgery. He returned less than two years later and was appointed head of the state hospital in Tehran in 1887. He produced eight books, on anatomy (*tashrih*), therapeutics, surgical pathology (*pâtology-ye jarrâhi*), preservation of health (*hefz-e sehhat*), percussion and auscultation, ophthalmology, and physiology, but all of these were translations from French sources.[36] While the French government agreed that the students should skip parts of the course in order to complete their studies in two years for political or diplomatic reasons, Dr. Schneider was concerned about their quality. This does not mean that

Schneider completely ignored France's commercial and political interests.[37] There is, however, no doubt that per his profession and intellectual interests he was concerned with the scientific quality of French education in Iran, which also constituted a goal on its own right. He confided to the Dean of the Faculty of Medicine in Paris that his British colleagues in Tehran openly criticized the "ease of obtaining a diploma in France."[38] In Schneider's view, the way the medical students undertook their studies harmed the reputation of French higher education. He therefore recommended that new students be sent to the recently established *Ecole du Service de Santé Militaire de Lyon*, and tried to keep himself informed of their progress.[39]

At times, humanitarian activities, such as the construction of dispensaries or hospitals by the missionaries who provided medical services to the local population, were deemed a channel of influence by Western governments. But, this was not the sole or the main purpose of these missions or their members. Economic and commercial interests sought by the Europeans in gaining concessions were also justified by the idea of modernizing the country. As to the members of the Iranian elite, such as Mirzâ Hossein Khân-e Sepahsâlâr, the granting of concessions was perceived by them to be in the interest of the country considering lack of local finance and expertise for modernization.

In 1882, Prime Minister Amin al-Soltan offered about 11,500 square yards in Tehran to Dr. W. Torrence, the physician of the American Presbyterian Board, for the construction of a hospital in Tehran, in acknowledgement of the medical service provided by a dispensary that Torrence had established in 1881. Dr. Torrence was also given full responsibility for the construction and management of the hospital. According to the plan, the hospital was to eventually accommodate 80 to 100 patients, irrespective of caste, nationality, or religion, and would begin with six to eight. About $5,000 of charity money was collected, and to raise more funds, the US Department of State intervened and made an announcement to the American public. Construction began on August 6, 1889, and the hospital became fully operative in 1892.[40] Four years after the grant of this concession, Spencer Pratt, the head of the American Legation, tried to woo the Shah in order to obtain others. Pratt reported that the Shah had given him the authority "to inform our government that through the Shah, extraordinary concessions will be granted to American capitalists to construct railway lines, cut canals for irrigation and draining purposes, open up mines, establish manufacturing and so on, and he assures me of his full support and protection for such undertakings."[41]

For the American government, commercial and economic interests were behind these humanitarian missions, but the Shah and his prime minister, Amin al-Soltan, believed that, unlike the Europeans, who "had merely sought their own advantage [in concessions] without doing anything in return, the United States have nobly taken the lead in the march of civilisation."[42]

The medical services provided by Western countries might have been aimed at gaining an economic or political advantage or for religious purposes, but, as D. Kumar pointed out, "to dismiss the colonial doctors reductively as the handmaidens of colonialism or capitalism would be to ignore a more complex and more interesting reality."[43] Provision of medical services or the introduction of modern medicine as a means to attain political, religious, or economic ends did not remain a means, or a passive agent, but generated its own dynamic and value. In the introduction of modern science and its development, nonmaterial factors or, in Parsonsian terms, transcendental aims were involved. It was not only objective/empirical but also subjective/intellectual benefits that motivated individuals in bringing research and education to non-Western countries.[44]

In order to better understand the mechanism of modernization, it is important to distinguish between the different goals of medical services generally. They included the promotion of modern science, furthering evangelical activities, enhancing the political or diplomatic presence of the country in question, or gaining economic benefits. Dr. Schneider pointed to the "advantages for France and its representatives in Iran in the constant presence of French doctors at the Qâjâr court."[45] However, to attain this, he insisted on a guarantee of higher quality from French medical education, and on creating properly qualified doctors. The French government's concern, on the other hand, was to have as many French-educated doctors at the Persian court as possible, which would compensate for their lack of a strong political presence. In the case of the American mission hospitals, we find a similar difference between objective and subjective motivations. The United States, as a new power seeking influence, used its missionaries as channels of influence and as a means to secure commercial and economic advantage. Even if in providing medical services, the ultimate goal of the missionaries themselves was the spreading of the gospel, the means to attain that end enhanced medical development. If we view the missionary hospitals or medical services as a social action, it involves the interrelationship between means and ends, just as Parsons argued.[46]

Parallel to the material aims of the American mission, we have Dr. Joseph P. Cochran, a doctor of the American church mission, who devoted his life in Iran to his hospital and patients. Between 1880 and 1882 he built a hospital two miles from Urumia. "There was no drug shop in the country, that is, he must prepare all his own medicines. He had absolutely no assistant... When an [surgical] operation was over, he himself would carry the patient to bed, as even for that work he could not trust his kind but rough helpers, and if it was a serious case, he would sit up all night, or at least come in several times during the night."[47] The hospital was initially funded by only $1,000, which later increased to $1,500. Even so, Dr. Cochran had to make personal contributions to be able to run the hospital.[48] Both Dr. Schneider and Dr. Cochran saw the goal of their missions as twofold: to secure French or American influence, as well as transfer modern medicine, and to provide medical service. In other words, there was an intrinsic relationship between their aims and their means. And the rationale behind this "system of action" was "efficiency," central to the works of both Schneider and Cochran.[49]

Western influence in Iran had two facets, political and intellectual, with much tension between the two. The relationship between the means and ends of the medical services depended on the kinds of political presence of the respective countries. Although the ideology of progress subjectively bonded Iranian intelligentsia and their Western counterparts in a "community of intellectuals," this bond was affected by the nature of each country's influence. For the Qâjâr elites, intellectual influence was not always subject to political weight and could even go against it, as in the case of the priority given to the French medical presence, in spite of its declining geopolitical influence. According to de Balloy, all Western powers in Iran used their medical assistance for political or commercial purposes, including Germany, "with the exception of France, whose interests in Persia are less political than moral. Its colony in Tehran is small and includes engineers, physician and industrialists."[50] Nevertheless, this low profile was dictated more by the geopolitical games with other powers than by the moral or civilizational goals of French diplomacy. Otherwise, there was no fundamental difference between France and other Western governments in their use of medicine for political purposes, but the disposition of the Iranian elite in favor of French cultural presence was more intellectually founded although it was not completely devoid of any political motivation.

In fact, a low political profile allowed the French medical service to gain more influence both because it was deemed politically neutral and because it was sometimes used by the Qâjârs to counterbalance the influence of Britain and Russia. On the other hand, Russia and Britain, whose influence was far greater and older, used their medical position to strengthen their military and political presence. To counteract the road construction concession granted to Russia between Tehran, Hamadan, and Rasht, Britain created a vice consulate in Kerman under the pretext of having an observation post on the borders with India.[51] A year later, when a British physician was appointed in Sistân, Russia obtained authorization from the Qâjâr government to create sanitary stations in Astara, Bandar Anzali, and Dar-e Gaz (in the northeast) in order to fight a potential plague outbreak.[52] The real aim of Russia in creating the sanitary cordon north of the Iranian border was to maintain its troops there;

and Britain clearly intended to reinforce its presence in the south by creating the consulate and providing medical services in Sistan and Kerman.

Regardless of differences between the influence of the imperialist powers, we might also view their presence as having multilayered purposes and projects, spanning military domination and economic and commercial profits, to humanitarian and cultural purposes, to purely scientific or intellectual interests, and all these existed in tandem and often in competition and opposition with each other.[53] Thus alongside their cooperation, there was also a tension between the physicians, who provided the medical services, and their governments, which used them for political ends.

Governments that sought a political or economic foothold in Iran were more active in creating mission hospitals. One of these hospitals was the Morsalin Hospital, created by the British Church Missionary Society (CMS) in Kerman in 1901. The British government had a medical center for the South Persian Rifles (SPR) in Kerman but this was just for British military forces in the south of Iran and was looked upon with antipathy by the local population. The CMS therefore created the Morsalin hospital.[54] Whether for evangelistic or purely humanitarian purposes, Church Mission Hospitals like Morsalin gained a great influence among the population. During the cholera of 1904, when the authorities and the rich (and many physicians) of Kerman fled the city, the Morsalin hospital, staffed by physicians and religious nurses, received cholera patients and provided medicaments and treatment to the poor free of charge.[55] While the aims of the foreign governments were political, mission hospitals played an important role in the medical transformation in Qâjâr Iran.

Waqf and Modern Hospitals

In addition to introducing modern medical practice, mission hospitals also inspired the local *waqf* founders to use their *waqf* for secular purposes such as madrasas and hospitals. The *waqf* is a multifaceted institution of social, economic, legal, and religious significance. The purpose of *waqf* (or *habs*), as originated in early Islam, varied, from eliciting tax from wealthy individuals, to turning a large amount of the conquered land into *waqf* in order to "purify" them.[56] Generally, however, setting aside part of a property to "purge" that property, or legalize it, was a common occurrence among the wealthy and the nobility. But faithful individuals also allocated part of their property for use by the public. This could be religious (such as the creation of a mosque) or secular (such as the construction of a hospital or bridge). In both cases, the creation of *waqf* was deemed to secure salvation for the soul of the founder.[57] We can, however, identify a change in the concept or interpretation of the principle of "purging" wealth in Islam. According to

Imam Fakhr al-Din Râzi (1149–1209), the "duties" of a Muslim's wealth consisted of taxes, such as *zakât*, *khoms*, expenses for *Hajj*, or Friday prayer, and other religious purposes. In a modern commentary on the Koran, ʿAllâmeh Tabâtabâyee (1902–1981) referred to the need to use wealth so that society's economy grows in the public interest.[58] According to "modern" jurisprudence (*fiqh*) the aim of *waqf* and its administration by the state organization of *owqâf* is economic management, which allows wealth to grow both in income and in expenditure, while the popular notion of *waqf* consists of "donations" for the purpose of (*taqarrob*), feeling close to God.[59] Although in both legal (*shar'i*) and secular (*'orfi*) *waqf*, or in the past and the present *waqf* systems, the aim was God's satisfaction (*le-morzât-allâh*) or approaching God (*taqarrob*), the administration and management of *waqf* changed toward using endowed wealth for public welfare and for the growth of wealth, rather than for individual "salvation" or purely religious purposes.[60] This indicates the evolution of the "legal" aspect of *waqf* in history, alongside sociopolitical developments.

Despite its legal support, a *waqf* deed was always subject to interpretation in line with social, economic, and political interests.[61] The *waqf* has consequently a double face: legal and profane. Its legal aspect was always referred to and used as a means for profane and mundane purposes. The inventory of endowments during the last decades of the nineteenth century and beginning of the twentieth century in Iran indicates that most *awqâf* (plural of *waqf*) were created for religious purposes, such as financing different religious festivals and funding ceremonies of mourning (*rowzeh-khâni*) for the imams. The madrasas were mostly *waqf* endowed, but given that they were for religious studies, such *waqf* had a religious rather than secular aim. Most *waqfs* were established for the creation of mosques and madrasas.[62] There were exceptional cases where the *waqf*'s were created for public purposes, such as the creation of water wells and cisterns.[63] From the end of the nineteenth century, there was a tendency to use *waqf* for public welfare.[64] Before his death (ca.1829), Hâji Bashir, the *Khwâjeh* (eunuch) of Prince HosseinʿAli-Mirzâ Farmânfarmâ Governor of Fars, endowed a large number of his properties for the expenses of *taʿzieh* (the annual commemorations of the tragic death on 10 Moharram 61/October 10, 680, of Imam Hossein).[65] This was the typical destination of *waqf* chosen by the nobility at that time, whereas in the first two decades of the twentieth century we find wealthy individuals or the nobility creating *waqf* for the construction of schools and hospitals. The Vaziri hospital, built by Mirzâ ʿIsâ Vazir, the Governor of Tehran in 1899, the hospital Amini in Qazvin, built in 1909, and the hospital Najmiyeh in 1926 by Najm al-Saltaneh, which were all funded by *waqf*, illustrate this shift.[66]

In Chapter 3, we referred to Cabanis' observations on Islamic hospitals and charitable hospices in the West. Mission hospitals in Iran represented a combination of modern hospitals and charitable hospices. In Islam, the "confinement" (*habs*) of a property for a charitable purpose is considered one of the good deeds that last "to the Day of Judgement" (*al-bâqiyât al-sâlehât*).[67] In Islamic history, the *owqâf* were sometimes endowed for the construction of secular establishments like hospitals. The *Robc-e Rashidi* (Rashidi Quarter), built by Rashid al-Din Fazlollâh, Minister of Qâzân, also included a hospital. The hospital cAzodi, created in ca.972 by cAzod al-Dowleh Daylami (r. 949–982) in Shiraz, was also endowed.[68] Nevertheless, these hospitals were quite rare, and most were established by statesmen. It is with the second half of the nineteenth century that we find an interest among the founders of *owqâf* in building hospitals as a result of sociopolitical movements at the turn of the twentieth century. It is also significant that these *waqf*-endowed hospitals, unlike the traditional *Dâr al-shafâs*, all hosted modern medicine, reflecting the movement for progress and modernity.

The author of the manifesto on the state and military hospitals (Document 505, discussed in Chapter 3) maintained that the construction of hospitals was the greatest of the *bâqiyât al-sâlehât*, while in Islamic law the *al-bâqiyât al-sâlehât* consisted of things like prayer, charity, and endowments for the construction of mosques and madrasas, among other things. However, creating a hospital was not the greatest act of charity in Islam. The author of Document 505 hammered home the idea that charitable donations for hospitals were an age-old tradition inherited from prehistoric Iran. There was indeed some kind of charitable donation in prehistoric *Pishdâdiyân* in Iran, but there is no specific indication of charitable hospitals in the Avesta.[69] It is nevertheless significant that for our anonymous author, the origin of charitable hospitals in Iran was in pre-Islamic Iran (but not in Islam). Whether such charitable hospitals in prehistoric Iran existed or not, reference to them as a model was obviously an attempt to justify a project that in fact, as he observed, was widely implemented in the West but not in Iran. Obviously, this author in advocating that devoting funds, which had previously been used for religious institutions, to secular and modern establishments was inspired by the Western tradition of charitable hospitals.

Modern Institutions and "National" or Private Funding

The new elite in Iran considered "illiteracy" the source of all economic and social problems, an idea expressed through new journals, including the *Tarbiyat* (1897), by Mohammad Hossein-e Zakâ'ul-Molk-e Forughi. Following the success of the Roshdiyeh School in Tabriz, and despite the opposition of the ulama, the reformist prime minister Amin al-Dowleh invited Mirzâ Hassan-e Roshdiyeh to build a branch of the school in Tehran.[70] What the ulama opposed in Roshdiyeh School was its new teaching method that Mirzâ Hassan had adopted from modern schools in the Ottoman Empire. A year later, a society called *Anjoman-e Maʿâref* was formed with the specific task of creating and expanding "modern" schools. The members of the society gained the support of Mozaffar al-Din Shah by having him as its head, while its management was entrusted to Nayyer al-Molk, the Minister of Education, through a royal order (*dastkhat*).[71]

The Society for Education (*Anjoman-e Maʿâref*) created several schools in different parts of Tehran, and drafted a constitution (*nezâmnâmeh*) that made members responsible for the collection of funds for new schools. Funds were usually collected from the nobility, whose members tended to follow the Shah's example.[72] The schools were named "National Mozaffari" (*Melliyeh-ye Mozaffariyeh*), to pay tribute to both the Shah and the private contributors. Initially, most of the students were the children of the nobility.[73] But some schools, like Sharaf, were built for children of the destitute.[74] At the Madresseh-ye Eftetâhiyeh, the first of the society's schools in Tehran, a section was also created for free primary education (*maktab-e majjâni*).[75] The establishment of new schools by private donations became popular and as ʿAbdollâh Mostowfi noted, each month one or two new schools were opened.[76] This trend gathered momentum under the Constitutional Revolution, and new methods of fund-raising, including the organization of concerts, were introduced—an entirely new phenomenon in Iran.[77]

The money collected by the *Anjoman-e Maʿâref* was of two kinds: capital for investment and funds earmarked for new schools and current expenses. The capital was managed by the trustees (or members of the society), who in turn entrusted it to a company. The profit made by this investment was paid monthly to cover expenses, which included teachers' and other employees' salaries.[78] The creation of the Society for Education began a new era in education through the foundation of secular funding rather than *waqf*, with management based on a kind of capitalist investment, the legal aspect of which was outside the control of the ulama. This was aimed at taking education out of the control of the clerics and basing the curriculum on modern science. Article 17 of the society's constitution said that "anyone who wants to create and construct a madrasa or *maktab* must get the permission of the Ministry of Education (*wezârat-e ʿolum*) and approval of the Society for Education."[79] The movement continued into the twentieth century up to the eve of the Constitutional Revolution. In 1904, for example, Weqâr al-Molk, with the association of others, created the *Madrasa-ye Soltâni* in Tehran.[80] The new charitable (secular) institution for the creation of modern schools inspired the *waqf* founders to devote their *waqf* to the creation of modern schools. The *Madresseh-ye Saᶜâdat* in Kerman built in 1908, for example, was financed by the endowments (*mawqufât*) of Hâj Mohammad Karim Khân-e Rashti.[81]

Under the Constitutional Revolution, modern schools underwent further change. The Society for Education was now consisted of not just the nobility but mostly revolutionaries, who were at the forefront of the movement, and who created schools or hospitals by collecting funds from individuals. Initially, most pupils were recruited from the family of the donors (most of whom were part of the nobility), but later the new schools increasingly recruited poor children free of charge. In fact, the constitutionalist political parties advocated the creation of compulsory and free education for all children, the creation of hospitals, and a "salvation army," presumably for the lower classes and the destitute.[82] After taking refuge in the mausoleum of ᶜAbdol-ᶜAzim in Rey, the constitutionalists, being opposed to the reactionary government of Mohammad ᶜAli Mirzâ, distributed articles saying that money paid for *rowzeh-khâni* (sermons by the mollahs), or for expenses incurred on pilgrimages of the saint imams, should rather be spent on schools for girls, the establishment of factories, and for the construction of roads and railways.[83] The Society for Education was represented in other cities as well. In Tabriz, the society was called *Anjoman-e Dabestâniyân* (Society for Primary Education), with the aim of integrating the existing traditional *Maktabs*.[84]

It was within the context of the new system of education, which moved away from traditional schooling toward modern science and from *waqf* funding toward private endowments, that medical modernization entered a new phase. This phase saw modern medicine overcome Galenico-Islamic medicine and found new institutional spaces and financial support.

The Ideology of Modernity

The new phase in the development of modern medicine originated in the idea of modernization for progress partly implemented under Mirzâ Hossein-Khân-e Sepahsâlâr.[85] This ideology gained momentum under the Constitutional Revolution, and found executive bodies in the newly created *anjoman*s (constitutional societies) and the *Majles* (Parliament).[86] The term "ideology" is employed here in the sense of "belief" in modernity. It was not, in Marxist terms, an ideology of the dominant economic class; it was born of a new era in historico-intellectual development, but it was also a utopia aimed at transforming Iranian society.

Modern science and technology, which were considered necessary to create a strong state and nation, also provided the model for new political organization. In its inaugural edition in 1906, the journal *Majles* wrote:

> Thus the members of the Parliament on the one hand and the ministers of the state on the other, should consider themselves as positive and negative poles of an electrical system, in which, if there is no collaboration between these two poles, within the very framework of their opposition or conflict they cannot illuminate the world or to operate telegraph, telephone and other instruments.[87]

Acquiring (modern) science and knowledge had become a social value because it was considered a source of intellectual wealth and material comfort. Upon the suicide of a young nobleman, Hassan-ʿAli Khân-e Eʿtezâd al-Molk, the state journal published the opinion of one of his relatives on the cause of his death. The article said:

> This young man killed himself because he was without knowledge. He was beloved of his parents but deprived of any culture. Once he [over] spent all his money and, out of "nobility," refused to be modest in his spending, he considered that he had nothing else of value in this world. He did not know French, or history. Had he known history and French so that in his spare time he could read [foreign] newspapers and could converse with cultured and learned people, he would not have [felt empty and therefore would not have] committed suicide.... It is therefore necessary that noble people force their children to study and educate them in order to avoid peril.[88]

By 1906, the modern science and technology that had been introduced into Iran by the wave of military modernization had gained wider meaning and was aimed not only at strengthening the state, but also at enlightening citizens, who could then create the foundation of a strong state. The military journal *Merrikh* (*Mars*), founded in 1879 by Sepahsâlâr, set out its aim as disseminating modern science as practiced in the West and other parts of the world, and discussed subjects relating to human rights (*hoquq-e ensâniyat*). The idea of human rights was taken from the West, but gained a different connotation in Iran. This journal emphasized that "the aim of an army is not to wage war against weak neighbours, but to establish and maintain order inside the country, to protect its borders and to take care of the King by enforcing his right over his subjects."[89] Thus, for *Merrikh* the introduction of modern science was to strengthen the state, thereby protecting the rights of the sovereign over his subjects and not vice versa. Significantly, by praising the efficiency of the modern army "to enforce the King's right over his subjects," the journal followed the traditional anthropomorphism that compared the country to the human body, whose head was the Shah and whose body was the nation. So, naturally, protecting the head was vital. In the early twentieth century, the anthropomorphist vision of the country evolved and it was by focusing on the education of its subjects that a strong state could be formed.[90]

The education of the people would be impossible without modern science. According to Malek al-Motekallemin, one of the constitutionalists, knowledge and science were the doors to progress and civilization.[91] In its first issue, the journal *Omid* (*Hope*), which began publication after the constitution was decreed, stressed that "science is a source of strength. No ignorant nation can have a strong state or ever reach fame. It is upon us to acquire science and reap its fruits ... we must create modern schools and encourage people to read."[92] According to *Omid,* the sciences and technologies that benefited human beings (*ensân*) and the nation (*mellat*) included modern medicine, engineering, agriculture, law, political science, commerce, mechanics, physics, chemistry, and mathematics. But for these to develop, schools were needed.[93] Scientific discourse also drew on existing religious resources, the Koran, and the tradition of the Prophet, which advocated the acquisition of knowledge.[94] "It is science," continued *Omid*, "that made the British sovereign over half of the world. It is their enlightenment that brought about their authority over 400,000,000 individuals.... Courage and happiness is the result of science, and trouble and poverty the consequence of ignorance."[95]

The "ideological" faith in modernity that propelled the constitutional movement also informed the second phase of medical modernization. The creation of a constitution to limit the arbitrary and authoritarian power of the Shah went hand in hand with the universalization of education through the creation of modern schools, the development of public health by cleaning the streets, increasing security by lighting the streets, preserving potable water from contamination, increasing the population through the improvement of hygiene, and the application of modern medicine. These measures were all part of the "ideology of modernity" taken from the West.

One can see this change of vision, from the traditional to the modern, in political organization and in the perception of air and water. The role of water as the vector of the cholera germ (*mâddeh-ye wabâ*) was mentioned earlier in the nineteenth century in state journals, but in medical discourse miasma was still considered the major cause. In the early twentieth century, the importance accorded to water as the vector of the pathogen agent gained weight only through the ideology of progress and modernity and its message for sociopolitical change. Even in 1893, the government's only remedial measure was to clean public cisterns. In the journal *Akhtar* (run by Iranian dissidents in Istanbul), intellectuals blamed the government for not paying enough attention to public baths, whose "water was rarely changed and generated a thousand kinds of disease."[96] After the epidemic of 1904, however, it was acknowledged that water, rather than miasma, was the transmitter of cholera, and measures were taken to prevent the contamination of potable water. Wash-houses or public laundries, equipped with a water tank with several taps, were built so that people do not use the running potable water for washing clothes. Sewage was channeled to deep pits dug nearby. The state journal was aware of the deficiency of this system because the sewage could infiltrate the surrounding land. It recommended the use of metal pipes, which were not affordable at that time, as an alternative. Initiated by the police, or *nazmiyyeh,* for further implementation of this measure, an appeal was made for private funds by the wealthy individuals to construct more public laundries.[97]

The Constitutional Revolution incorporated the hygienist movement and gave it a political and ideological dimension. Lighting the streets, cleaning public baths, and the management of water, in line with the modern theory of contamination, were on the agenda of the constitutionalists.[98] The City Council (*showrâ-ye baladiyyeh*) was advised that its main duty was to prevent water contamination. The old regime's customary official for distributing water, or *Mirâb,* was, according to an article in *Tamaddon,* oppressive and discriminatory as he did not distribute water equitably. The first task of the City Council was to stop unjust water distribution, so that everyone was supplied with clean water at least once a week, and the second was to cover streams in order to avoid pollution.[99]

We find a similar change of vision in the concepts of "nation," "nationalism," and "state power." State power had been defined in terms of the army, but was now extended to the nation. Nationalism, which had been vaguely based on traditional Persian literature and the unifying role of the Shah, gained a new conceptual depth thanks to new archaeological discoveries. Referring to the works of the Dieulafoy couple, the journal *Tarbiyat* noted that "they excavated the ruins in Shushtar (Shush) and have taken the objects that we consider the remnants of the period of ignorance and lethargy (*nâdâni va bikâri*), to France for the Louvre Museum; these materials are the testimonies of the forgotten or ignored potentials and capacities that the Iranians have."[100]

Ernest Renan was praised by the editors of *Tarbiyat,* when he distinguished Iranians from other Muslims for their "intelligence" and "the importance they accord to science," maintaining that the Iranian elite of the Sassanid period, by serving the Abbasid caliphate, introduced Persian know-how and civilization into Islam.[101] The Iranian intelligentsia built the idea of nationalism on the pre-Islamic power of Iran and claimed it as a source of national modernization and progress. The periodical *Omid* devoted sections to reminding readers of their country's bygone power in past eras. But it argued that the revival of past glories was only possible through (modern) "science" and that it was due to the lack of science that the country had become weak, and had, for example, granted railway construction concessions to the Russians.[102]

Just as the ideology of *mashrutiyat* (constitution) promoted modern education, science, and patriotism, it also sought respect for the rights of citizens. This respect helped meet "human rights," one of the principles of the Fundamental Law.[103] Increasing attention was paid to hygiene and public health, cleaning, and street paving. Shortly after the establishment of Parliament in 1907, *Omid* suggested to the deputies that major projects should be prioritized, including the establishment of schools and a polytechnic in each city, and the enlargement and cleaning of the streets.[104] No doubt the principle of the "rule of law," order and discipline were going to affect the nation's lifestyle.

A. Williams Jackson, an American professor of Persian studies, who visited Tehran during the Constitutional Revolution, shortly after the creation of the National Assembly, pointed to the watchword of dozens of newspapers, "liberty, equality, and fraternity." He observed: "Signs of progress were equally noticeable in the introduction of municipal regulations and the improvement of matters of domiciling.... houses were numbered; the names of streets were indicated.... The main thoroughfares were occasionally sprinkled and cleaned."[105]

Modern Science, a Model for Medicine

The scientific discourse that gained prominence during the Constitutional Revolution also represented a watershed in the development of medicine. From this period onward modern Western medicine overcame the dominance of traditional medicine. While earlier the public press published on both modern and traditional medicine, constitutionalist journals published articles only on modern medicine, despite the fact that no fundamental changes occurred in the medical landscape of the country. Contemporary sources tell us about physicians who boasted of their practical experience of hospitals and their surgical skills, alongside their traditional theoretical knowledge. Modern medicine became a fashion, prompting traditional physicians, who had not properly studied modern medicine, to begin claiming to be skilled.

Dr. Schneider, whom we met earlier in Chapter 3, considered this situation perilous for public health. In the course of its activities, the Sanitary Council had acquired various responsibilities, including the examination of practitioners and the appointment of both traditional and modern physicians for different parts of the country. Under the influence of modern science and the constitution, however, the Sanitary Council proposed that a diploma in modern medicine be made compulsory for the practice of medicine, with harsh penalties for noncompliance:

> We have been informed that the Sanitary Council has established that only those who hold a diploma of medicine from the Dâr al-Fonun approved by the Ministry of Education (*wezârat-e maᶜâref*) and five years of experience in the hospitals in London and Paris, approved and sanctioned by the medical universities there, has the right to practice. And if a person who has no such diploma or certificate treats a patient, even if that patient has recovered, that doctor should be imprisoned for having practiced without appropriate documents. And if the patient dies of the treatment, the doctor must be given a life sentence.[106]

This project, which answered Schneider's concerns, was not, however, accepted by the physicians. They objected that it would take several generations to provide enough physicians trained in modern schools in Iran or in Europe and that it required 30 years for the realization of this project.[107] The opposition to the project was by no means unreasonable, as in 1932, the head of the *Sehhiyeh* (Medical Centre) of Yazd wrote a report to the local authorities about Hâj Esmâᶜil Hushmand, advising them to prevent this traditional doctor from practicing modern medicine. Despite having permission to practice humoral medicine only, Hushmand had injected a patient. The patient developed gangrene in his hand and died two days later. While a report was on its way to the authorities, Hushmand had already left Yazd for Isfahan or Tehran to take his exams in order to obtain permission to practice modern medicine (Figures 4.1 and 4.2).[108] In 1933, the local Office of Education (*edâreh-ye maᶜâref*) in Azarbaijan sought advice from the Ministry of Education (*wezârat-e maᶜâref*) as to which measure to adopt regarding untrained

midwives: "There are many midwives without permission who practice. But at the same time there are not enough trained and skilled midwives here."[109] The ministry answered that "as long as there are not sufficient numbers of educated and trained midwives in the provinces, you should use the untrained but experienced ones."[110]

Despite the enthusiasm for modern medicine, traditional medicine continued to be taught, and at the Dâr al-Fonun both were studied (Figure 4.3).

The expansion of modern medicine through university education under Rezâ Shah could not prevent the practice of traditional medicine for the decades to come. Many doctors who graduated in modern medicine still practiced traditional medicine as late as the 1960s. Dr. Ahmadiyeh, who graduated from the Dâr al-Fonun in 1915 and also studied in Paris, used the medicine of Avicenna and Gorgâni when he felt he could not cure his patients according to modern medicine. But the medicine of Avicenna was only used "alternatively," and not as an initial treatment.[111] Therefore, if in the nineteenth century traditional medicine was practiced alongside modern medicine by conviction, or for institutional reasons, in the postconstitutional period it was practiced in response to market requirements.

In the same way that mastering modern science and technology as one of the ingredients of constitutional discourse became a norm and received social esteem, Western physicians, who presumably embodied modern medicine, were accepted in the local market despite the fact that they represented a potential threat to their Persian counterparts. From the turn of the twentieth century, Western physicians opened practices in Tehran, and newspapers advertised their activities: Dr. Haaz from Germany, who received patients daily from 7 to 11 AM and 3 to 8 PM,[112] and Dr. Trafus, the Greek physician, who received patients daily from morning to 12:00.[113] They were also active in advising on public health, sending instructions to the provinces

Figure 4.1 Kholâsat al-hekma (late nineteenth to early twentieth century), a modern medical text used by traditional physicians

۲

جهان مظفرالدین شاه و خلّدالله ملکه وسلطانه وایّدالله جیشه
واعوانه و بین نابه بتخصّص شاخص کافی و رادا بمجد وافی و نیر روشن ضمیر و اعلم حضرت
مستطاب معظم اتابک اعظم مدّظله العالی
لم
دیدکه ریاست جناب مستطاب اجلّ اکرم و زیر علوم و معارف صاحب مقامات تعا
و محامد و اوصاف متعالیه ضاعف الله اجلاله و ادام ایّامه
ل
در نشر و انتشار رشته از فنون متنوعه و علوم و تشعیب از شعب متکثّره صنایع بذل
جهد مینمایند و در تشکیل مکاتب و مدارس عالی و صنعتی مساعی جمیله مبذول
میدارند این بنده ... امیدواری و از آرزومنده دولت جاوید آثار حاجی بن العابدین خان
دکتر ولد حاجی شکرالله خان تا جارقوانلوازسلسله وطایفه برجی اعام سلطنت
نیز بهمراهی پاجمع مصمم گشت که محض خدمتی به دولت و ملّت ایران و برادران
وطنی کتابی مشتمل بر معالجه امراض از کتب معتبره فرانسه و فارسی ترجمه نمود
و بخصوص ادر عبارت سعی کرده که سهل و آسان و خالی از اشکال و دشواری باشد
تا خاص و عام از آن بهره برند و در و نزد یک بدون زحمت از وفاء آنچه منظور
منتفع شوند در حقیقت تذکره بقصد یادآوری برای محل عمل مدقّن سنه
از خلاصة الحکمة ناامید رجاء واثق آنکه در حوزه نافع و مقبول گردد و در
پیشگاه اقدس و دایره قدس پسند و قبول سجود یمتد الحبیب و کبد العیم
واین کتاب شریف لطیف بهمراهی جناب مستطاب اتابک میرزا محمد رضا خواهی که در علوم تعالی
عربیّه و حکمت و طبّ عتیق و جدید خطی وافر دارند خاصّه در طبّ جدید و باقی علوم معقل
و از قبیل ریاضی و فیزیک و شیمی و تشریح و کتب طبّی آلمانها سالها است نزد ایشان
تحصیل نموده و همه را بوجه اکمل دارا شده توجّه شد و به طبع و جلیه طبع و استر گردید
و بالله التوفیق و علیه التکلان

مقصود

Figure 4.1 (Continued)

and cities on preventative measures.[114] On behalf of the French embassy, Dr. Schneider, who was appointed head of the Central Sanitary Council (*Majles-e Hefz al-Sehheh*), provided all physicians in the capital with smallpox vaccine, vaccine against cholera, and diphtheria serum free of charge. The

Medical transition under the Constitution • 143

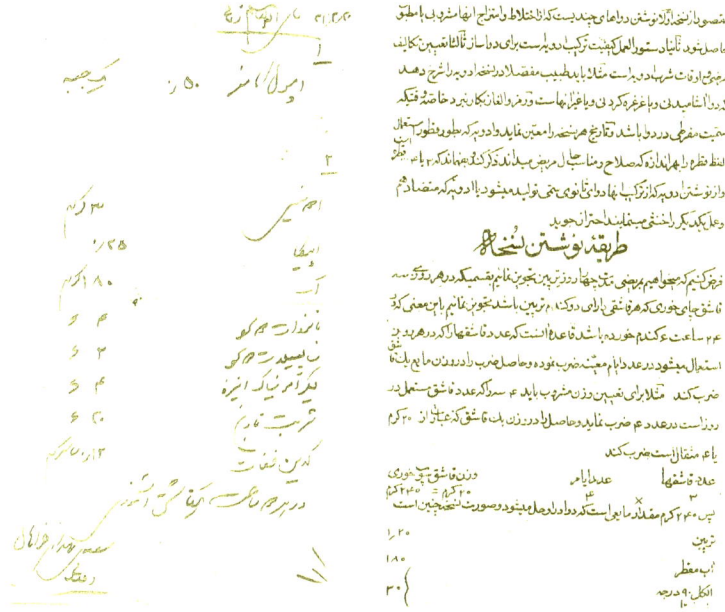

Figure 4.2 Prescription according to Kholâsat al-hekma and a sample prescription in 1332 (1953)

Figure 4.3 Traditional and modern educations

vaccine and serum were imported monthly from France and were sent to other provinces on request.[115] This is all the more significant because at the same time, under the Constitutional Revolution, the consumption of foreign products was criticized, whil locally made products were encouraged.[116]

When, after World War I, Western physicians left the country, the Committee of Advocators of Medical Reform strongly criticized those responsible for their departure. The Committee also complained that "physicians, who had studied in Europe about twenty years earlier, should no longer be permitted to teach medicine, which has advanced in leaps and bounds. Instead, skilled and knowledgeable physicians should be invited from Europe to replace them."[117] The number of Western physicians increased under Reza Shah, despite competition from local physicians. For instance, we find Dr. Schtrong, who taught pharmacology as the head of the state School of Art.[118] At the request of the German embassy, Dr. Gustav Frank, who had worked for about ten years in Iran (seven in the provinces and three in Tehran), was allowed to continue practicing by the Ministry of Foreign Affairs, along with his wife, who was a midwife.[119]

While Western doctors were allowed to practice in order to supply "skilled" medical services, many local physicians who wanted to study at the school of medicine faced difficulties. In 1926, Mirzâ Assadollâh Esfahâni was denied admission to the school of medicine despite the fact that he was from a family of physicians and had finished the fourth year of the Dâr al-Fonun, after obtaining his diploma from a secondary school (*motewasseteh*) in Tehran. So, he wrote a petition to the minister.[120] Likewise, Mostafâ Falâti, who had studied medicine with a certain Dr. ʿAli-Khân-e Falâti in Tabriz and had prepared to take his exams in order to study at the school of medicine in Tehran, was accepted only as an "unofficial" student,[121] which meant that he could not obtain a diploma. These difficulties were apparently due to a lack of buildings, laboratories, professors, and equipment. We are told that students were poor and requested financial help or worked outside the profession to make a living. Such shortages naturally generated irregularities and nepotism, but it indicates the degree to which the study of modern medicine was in vogue.

Modern Public Health and the Rebirth of Iran

In both pre-Islamic and Islamic Iran, bodily health and hygiene, whether in ritualistic, religious, or secular forms, was part of the cultural heritage. In premodern Iran, wrestling and sport in *zurkhâneh* (a kind of palestra, or Persian gymnasium) were advocated as instruments to enhance moral qualities such as chivalry, courage, loyalty, and generosity.[122] In the legendary history of the Shâhnâmeh, Rostam, who symbolizes the strong man, is hailed for having saved his king, Kay Kâvoos, from the demons (Divs).[123] In the nineteenth century, both hygiene and physical strength became more consciously linked to the strength of the nation and revival of Iran.

The association of health, population, and nation became more articulate after the constitution. In a conference on moral renaissance (*now-zohur-e akhlâqi*) organized on June 8, 1908, by the Society of Students (*Anjoman-e Mohasselin*), a lack of science, knowledge, and social consciousness was associated with a history of "despotism and tyranny that reigned for thousands years."[124] The question of moral (or national) degeneration was also linked to physical illness and weakness. Traditional education, in which the custom of punishment was a cornerstone, was deemed responsible for moral degeneration. Published from 1910, the journal *Dânesh* placed emphasis on the physical health of children: "Weak and ill children are often bad tempered, obstinate and pernicious. The solution is not punishment; a system of education borrowed from tyranny; but in medical treatment."[125] The contrast was depicted in a comparison of the two branches of education. In the modern school, students sit on benches in a French class in an orderly manner; there is no punishment, but comfort. In the traditional school, a lack of hygiene is illustrated by a water container with a bowl that everyone uses. In the modern school, the students use tap water.[126] Significantly, the teacher in the modern school says:

> What causes progress, wellbeing and reputation is science. The relation of an ignorant to a knowledgeable individual is that of a blind man to a person who has eyes. The human value of an individual increases with the number of sciences he masters. If you know French and German, you are two people instead of one.[127]

Note that here the subject of the class, French, which represents "modern Western science," is particularly significant.

The birth rate in nineteenth-century Iran was not probably low, but infant mortality was high mainly due to smallpox and diphtheria. The military depended on the population, which was an indicator of the strength of the state. The importance of medicine and physicians for the army was underlined in medical texts.[128] With the constitutional reforms, the relationship between public health and population strength grew even clearer. The journal *Shekufeh* (*Blossom*), established in 1912 by Maryam-e Mozayyen al-Saltaneh, the daughter of the late Mirzâ Seyyed Razi Ra'is al-Atebbâ,[129] was aimed at women and published articles on *akhlâq* (ethics) and *adab va tarbiyat* (discipline and education), teaching women to raise their children in line with modern public health and medicine to increase population.[130]

Parallel to the revival of the past came a new interest in science and race. This revolved around the idea, popular in the West, of the superiority of the Aryan race. These ideas were all the more influential beyond the West, as they were fresh and vigorous and received with conviction and faith in their places of origin. Archaeological explorations in Iran had reinforced the theory that Iranians were cousins to Europeans. Gobineau and Croizier[131] were explicit on this issue, while others like Ernest Renan focused on the intellectual superiority of the Iranians over other Asiatic races, particularly the Arabs. This discourse, while satisfying the age-old anti-Arab bias, further encouraged the Iranian intelligentsia in their modernist ideas. Anti-Semitic and anti-Islamic ideas characterized the writings of Mirzâ Aqâ Khân-e Kermâni, who, praising the sociopolitical system under the Sasanians (224–651 AD), believed Iran had actually become sick since the arrival of Islam. Referring to European physicians who believed that the dreams of pregnant mothers influenced the character of their children, he pointed out that lack of care for the education and material well-being of women had caused the degeneration of their children. "While the Europeans devote great attention and care for the rights and education of women, in Iran women are not even counted as human."[132]

According to Kermâni, this moral corruption was illustrated in the character and behavior of the mollahs, who "considered lies or dissimulation (*taqiyya*), hatred, ugliness, quarrelsomeness, superstition, and slander, as necessary (or religious) qualities."[133] It was the Arab domination that had corrupted Iranians, who were now characterized by a lack of stamina and poor demeanor and posture: heads forward rather than up and with their backs hunched rather than straight.[134] One can see the influence of theories of hereditary and neo-Lamarkian genetics in Kermâni's ideas.[135]

The revival of Iran was deemed to be best achieved through regaining its past greatness and through strengthening its moral and physical health. Schooling did exist in Iran, but only modern schooling was associated with hygiene, health, and physical strength. In the early twentieth century, the journal *Adab* listed several conditions for schools, drawing a direct relationship between modern education and patriotism.[136] One of the constitutional societies—the *Anjoman-e Farhang* (culture)—claimed that its ideology was threefold: constitution, patriotism, and dissemination of culture and science. To this end, it created a free evening class for adults in the *Anjoman*, with a curriculum consisting of French, English, Arabic grammar, history, geography, mathematics, and administration.¹³⁷ The term *vatan* (homeland) coupled with *maᶜâref* (education) became the leitmotif of the modernists in the years following the constitution.

Constitutionalism and Modern Medicine

In its inaugural issue, one of the periodicals newly created under the constitution stated that:

> If a nation (country) is without a National Assembly (*anjoman-e melli*) that nation is known as savage and uncivilised. Just as a callous and mean-spirited physician (*tabib-e bimorowat*) wishes that people are always sick, the men of power prefer the nation to be without knowledge and science, so that they can impose their will on them.¹³⁸

Likewise, in its inaugural edition in 1906, the periodical *Majles* (*Parliament*) wrote:

> The society of the National Council [Parliament] is like the brain, which makes decisions and the ministries of the state operate like hands and legs and other limbs of the body, which execute the ideas of the brain. One can even say that Parliament is the centre of the understanding and perception-nerves, while the state is the centre of the locomotion-nerves. So long as both are healthy, the body of the state stands up. However, once one part falls ill, the body is deprived of its vitality.¹³⁹

The use of such medical metaphors by the constitutionalists is revealing. They aimed at illustrating the transition from an old and sick society, where traditional science prevailed, to a modern and healthy social and political system, with the help of modern science. The comparison between an unjust physician and an unjust ruler, and the equating of the lack of (modern) science with illness, clearly contrasts traditional medicine as a symbol and cause of illness with modern medicine as representing and creating a healthy nation. In 1905, the Head of the Sanitary Council was Dr. Schneider, who selected medical officers or physicians to work in SC branches in the provinces.¹⁴⁰ In 1907, Octave Le Comte, a pharmacist to the French army, was sent to Iran as the Pharmacist of the Shah, through the recommendation of Nazar-Aqâ, the Persian ambassador in Paris (Figure 4.4). Le Comte also taught at the Dâr al-Fonun.¹⁴¹ By taking the side of modern medicine, the constitutionalists aimed to formalize its triumph over traditional medicine, which was still sponsored by the state and court.

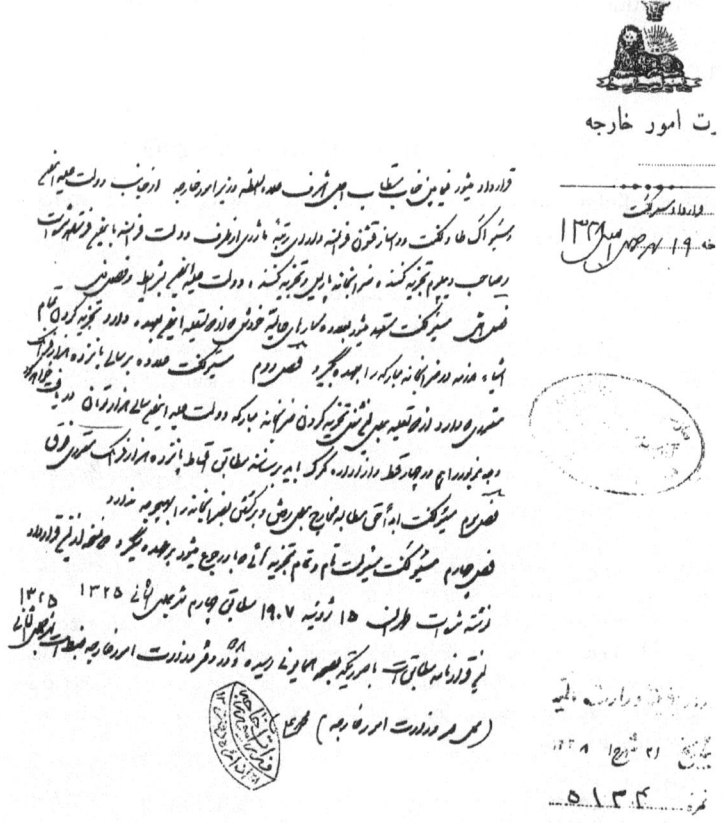

Figure 4.4 Pharmacist Octav Le Comte

By the time the constitution was established, modern medicine was considered as "a unique know-how" in the possession of the Western powers. Under the constitution, the reformers, for the sake of progress, accepted modern medicine in its "quasi-colonial" status, considering it as a knowledge superior to "national," traditional medicine. In December 1909, the American hospital issued licenses to three physicians Mirzâ Qawâm al-Din, Mirzâ Mozaffar Khân, and Mirzâ Amir Khân, specifying that "they have studied anatomy, physiology, pharmacology, practised medicine, surgery and the use of antiseptics, ophthalmology and midwifery for six years; they have always been present in the medical and surgical clinic at this hospital." The

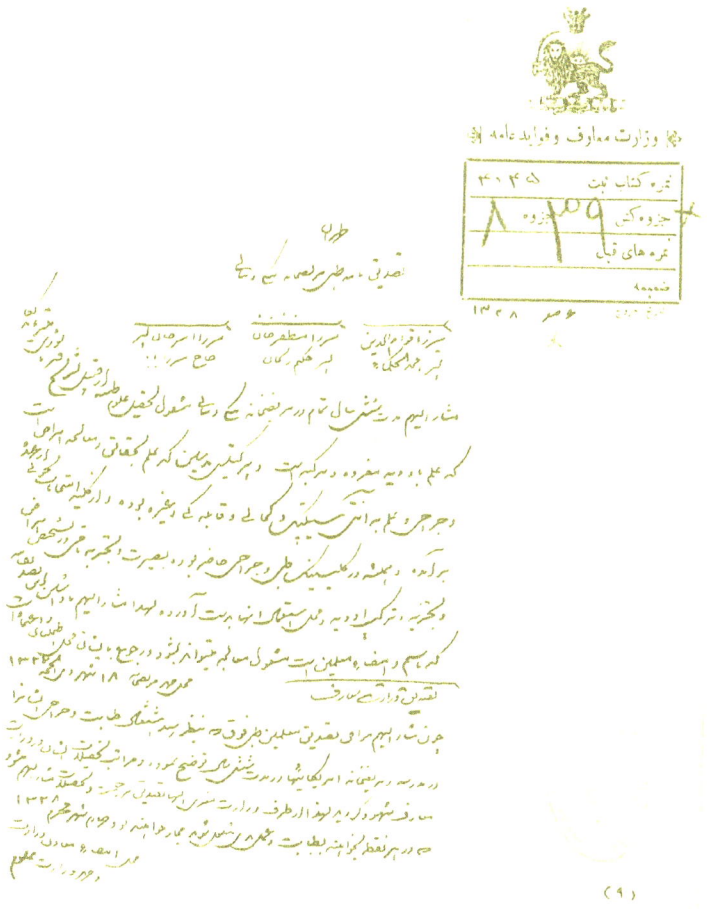

Figure 4.5 Certificate issued by the American hospital

hospital issued this certificate in December 1909 and the Ministry of Education (*maʿâref*) approved it two weeks later, in January 1910 (Figure 4.5).[142] Significantly, all those who applied for a license to practice or for a scientific order (*neshân-e ʿelmi*) flagged up their theoretical and practical skills, which they had acquired at modern schools, the state hospital, and at American or British establishments.

The regime change was another reason for physicians to study modern medicine. When the ideology of modernity was supported by a political regime, it is likely that physicians moved to modern medicine not only from conviction, but also because modern medicine had political weight. References to traditional medicine were still made after the constitution,[143] but these were fewer and most applications for practice at that period referred to modern medicine, surgery, and practice at hospitals, and not one application boasted of skills in traditional medicine, despite the fact that it was still active at the Dâr al-Fonun until 1911, when it was abolished.[144]

The constitutional government became directly involved in the matter of public health. In 1907, it "required the governors of the provinces to 'pay strict attention to... public health... and [to] act with utmost speed in the case of epidemics. In 1910 the government took the first step toward introducing vaccination in Iran by setting aside '10 per cent of the taxt on transportation... for improvements in public health and particularly for general and free vaccination against smallpox and diphtheria.' "[145] However, the revenue of transportation tax, if it was leveied at all, was insufficient for the remuneration of physicians. In 1909, the inhabitants of Quchân requested a medical center, including a physician and a pharmacy. At an annual cost of 2,000 tomans, they asked Customs to pay its expenses by levying 0.5 *shâhi* from each postal package. It was then agreed that 1,000 tomans should be paid by Quchân Customs, 300 from central government, and 700 from the Russian consulate and Russian bank plus contributions from the large companies in Quchân.[146] Likewise, the salary of the above-mentioned Octave Le Comte was paid by Customs in four installments.[147] Physicians were paid by different organizations, the money raised in different areas. For example, some, like Mirzâ ʿAbdollâh Loqmân, were paid by the Dâr al-Fonun, while another physician, Mirzâ Yahyâ Khân-e ʿEmâd al-Hokamâ, was paid by the Royal Treasury, apparently because he often accompanied the Shah (*dar rekâb*).[148] The constitutionalists failed to set up an efficient system to finance medical services. However, private funds, which emerged as one of the results of the constitutional movement, came in the form of donations and in secularizing the *waqf* (Figures 4.6 and 4.7).

The development of medicine reflected sociopolitical changes. If previously the state and its self-strengthening strategy informed the creation of modern medical institutions and how they were structured and organized, it was now the emergence of an alternative voice (civil society, or, in the words of *Iran-e Soltâni, tarvij-e mellat*) (development of nation), that decided how medicine was going to change.[149] Throughout the nineteenth century, few military or state hospitals were built in Tehran. During the cholera outbreak of 1904, the Mozaffari hospital, near the Qazvin Gate, was the center of distribution for European medicines. In 1905, the homeless and insane, in Tehran, were transferred to the state hospital (*Bimârestân-e dowlati*). This

Medical transition under the Constitution • 151

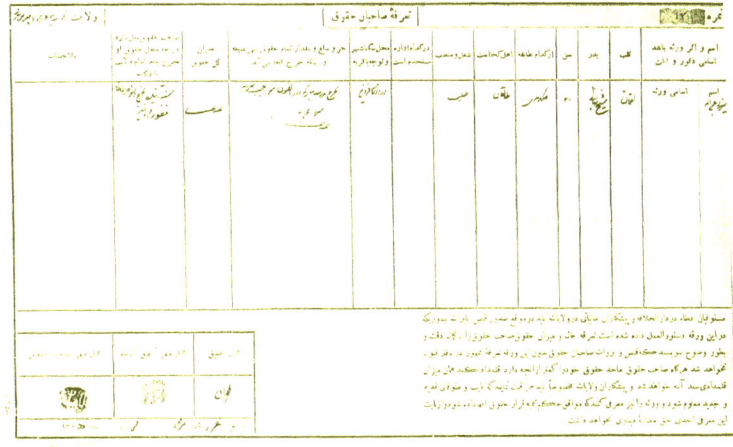

Figure 4.6 Source for the salary of Mirzâ ᶜAbdollah Loqmân (*Asnâd*)

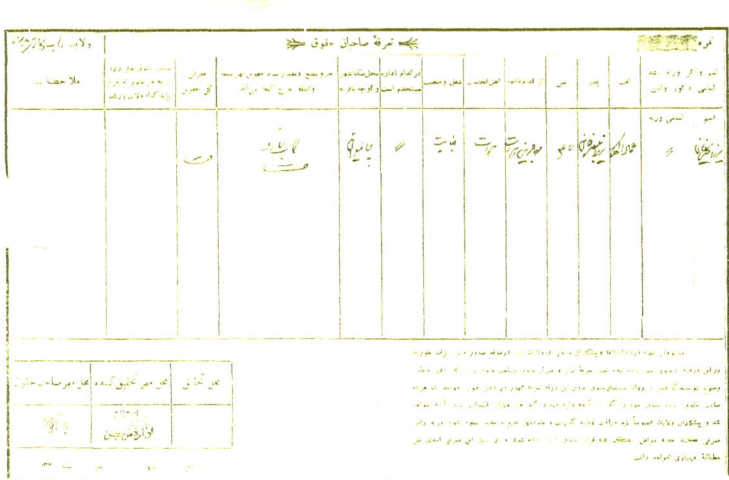

Figure 4.7 Source for the salary of Mirzâ Yahyâ ᶜEmâd al-hokamâ (*Asnâd*)

hospital may be the one that was initially built in 1853, outside the Gate Dowlat.[150] Even though the Central Sanitary Council planned the construction of new hospitals to tackle the cholera epidemic, there is no evidence that this was implemented.[151] The sole charitable *Dâr al-shafâ* was in Mashhad, while charitable money, largely available through *waqf*, did not contribute to its expansion. On the other hand, in the twentieth century while no additional state hospitals were built, several hospitals were created by *waqf* foundations and/or charitable collections through the driving force of the constitutional movement.

The tendency toward the secularization of *waqf* was in tandem with ongoing medical reforms. In the 1870s, Mirzâ Hossein Khân-e Sepahsâlâr, who initiated radical changes in state administration and contributed to the introduction of modern science, established a complex called Madrasa-ye Sepahsâlâr, which contained a madrasa, a mosque, and a *dâr al-shafâ*, in which traditional medicine was practiced.[152] The use of *waqf* for such an establishment by a "modernist" occurred when *waqf* was for a religious purpose. Under the constitution, when the secularization of *waqf* was at work, Hojjat al-Eslâm Sadr'al-Ulama, a constitutionalist cleric, created a *marizkhâneh-ye melli* (national hospital) in Guilân. In his letter of November 1907, he requested that Nâsser al-Molk, the Minister of Finance in the first constitutional cabinet, announce it publicly, so that others would follow.[153] Through the impact of the constitutional movement and the growth of the *anjomans* (societies), similar, charitable societies, *anjonam-e kheiriyyeh*, also appeared. These provided a new institutional frame for the creation of hospitals and schools, such as the national hospital and madrassa-ye Roshdiyyeh in Guilân.[154] The charitable associations did not replace *waqf* but inspired *waqf* founders to use *waqf* for secular establishments. One of the charitable hospitals created by the *anjoman-e kheiriyyeh* (charity society) of Rasht on the eve of the Constitutional Revolution was the Bimârestân-e Bistun. Hâj Seyyed ʿAli Aqâ Fumani donated his house to the *anjoman-e kheiriyyeh* of Rasht to be used as a hospital. Such hospitals were called "national" hospitals in the same way that new schools created by the *anjomans* were known as "national" madrasas.[155] The hospital Vaziri built in 1899 appears to be the first *waqf*-endowed hospital. From the constitutional period onward, other such hospitals were established at a pace that depended on sociopolitical conditions and led to hospitals of Qazvini in 1909 and Nuriyeh in 1917. The latter was established by Nurollâh-Khân-e Zahir al-Mamâlek, a wealthy man from Kerman, and the grandson of Ebrâhim-Khân-e Zahir al-Dowleh, the Governor of Kerman (1804–1825). In the *waqf* deed, Zahir al-Mamâlek specified that two doctors (educated in modern medicine), one traditional physician (*tabib-e irâni*), and one midwife should be employed in the hospital. The two doctors should preferably be graduates of European faculties. Failing this, they should have studied in the medical schools in Iran but have a license from the Ministry of Public Instruction (*Wezârat-e Maʿâref*). This endowed hospital also featured a medical class, apparently inspired by the American mission hospitals.[156]

The most significant medical establishment partly endowed by *waqf* was the Pasteur Institute in Tehran. From the early twentieth century, vaccines, sera, and glass ampoules were imported from France under the auspices of Dr. Schneider. However, this proved unsatisfactory, as, for example, given the long distance, calf lymph arrived in an inert state.[157] During his visit to Paris in 1919, where he attended the peace conference, Prince Firuz Mirzâ Nosrat al-Dowleh, Minister of Foreign Affairs and son of ʿAbdol-Hossein Mirzâ Farmânfarmâ, visited the Pasteur Institute there, and proposed creating one like it in Tehran. The French Council of Ministers approved his suggestion and Dr. Mesnard was appointed as the first Director of the Pasteur Institute of Persia, which opened on August 24, 1921, in temporary premises in Tehran. This became the foundation of biomedicine in Iran, and involved activities such as microscopic and chemical examinations, the inoculation of animals against anthrax, and the preparation of vaccines and sera.[158] Prince ʿAbdol-Hossein Mirzâ Farmânfarmâ donated more than 10 square kilometers for the construction of the institute and a small hospital, and 10,000 tomans (£2,500) toward the cost of construction. The government supplemented this with 15,000 tomans (£3,750). Due to the dire economic conditions of 1921, building work stopped, however, as the government could not contribute further.[159]

In 1903, prior to his involvement in the Pasteur Institute, when he returned from exile in Iraq and was appointed to the government of Kermânshâh, Farmânfarmâ had established a hospital there and appointed Dr. Mahmud-e Moʿtamed, a graduate of the Paris medical school, as its director.[160] He had allocated part of the tax levied from the transportation of bodies to the ʿ*Atabât* (Karbala and Najaf) to its expenses.[161] The creation of a *waqf* by Farmânfarmâ and his contribution toward the Pasteur Institute should be viewed within the framework of political, social, and economic developments.[162] Farmânfarmâ's support for the constitution and the Pasteur Institute indicates that the constitution and biomedicine, as epitomy of modern science, were associated, at least in the mind of members of the Qâjâr elite such as Farmânfarmâ. Whether this was for reasons of conviction or politics, such actions can often be defined as "conformity to norms."[163]

Although medical modernization was on the agenda of the constitution, as part of the enlightenment project, it did not reflect the current practice as "people preferred old-fashioned doctors."[164] The prevalence of traditional physicians was primarily due to a lack of modern trained doctors, to the extent that constitutional law licensed many traditional physicians. However, if the inclusion of humoral medicine in the Dâr al-Fonun's curriculum had been justified by traditional education, after the constitution the practice of traditional medicine was judged according to need rather than arising from theoretical necessity. Paradoxically, a constitution that advocated political independence and nationalism appointed a Sanitary Council that was led by Western physicians. These included Dr. Schneider (until 1907), Dr. Coppin, and, until 1911 when he was replaced by Dr. Gachet, a French naval surgeon, Dr. Georges. In time, Dr. Neligan, who had been vice-president of the Council, succeeded Dr. Gachet and became acting president.[165] The Health Council that sat on June 5, 1906, at the Dâr al-Fonun was headed by Dr. Lindley, who was acting president (replacing Dr. Schneider) and its members included nine Europeans and six Iranians: Drs. Bâqer Khân, Bongrand, Cormick, d'Obermayer, Galley, Georges, Hakim al-Molk, Mirzâ Ebrahim Khân, Mirzâ Mahmoud Khân, Nezâm al-Hokamâ, Regling, Sadowsky, Wishard, Rezâ Qoli Khân, and Wibier.[166] Dr. Odling, the physician to the British Legation, was also a member of the Sanitary Council until his death from typhoid fever on February 17, 1906.[167] This represented a radical change, as throughout the nineteenth century, traditional physicians always outnumbered Western colleagues in the Sanitary Council, which was always headed by Iranians.[168] Such a radical change, on the other hand, did not so much reflect, nor respond to, practical need as it was in tandem with theoretical transformation toward biomedicine.

Throughout the nineteenth century, a model of medicine that consisted of the association of traditional and modern medicine was shaped for both theoretical, as explained in Chapters 1 and 2, and practical and political reasons, as discussed in Chapter 3.[169] The sudden manner in which traditional medicine gave way to modern medicine appears to be a consequence of the Constitutional Revolution and the *political* ideology of progress and the social and professional values it promoted. Constitutionalism and its politically loaded ideology of progress gave social value to modern medicine and public health. It was not only in terms of practice—as it was in the previous decades—but also in ideological or epistemological terms that modern medicine was considered superior in sociopolitical and intellectual discourse. As Cyril Elgood noted, it was due to the "change of outlook" that "a series of laws regulating the practice of medicine and pharmacology approved by the Sanitary Council and ratified by the *majles* drove the last nail into the coffin which held the body of Greek and Arab medicine."[170]

A turning point came in 1911, with a law that aimed to regulate the practice of modern medicine. This law forbade the "practice of medicine or its different sections and branches in any part of Persia unless a permit be obtained from the Ministry of Education and registered at the Ministry of the Interior."[171] The only documents recognized by the ministry were issued by state medical schools or by the schools of foreign countries. "Those who did not possess any of these diplomas and have practiced for more than five years but less than ten years have the right to appear before a special committee and pass an examination" within three years of the promulgation of the law. The examination committee was composed of "professors of the School of Medicine, four well-known physicians and a representative of the Ministry of Education." After 1911, the only documents authorizing the practice of medicine were diplomas from schools of medicine in Iran or Europe. Those who had practiced medicine continually for more than ten years had the right to obtain a permit.[172] Following the constitutional law of 1911, the Ministry of the Interior appointed skilled physicians via the Central Sanitary Council. These physicians were to examine "licensed" physicians by asking them to write a report on current diseases and their patients. For instance, in 1913, Dr. Dâvood Khân-e Afshâr was sent on a mission to Hamadan, to report unlicensed practitioners to the local government, and to test the practice and knowledge of those who were licensed by examining some of their patients in order to see if they were diagnosed and treated accurately.[173] However, nepotism enabled many unskilled physicians to acquire a license to practice modern medicine. In 1924, Gilmour reported that out of a total of 905 doctors with permits across the country, only 253 had diplomas and the rest had received their permit after the 1911 law was promulgated. According to Gilmour, those who had a permit but no diploma were "very old-fashioned and knew little of modern medicine."[174] The door was thus left open for experienced practitioners of traditional medicine.

Although restrictions had been imposed on selling chemical drugs for many years, it was only in October 1919 that a decree required every pharmacist to have a permit. This permit was given to those who had a diploma from a European school of pharmacology, or those who had passed an examination in practical pharmacology, chemistry, and toxicology.[175] The pharmacy law did not exclude traditional herbalists from working. Its article 20 allowed them to sell certain chemical drugs, including aspirin, vaseline, quinine sulfate, magnesium carbonate, potassium chlorate, and phenacetin.[176] Just as for traditional physicians, doors were left open for traditional pharmacies, as the market required. However, the law set strict regulations to check their activities. This was done via a committee of five: "a doctor knowing chemistry or a druggist nominated by the Ministry of Education; one druggist from those of the town; one person from the Ministry of the Interior, and one from the Ministry of Justice."[177] The principles of modern medicine were the main criteria in providing licenses to traditional doctors and herbalists. In 1911, the post of lecturer in traditional medicine at the Dâr al-Fonun was abolished, and the new government employed professors from the West to infuse life and new ideas into the Dâr al-Fonun. These included Dr. Gachet, professor of physiology; Dr. Lattes, of chemistry; Dr. Le Blanc, of natural history; and Dr. Stump, of dentistry.[178]

From time to time, there was resistance to the European control of medical services or quarantines. In March 1906 a "fanatical Persian doctor named Hukema incited the population of Sistan to rise up against the European doctors who had begun to take plague sufferers from their houses to the hospital. The crowd demolished the hospital, destroyed the medicines and surgical appliances and attacked the British Consulate."[179] At the same time, the dispensary of the British Legation in Tehran gave medical advice and medicines "free of charge to Persians of all, but chiefly the lower classes."[180] Likewise, the dispensary of the British Consulate in Esfahan was frequented by local patients.[181] The reaction of the population at large for or against modern medicine was somehow emotional and a matter of circumstance and need, while the position of the constitutionalist elite toward modern medicine was conscious and consistent. The elite's adherence to modernity and progress, favoring medical modernization, went hand in hand with the quasi-colonial domination of Western physicians and medicine in Iran from the early twentieth century onward. This was a trend that mirrored the constitutionalist *bast* (refuge) in the British Legation in order to gain British support against the anticonstitutionalists in the government. Whether the constitutionalists were aware of what was behind the support of the British Legation or not, either they had accepted that modernity and progress was making its way to Iran through Western domination or they wanted to use Western support to put their reform projects into effect at any cost, an attitude that we meet later in Reza Shah's policy toward the Western powers.

Medicine and Public Health Under Reza Shah: An Overview

The constitution and its reform projects were far from being achieved by the time Reza Khân rose to power in 1921. The functions of the constitution were interrupted not long after it was reestablished in 1911, as throughout World War I there was no central power to establish security and order, let alone implement reforms. The country was the scene of conflicts between the occupant armies, and food shortage caused by the war turned into famine and mortality from 1917 onward.[182] It was after the restoration of a central power and the new administration set by Reza Khân that the constitutionalists found opportunity to resume their activities. In fact, Reza Shah's reforms in education and medicine were a response to the modernist ideas of the constitution, and, as Ali Ansari pointed out, "it was the rise to power of Reza Khân which brought the modern to Iran in a material sense, which affected every section of the population."[183] Nevertheless, insofar as the constitution was not only an ideology but also a utopian vision, the failure to create a "constitutional" monarchy or to respect its tenets cannot be attributed to Reza Shah alone. Before him none of the Qâjâr Shahs wanted, or were able, to realize it.

On the other hand, Reza Shah's reforms, mainly designed by his three ministers, Teymourtash, Prince Firuz, and Davar,[184] were those to which the Qâjâr reformist elite aspired since the end of the nineteenth century, but new devices and regulations were put in place for implementation. The inherent link established between patriotism and public health was provided with a more efficient institutional framework in the aftermath of the 1921 coup d'état. The size of the population that Dr. Amir Aʿlam, Minister of Public Health under Reza Shah, considered as "one of the main sources of wealth, power and greatness of the nations"[185] had featured in medico-political discourse since the mid-nineteenth century but was given fresh expression in the project of creating maternity centers (*dâr al-welâdeh*) and schools of midwifery. Likewise, what, according to Kashani-Sabet, emerged as the "maternalism" from the second decade of the twentieth century to tackle the issue of population decline[186] originated under the constitution, when this theme was borrowed from the West, together with nationalism, love of the homeland (*vatan*), modern science, and the concept of a healthy body as tools to construct a strong nation. In 1927, recommending a school of midwifery,[187] the journal *Ettelâʿât* took up the argument of the journal *Shekufeh* in 1912, which reported that despite polygamy and a young age of marriage, Iran had declined in population, while in Europe, where women often married after the age of 25, and men after 35, the population increased. This, it said, was because "we are ignorant in the science of rearing our children and we do not have educators [of midwifery]. We have no public health and children are raised like herbs in the desert."[188]

Under Reza Shah, from 1921, public health was organized in line with French models. Each city was controlled by a municipality, directly under the authority of the prime minister, which collected its own taxes. Each municipality had accounting, engineering, and medical departments. This was meant to increase efficiency by avoiding the labyrinth of ministries or other institutions previously involved in medicine and public health. The medical department was divided into two sections: the sanitary section and public assistance. The sanitary section was responsible for the cleanliness of the city and public assistance was provided through a hospital and a lunatic asylum.[189] The municipal hospital in Tehran was built outside of the city, and had 72 beds. Its staff consisted of 1 surgeon, 1 physician, 3 assistant doctors, 1 oculist, 1 laboratory director, 3 medical student interns, and 12 male and 6 female nursing orderlies alongside 2 pharmacists.[190] The Sanitary Council that had been formerly associated with the Ministry of the Interior was reinvented as the Ministry of Public Health.[191]

The employment of Western advisers under Reza Shah was a continuation of the policy that had been established under Mozaffar al-Din Shah in 1896. The head of one of the state hospitals built in 1868 was Mirzâ ʿAli-Akbar Khân-e Nâzem al-Atebbâ, a graduate of the Dâr al-Fonun. He was its director probably until 1883, after which the monopoly of local doctors ended when Dr. Albo, a German, was appointed as director.[192] In 1896, Mozaffar al-Din Shah ordered the hospital to be organized by the German Legation.[193] From this point onward, the state hospital was run by Western physicians, except between 1915 and 1919, when most Western physicians left the country. The government was unable to meet its expenditure, and Dr. Ilberg, its director, paid 13,372 tomans from his own pocket to meet its costs.[194] This policy continued under the constitutional government from 1909 onward, with the employment of several advisors such as Belgian M. Mernard as supervisor of customs in 1910 and Morgan Shuster as finance advisor.[195]

The changes implemented by Reza Shah were not so much to employ Western advisers for medical modernization as to bring everything, whether under Persian or Western direction, under state control, partly as a reaction to the increasing influence of European countries during World War I and in line with the national independence dear to the Iranian modernist elite. When the state hospital was under the direction of Dr. Ilberg, it was sometimes called the German hospital (*marizkhâneh-ye âlmâni*).[196] Likewise, after the 1919 Anglo-Persian Convention, which was to transform Iran into a protectorate of Britain, when the state hospital was taken over by British physicians and financed by the British government, it was known as the *marizkhâneh-ye englisi* (British hospital).[197] After 1919, there was a period of stiff opposition to the employment of foreign and particularly British physicians.[198] Iranian physicians trained in the West wished to direct medical services themselves rather than work under Western experts. Nevertheless, they could not avoid the domination of Western medical services, via military, civil, or missionary hospitals, in several cities that were also used for medical education. In the 1920s, Western hospitals in Iran still outnumbered those built by the government or by charitable foundations, and they also had greater accommodation for patients. Iranian hospitals were limited to the Imperial (or state) hospital, which could accommodate 50 patients, the Vaziri hospital (established in 1899 by Mirzâ Mussâ Vazir, but taken over by the government in 1921 and reorganized under the control of the Public Health Administration, *Sehhiyeh-ye Koll-e Mamlekati*) with 30 beds and a hospital for women with 20 beds. The *waqf*-endowed hospital of Amini, created in 1909 by Mohammad-ᶜAli Amini in Qazvin, became practically inoperative after 1919, due to mismanagement by his sons.[199] There were some military hospitals in Tehran and other cities, including the 50-bed Ahmadiyeh hospital in 1914, which was for the Gendarmerie, and other small or temporary hospitals, set up mostly as a the result of the war.[200] There were also three municipal hospitals. The sporadic movement for building privately funded or *waqf*-endowed hospitals had petered out under the economic crisis during and after World War I. Only a few hospitals kept their doors open during this period. In 1913 the Gendarmerie established a small hospital in the south, with a budget of only 1,000 tomans.[201]

On the other hand, there were more than ten foreign hospitals in the country. The British CMS had four: one in Isfahan (120 beds), one in Kerman (60 beds), one in Yazd (60 beds), and one in Shiraz (50 beds). The American Missionary Society maintained six: one in Mashhad (50 beds to be extended to 100 beds), one in Tehran (45 beds), one in Rasht (25 beds), one in Hamadan (25 beds), one in Kermânshâh (25 beds), and one in Tabriz (100 beds).[202] It is possible that the figures given by Gilmour are underestimated, as according to a Persian document, the British hospital in Esfahan under Dr. Carr had 120 beds for men and 70 beds for women.[203] There was also a British hospital in Sistan, which is not mentioned by Gilmour.

The demise of the state hospital between 1915 and 1919 was due to lack of money and particularly due to decentralized administration. Three establishments were involved in the running of the hospital: the Ministry of *Maᶜâref va Owqâf* (Public Instruction and Endowment), which oversaw the hospitals, the medical schools, and the sanitary councils. The budget of the hospital, about 1,200–1,300 tomans per month, came from customs revenues (*Gomrok*), while the Ministry of Finance (*vezârat-e mâliyeh*) controlled the budget of all governmental establishments. In 1918, the state hospital could not maintain all 30 of its beds, a condition for receiving full funding.[204] But this was apparently because it was without physicians and surgeons.²⁰⁵ The budget of the hospital was often in arrears and there was no money to purchase food and clothing for patients, let alone a salary for the employees.²⁰⁶ The ministry of *Maʿâref* and *Owqâf* that managed the budget did not consider itself accountable to the Ministry of Finance for any expenses and Customs did not pay expenses regularly. Finally, the Ministry of Finance requested that Customs pay at least 600 tomans at the beginning of each month for food with the rest of the expenses to be paid whenever possible.²⁰⁷

Such chaotic administration could only be overcome by central authority. The major work of Reza Shah was to bring all these under a new institutional framework. For example, the *baladiyeh* (town hall) of Qazvin took over the Amini hospital, and used it to host sick poor "who, unable to pay for a doctor or treatment, died every day in the alleys and streets."²⁰⁸ A similar transfer was made with endowed properties, such as one belonging to ʿAziz-Khân that was earmarked for the expenses of a house for orphans.²⁰⁹ The *Sehhiyeh-ye Koll-e Mamlekati* (Central Public Health of the Kingdom), later renamed as *Edârehye Koll-e Behdâri* (Central Administration for Health) that managed medical institutions such as hospitals and medical schools, came under the authority of the Ministry of the Interior. In 1941, this administration was transformed into the Ministry of Public Health (*vezârat-e behdâri*).²¹⁰

By the time Reza Shah took power, the outlines of modern medicine in Iran were fixed, through a process that began at the end of the nineteenth century and gained momentum under the constitution. Had new ideas not been brewing during the constitution, Reza Shah could not have undertaken his project of modernization. The foundations of medical transition had been set and modern medicine had gained the legitimacy it needed. What hindered the reform projects of the constitution was the state of chaos and lack of order that prevailed after 1910.²¹¹ What remained to be achieved was the expansion of modern institutions, medical schools, hospitals, and health centers under the Ministry of Public Health, which Reza Shah endeavored to realize.²¹² All these reforms that responded to the concern of efficiency were fundamentally driven by the combined strategy/ideology of centralism and nationalism. It was such ideology that also informed the creation of the University of Tehran by bringing together six higher education institutions in 1934, the efforts for nationalization of the banking system, Persianization of the state, and attempts to nationalize the Anglo-Iranian Oil Company.²¹³

Fundamentally anti-British and anti-Russian, Reza Shah endeavored to seek other allies. After being frustrated by the Americans' refusal to provide Iran with assistance in education and modern science, as they believed it was "not worth the price of antagonizing the British," Reza Shah looked to Germany.²¹⁴ He did not woo France, considering perhaps "that difficulties over the Syria and Lebanon Mandates left that country with little inclination for involvement further afield."[215] Nevertheless, France, as under the Qâjârs, preserved its dominant place in medicine in Iran all the more so as Britain continued to conspicuously exploit its medical services to political ends.[216] Once the post–World War I political storm subsided and the 1919 Anglo-Persian Agreement was definitively buried by the veto of the *Majles*, a trend similar to that which occurred in the second part of the nineteenth century resurfaced: The heads of the Tehran Pasteur Institute, with a short interval in 1926, when Dr. Bahrâmi was director, were French: first Joseph Mesnard and then Jean Kérandel. Following Kérandel's death in 1834, Dr. Hosseine Mashʿuf was director for two years and then Dr. René Legroux was sent out to Tehran as honorary director of the Institute, even though the effective administration was under the Iranian doctors Abol-Qâssem-e Bahrâmi and Mehdi-ye Ghodsi.[217] During his work as director of the Pasteur Institute, Dr. Kérandel was also entrusted with the reorganization of the Public Health Administration, while Dr. Charles Oberlin was appointed Dean of the Faculty of Medicine, Dentistry and Pharmacy of the university in 1939, although some newly established institutions were also to be run by nationals of other Western countries in particular the Germans.[218] This situation reflected the tension between the Iranian struggle to become politically independent and its need to modernize with Western help.

Conclusion

This chapter has followed the conception and implementation of the projects of medical modernization under the constitution, where the ideological foundations of reforms undertaken by Reza Shah and his ministers were laid. Nevertheless, the medical reforms under Reza Shah ushered in a new period of development, where one can see signs of maturity in relation to the practice of modern medicine, now that modern medicine had asserted its authority and could proceed with confidence and self-reflection. After a period of enthusiasm for modern medicine and belief in the efficiency of biomedicine, physicians began realizing that their zeal for modern medicine had overshadowed preventive medicine, ingrained in their traditional *Hefz al-Sehheh* (health preservation). In 1924, referring to the military medical service, John Gilmour stated that its efforts "appear to be directed entirely towards the treatment of sickness and not towards its prevention." In the same vein, Dr. Charles Oberlin, who was in charge of reorganizing the Ministry of Public Health (*Vezârat-e Behdâri,* literally Ministry of the Preservation of Health), pointed out that he was surprised by the privileged place accorded to hospitals by ministers and physicians in Iran, while, he continued, "the final conclusion of modern medicine [thus far] is that prevention is superior to, and better than treatment."[219] It is not surprising that from now on we find modern educated physicians who wrote on the theme of health preservation but from a new approach.[220] Modern medicine provided physicians with new insights into both the futile theories of Galenico-Islamic medicine and its wisdom, which was the fruit of thousands of years of practice. Reza Shah's period can be considered as the third phase of medical modernization with the overwhelming superiority of modern medicine, while it continued to coexist with traditional practice, as many traditional doctors adopted modern practice and knowledge. However, while in the second part of the nineteenth century there was symbiosis between traditional and modern medicines, the coexistence in the twentieth century was not of symbiosis but one of practical convenience that worked for the benefit of biomedicine and led to the definitive demise of traditional humoral practice.

Conclusion

In this book, I have tried to draw a picture of the conceptual and socio-institutional dimensions of medical transformations in modern Iran. One conclusion from this study is that the relationship and dialogue between old and new, traditional and modern medicines was at work throughout their respective developments. This relationship reflects a historical reality that I have endeavored to highlight and that has for a long time been overshadowed by the systematic and unqualified emphasis on a contrast between traditional and modern that has often informed the historiography of medical modernization. This book has examined five major aspects of this medical transformation. The first and fundamental step toward modernization was the creation of modern institutions where both modern and traditional medicine were taught and practiced and in the activities of which traditional physicians were directly involved. This medical modernization was embedded in political, administrative and institutional transformations mainly aimed at military reform. As a result, conceptual transformation in medicine took place within the framework of institutional and political reforms. Both institutionally and theoretically, modern medicine was indebted to traditional medicine for being assimilated and established. Finally, this pattern of transformation was informed by the competing geopolitical strategies of the European powers in Iran. In summary, it was institutional, administrative, and political factors that not only caused the introduction of modern medicine but more fundamentally helped traditional medicine and physicians to transform.

As to the mechanism of transformation, I have shown that far from being linear, the transition was made up of three parallel strands that, taken separately, appear to embody different medical schools—traditional, hybrid, and modern. However, in reality, these categories functioned and developed within the framework of interrelationships, cooperation, and conflicts of interests and ideologies at the same time, representing as a whole an overarching transformation. As a result, medical professionalization was under way before modern medicine gained currency. It is therefore not accurate to talk about the medical profession in a way that restricts it only to students and practitioners of modern medicine. In the same vein, we can grasp the transformation process only if we take into account the interaction and dialogue between all the components of the medical profession, including traditional and modern practitioners.

This pattern of medical transformation was a response to the modernization movement in Qâjâr Iran, which aimed at centralization. From this viewpoint, the important aspect of medical modernization was to bring medical education and practice under state control. This explains why in Iran, traditional physicians, alongside their modern educated counterparts, were involved in the activities of the newly established institutions. Modern institutions, including modern schools, hospitals, and sanitary councils, as well as modern journals provided new spaces for the activities of physicians and, as new networks, played an important role in medical professionalization. One cannot neglect the importance of the expanding state bureaucracy in Qâjâr Iran, even in its patrimonial format, as an institutional context within which modern science and technology were taught and pursued. It was, as explained in Chapters 2 and 3, this expansion that allowed the encounter between modern and traditional to take place. In a similar vein, it is true that due to the patrimonial nature of the state an independent merchant class did not develop in Qâjâr Iran. However, within the framework of the patrimonial system, some merchants (such as the Amin al-Zarbs), or some business-minded princes-turned-speculators (such as the Farmanfarmâs), emerged, and they played a key role in the creation of modern schools, the national bank, and the Pasteur Institute.

However, the coexistence of tradition and modernity within the new institutions did not last long, not least because this was only one phase or part of the long process of enlightenment, whose achievement was deeply influenced by Western science and ideas. The constitutional movement, though fragile, was rooted in this process of enlightenment, which was influenced by both internal and foreign developments, including the constitutional movements in the Ottoman Empire and Russia.[1] As the reform movement became more conscious and radicalized, intellectual/ideological consistency and institutional/social solidarity appeared on the agenda. In other words, intellectual and ideological distinctiveness was a requisite for sociopolitical integration and political action, and it was provided by the paradigm of modern science (*'elm-e jadid*) and mobilized the intelligentsia, as well as a wider public, for reform under the constitution. Modern science was the new paradigm in its "sociological sense," as Thomas Kuhn put it, which consisted of "the entire constellation of belief, values, techniques and so on"[2] shared by the members of the constitutionalist reformers.

Although modern science was first introduced in Iran as applied science through the polytechnic school (or Dâr al-Fonun), it was not entirely devoid of intellectual and conceptual lure. The question is, to what extent was the intelligentsia conscious about distinction between technology (applied science) and science? And when we talk about medical transformation, did this relate to technology (e.g., vaccination, dissection, surgery) or to new concepts? According to some historians, by this time, science and technology were fused, a fact that ensured the universality of modern science.[3] Was this fusion also at work in nineteenth- and twentieth-century Iran? As it was argued in Chapters 1 and 2, there is evidence that in the encounter with modern medicine, a conceptual transformation took place alongside the introduction of modern anatomy and surgery. This was partly due to the coexistence of traditional and modern medicine within the new institutional setting. There had to be some dialogue and negotiation, as modern medicine was not applied in an empty intellectual, cultural, and social space.

It is this space that was examined in Chapters 3 and 4. Chapter 4, however, described the radical shift toward modern medicine under the auspices of the constitutional movement. Paradoxically enough, with its encroachment to the detriment of traditional medicine, modern medicine before the 1930s remained deprived of its theatres of anatomy and dissection, under the pressure of religion. In India and Egypt, as in Iran, dissection was loathed, but it was nevertheless practiced as it was allowed and enforced by the state or colonial administration; however, in Iran, disallowed by the clerics, it was not practiced. Although the state did not formally endorse the clerical opposition, it did not oppose it either. The situation was not much different in India and Egypt insofar as religious prohibition is concerned. In Iran, however, the Qâjâr state, unlike Mohammad Ali's government in Egypt and the colonial administration in India, practically shared power and legitimacy with the clerics. Consequently, the relationship between modern science and religion in Iran should be examined from political as well as religious standpoints. Opposition of the clerics to dissection was determined not only by religion but also by politics. In other words, attitude toward dissection in Qâjâr Iran, despite its religious veneer, was loaded politically. Accordingly, the clerical opposition to dissection was seen more as an indication of political independence than as a fundamental principle that medicine should adhere to. This signifies that dissection for the purpose of research or medical application could be justified whenever politically suitable. It is in this light that we can explain the attitude of the clerical power in Iran today toward organ transplantation or assisted reproductive technique. This question further highlights the above-mentioned relationship between theoretical knowledge or scientific research on the one hand and application of scientific founding on the other.

The question that requires further investigation is whether biomedicine, after being fully developed in Iran, continued to be a science or applied technology, or it incorporates both? Are medical technologies now developing in Iran, such as cloning and embryo stem cell research, mere applied techniques or are they accompanied by innovative scientific research and further stimulated by a spirit of enquiry? After all, could the development of technology be viable without being sustained by research? The study of this question will require reflection on social and political factors, which have underpinned such technological development or research so far. It will be important to see how these factors inform or nourish institutional interface within which medical research is undertaken. As a matter of fact, while discoveries or initiatives of various assisted reproductive technique originated elsewhere, Iran adopted them with the endorsement of the clerical power. Despite his conservative views on social policy, Ayatollah Khomeini, the founder of the Islamic regime, issued a fatwa in 1984 that permitted artificial insemination of married couple but did not allow the involvement of the gametes of the third party. His successor, Ayatollah Khamenei, however, issued fatwa in the late 1990s, which authorised artificial insemination even with the use of a third party's egg or sperm, embryo or uterus, whether the donor is a close relative or not.[4] It is now important to see to what extent the research institutions, such as Royan Institute in Tehran, and Yazd Research and Clinical Centre for Infertility, allow leeway to, or condition, the spirit of research and inquiry. The experience of medical transformation in Qâjâr Iran indicates an inherent relationship between theory and practice, between intellectual inquiry and technological application, even though they were not given equal importance. However, the share of these components (technology and research) in science and medicine can be subject to change according to historical and sociopolitical contexts, as their value, nature, and function vary across time and space. Modern technology may not be viable without the spirit of research but has the spirit of research in Iran today a similar value and function as it had in the Ancient Greece or during the Enlightenment in Europe?

Notes

Prelims

1. Ibn Qayyim was aware of the difference while he was committed to both Greek medicine and Islamic principles. See Introduction by Penelope Johnstone to the *Ibn Qayyim al-Jawziyya Medicine of the Prophet*, p. xxiv.
2. Jamshid Khodâdâdi, *Kelid-e vorud be tebb-e qadim* (key to the traditional [ancient] medicine), (Tehran, Shahr Press, 1389/2010). This volume is published with the support of the Research Institute of Koranic Medicine (*Mo'assesseh-ye pajuhesh-e tebb-e qor'âni*). Koranic medicine is based on medical, usually nutritive, instructions of the Koran and on *shafâ* (healing by God or saints), and gives importance to prevention. See the Foreword by Mohammad-Hâdi Mo'azzen Jâmi, pp. 26–27. This is an illustration of the antagonistic perception of the Islamists with modern science. For more on "return to the tradition" amongst the Islamist scholars, see: M. Iqbal, *Islam and Science*, and Leif Stenberg, *The Islamization of Science: Four Muslim Positions Developing and Islamic Modernity*.
3. Hormoz Ebrahimnejad, *Medicine, Public Health and the Qâjâr State*, 2004.
4. The change of vision and concepts in Unani medicine in India was more perceptible and conscious than in Iran, and also took a different form than in Iran. The change in India was called "renewal" that can be understood in terms of the adaptation or assimilation of the theories and practices of modern medicine. See M. Harrison, "Science and the British Empire," p. 60.
5. In India, the contribution of the traditional hakims to introduce allopathic medicine was better organized under the colonial administration, particularly in the Punjab and then Lahore. Despite this organizational success, many "modern trained" doctors opposed the hakims' contribution, apparently for scientific or quality reasons, but they might also have been motivated for professional reasons. In the Punjab, the hakims were employed as vaccinators in 1857, and then in 1867 their employment became more in sync with the Mercer program, in which hakims or their sons would practice medicine in rural areas as they had the confidence of the population. In Lieutenant Colonel T. W. Mercer's view, "these men

would form the nucleus of a well-educated medical profession . . . and would take advantage of the educational opportunities offered by [this] system." John Hume, "Rival traditions: Western medicine and Unani Tibb in the Punjab, 1849–1889," pp. 219–222, 227, 230–31.

6. Tancoigne, who accompanied the mission of General Gardane in Iran, in his note of January 30, 1808, mentions that "Iran is the true name of Persia amongst the Orientals . . . it would be unintelligible to the [Iranians] to term it Persia." M. Tancoigne, *A Narrative journey into Persia, and residence at Teheran*, p. 147.

Introduction

1. Some recent works, however, have made a good start: Cyrus Schayegh, *Who is Knowledgeable Is Strong*; Kamran Arjomand, "The emergence of scientific modernity in Iran"; Maryam Ekhtiar, *Modern Science, Education and Reform in Qâjâr Iran*. Mohammad Tavakoli-Targhi has pioneered a critical study of the historiography of modernity in Iran. According to Tavakoli-Targhi, the historiographical narrative of modern Iran has been based on Western modernity ignoring the Iranian side of modernity or intellectual dynamism. As a result, a large number of texts have not found their place in the work of the historians on modern Iran. "A decolonization of historiography is therefore needed." See "The Homeless Texts of Persianate Modernity."
2. George Basalla, "The Spread of Western Science"; Donald Fleming, "Science in Australia, Canada, and the United States"; Roy MacLeod, "On Visiting the Moving Metropolis: Reflections on the Architecture of Imperial Science"; idem, *Technologies and the Raj: Western Technical Transfers to India 1700–1947* (New Delhi, Sage, 1995); Bernard S. Cohn, *Colonialism and its Forms of Knowledge*; Michael Worboys, "Science and Colonial Empire"; M. Harrison, *Climate and Constitutions*; idem, *Public Health in British India*.
3. Anne-Marie Moulin and Khaled Fahmy have produced valuable works on the process of medical modernization in nineteenth-century Egypt, which was quite similar to Qâjâr Iran in the sense that it was not formally colonized. See A-M. Moulin, "Révolutions médicales et politiques en Egypte"; idem, "L'esprit et la lettre de la modernité égyptienne"; K. Fahmy, *All the Pasha's men*.
4. See for example D. Arnold, *Colonizing the Body*, p. 14; W. Ernst, 'Beyond East and West,' p. 513.
5. Mohamad Tavakoli-Targhi, *Refashioning Iran*, pp. 4, 9, 17.
6. This thesis is particularly developed in Chapter 1, "Modernity, Heterotopia and Homeless Texts," and Chapter 2, "Orientalism Genesis Amnesia," of *Refashioning Iran*.
7. Edward Said, *Orientalism*.
8. *Selections from the records of the Bengal Government, no. XIV, Papers Relating to the Establishment of the Presidency College of Bengal*, pp. 10-11; quoted in Seema Alavi, *Islam and Healing*, pp. 57–58.

9. In this work, I am particularly inspired by Jan Golinski, *Making Natural Knowledge*. For other scholars using this approach, see Ludmilla Jordanova, "The Social Construction of Medical Knowledge"; Kapil Raj, *Relocating Modern Science*.
10. As L. Jordanova noted, "If social historians of medicine attempt more than anecdotal or descriptive history, they frequently adopt social constructionism in one form or another, even if they have been less explicit than historians of science about their conceptual manoeuvers." See Jordanova, "The Social Construction of Medical Knowledge," p. 361.
11. For instance, the Frenchman Philip Colombe was the Commander of Artillery in the army of Shah Soltan Hossein Safavid. At this time, Iran asked France to help build a foundry to construct artillery guns, but Louis XIV refused. Jahângir Qâem-Maqâmi, "Ravâbet-e nezâmi-ye Iran wa farânseh dar dowreh-ye Safaviyeh," pp. 122–124.
12. A similar picture of the introduction of modern science can be found in the Ottoman Empire. The state elites, eager to modernize the country and particularly the army, did not hesitate to introduce modern science and medicine by employing Western physicians to the detriment of local court physicians; as in the case of the Austrian Dr. Minas, who came to Istanbul to modernize the Quarantine. The Chief Physician Abdulhaq Molla was removed from his position at the Quarantine Council in July 1838. See N. Yildirim, *A History of Healthcare in Istanbul*, pp. 23–24.
13. For example, the work of Sâleh ibn Nasrollâh al-Halabi was translated in ca.1810 into Persian. *Tebb-e jadid-e kimiyâyee t'â Bereklus* (originally written in 1669), dated 1225/1810, Library of Oriental Institute, St Petersburg, manuscript no. C 1612, fols. 1b–69a. The intellectual interest in modern science was no new phenomenon in Qâjâr Iran. Under the Safavids, the Italian traveler Pietro Della Valle reported the enthusiasm of an astronomer, Mollâh Zein al-Din-e Lâri, in learning about new scientific developments in Europe; cited in Kamran Arjomand, "The emergence of scientific modernity in Iran": op. cit., p. 7. See also M. Tavakoli-Targhi, *Refashioning*, Ch. 1.
14. Controversy among physicians was particularly illustrated by "commentaries," a format in the Islamic medical literature, where physicians explained the difficult passages of their predecessors or criticized and completed them. There were also straightforward critical works: Râzi's *Shokuk alâ Jâlinus* (*Doubts on Galen's ideas*) is one example of such work.
15. ʿAbdol-Hossein Zenuzi Tabrizi (Filsuf al-Dowleh), *Matrah al-Anzâr*.
16. In writing his Matrah al-anzâr, Zenuzi Tabrizi was probably inspired by the *Nâmeh-ye dâneshwarân* (*Book of learned men*) initiated by ʿAli-Qoli Mirzâ Eʿtezâd al-Saltaneh, first published in 1878 in two volumes, consisted of the biography of men of science of all origins and persuasions, traditional and modern. See ʿAliqoli Mirzâ Eʿtezâd al-Saltaneh (ed.), *Nâmeh-ye dâneshvarân-e Nâsseri dar sharh-e hâl-e shshsad tan az dâneshmandân-e nâmi*, entesharât-e Dâr la-fekr, 1333/1954.

17. See, for example, the anonymous author of a treatise on the establishment of state hospitals, who often referred to Westerners for their endeavors in establishing modern hospitals. Edited in H. Ebrahimnejad, *Medicine, Public Health*.
18. "J'ai parcouru les deux volumes de pathologie [Chahâr-Maqâleh] et je déclare que leur lecture sera très utile en Perse. C'est un résumé exact et complet de l'état de nos connaissances sur les maladies internes. Le Rédacteur a fait preuve de tact et de beaucoup d'intelligence. Ce livre mériterait d'être imprimé aux frais de l'Etat. Téhéran, Octobre 1871, Dr. Tholozan." Tholozan's preface to *Chahâr Maqâleh-ye Nâsseri dar Tebb* by Nasrollah Mirzâ, National Library, Tehran, MSS 722.
19. Nasrollah Mirzâ Qâjâr, *Chahâr maqâleh*, pp. 9–10, 15, 78–79.
20. In the summer of 1862, a certain Fath'Ali wrote an article that he asked the state journal to publish. In the article, he criticized the chronicler Mirzâ Rezâ Qoli Khân-e Hedâyat (the author of the *Rowzat al-Safâ-ye Nâsseri*) because he scattered his narrative with poetry that "disturbed" the readers who were interested in the course of the events and not in poetry. In a fictional conversation between Fath'Ali and Hedâyat, who was the director of the Dâr al-Fonun at that time, Hedâyat acknowledged that he used many expressions for the purpose of rhyme, without any particular relevance to the narrated events. Fath'Ali then explained that "in our country if one does not understand a text, s/he believes that the author of this text was at the pinnacle of knowledge. But, if a philosopher expresses an important idea in a simple way so it is understood by everyone, he is deemed without knowledge because, it is said, 'any illiterate person can understand him'." See Fath'Ali, *Wasyyatnâmcheh*.
21. A. Mostowfi, *Sharh-e zendeqâni man*, vol. 1, p. 714.
22. This is made by Schayegh to indicate the "semicolonial" nature of science in Iran. *Who Is Knowledgeable*, op. cit., p. 197. This is also maintained by other scholars to whom Schyegh refers in support of his argument: Michael Worboys, "Science and the Colonial Empire, 1895–1940."
23. G. Attewell, *Refiguring Uniani Tib*, p. 91; on this question see also Claudia Libenskind, "Arguing Science": Unani Tibb, pp. 58–75.
24. Ann Marie Moulin, "Introduction: Se repérer dans le labyrinthe du corps et de l'Histoire," p. 7.
25. Ann Marie Moulin, "Introduction," op. cit., p. 7.
26. In October 2008, after India launched the Chandrayan I, an unmanned craft, to scan the whole surface of the Moon, the mission was considered a message of "Indian independence" and the Indian prime minister, Singh, repeated the term "fully indigenous" several times in his speech when describing the technology used for this mission.
27. The World Bank data, http://data.worldbank.org/indicator/NY.GNP.PCAP.Pp.CD date of access: 08/03/2013. The expenses for space programs in China and India are only an estimation given the ambitious projects of these two.

28. Mohammed Ghaly, "The Beginning of Human Life: Islamic Bioethical Perspectives," p. 177. For an account of the different opinions on the beginning of life, see pp. 180ff.
29. Farzaneh Zahedi & Bagher Larijani, "National bioethical legislation and guidelines for biomedical research in the Islamic Republic of Iran," p. 631.
30. Farzaneh Zahedi & Bagher Larijani, ibid, pp. 630–631.
31. Mansureh Saniei, "Human Embryonic Stem Cell Research in Iran," p. 330.
32. Farzaneh Zahedi & Bagher Larijani, ibid, p. 631.
33. Kiarash Aramesh, "Iran's experience with surrogate motherhood: an Islamic view and ethical concerns," p. 320.
34. Mansureh Saniei, "Human Embryonic Stem Cell Research in Iran." See also M. Saniei & R D Vries, "Embryonic stem cell research in Iran." On the history and development of stem cell research in Iran since 1991, see Y. Gheisari et al., "Stem Cell and Tissue Engineering Research in the Islamic Republic of Iran," 2012.
35. p. Johnstone, *Ibn Qayyim al-Jawziyya*, op. cit., p. xxiv.
36. M. Ghaly, op. cit., p. 181.
37. See, for example, Mark Harrison, "Medical experimentation in British India: the case of Dr Helenus Scott"; John Manton, "Making modernity with medicine: mission, state and community in leprosy control, Ogoja, Nigeria, 1945–50"; Kapil Raj, *Relocating Modern Science*.
38. Mohamad Tavakoli-Targhi, *Refashioning Iran*; idem, "The Homeless Texts of Persianate Modernity."
39. See, for example, Mirzâ ᶜAbdol-Karim Tehrâni, *Ketâb-e ᶜAlâyem al-Amrâz*, p. 2.
40. As it was claimed by Mme Jane Dieulafoy, *La Perse, la Chaldée et la Susiane*, p. 121.
41. Abbas Amanat, *Pivot of the Universe*, pp. 200, 265–266.
42. One can refer to the legal reforms introduced by Mohammad-ᶜAli Pasha of Egypt to reinforce state control particularly in the context of, and as a reaction to, its dependency on the Ottoman Empire. See Khaled Fahmy, "The police and the people in nineteenth-century Egypt"; idem, "The anatomy of justice: forensic medicine and criminal law in nineteenth-century Egypt."
43. Although this network was informal, under the strange name of Farâmushkhâneh, it was soon banned by the order of Nâsser al-Din Shah. Its successors, the Adamiyat Societies, were also clandestine under the Constitutional Revolution.
44. Clifford Geertz, *The Interpretation of Cultures*, p. 35. Geertz model provides probably the most appropriate methodological and theoretical framework for the analysis of the particularity of modernity and modernization in Iran, including the idea of constitutionalism and nationalism.
45. *Tarbiyat*, no. 3 (25 Rajab 1314 H/31 December 1896), printed in two volumes by Ketâbkhâneh-ye melli (National Library), 1376/1997, vol. 1, p. 10.
46. *Iran*, Year 59, no. 8 (June 28, 1906), rpnt, p. 438.
47. As J. Hodge indicated, "The collecting of statistics on everything from finance, trade, health, population, crime and much more was a necessary precondition for

the consolidation of a colonial state [in India]." Joseph M. Hodge, "Science and Empire: An Overview of the Historical Scholarship," pp. 6–7. For the wider context of imperial science, see Roy MacLeod, "On Visiting the Moving Metropolis." See also John Gascoigne, *Science in the service of empire, Joseph Banks, the British State and the uses of science in the age of revolution.*

48. This is illustrated in the secret Convention of 1907 between Britain and Russia, dividing the country into two zones of British and Russian influence and its effective occupation in 1915.
49. On Chamberlain and Curzon see Michael Worboys, "Science and the Colonial Empire"; on Curzon see Roy MacLeod, "Scientific advice for British India: Imperial perceptions and Administrative Goals, 1898–1923."
50. Joseph M. Hodge, *Science and Empire, op. cit.* p. 14.
51. As Ervand Abrahamian mentioned, referring to Marx, "radical ideologies could, at times, develop into forces capable of helping undermine the ruling class." See "The Causes of the Constitutional Revolution in Iran," p. 383. The frenzy around the constitutional movement was so strong that some people paid bribes to the constitutional societies to win their favor. See Fâruq Khârâbi, *Anjomanhâ-ye ᶜasr-e mashruteh*, pp. 286ff. A similar phenomenon occurred under the 1979 Revolution, and thousands of committees (*komités*) mushroomed across the country but their ideology was inspired by religion and Islam.
52. Cabanis, *Coup d'oeil sur les révolutions et la réforme de la médecine*, cited by A.M. Moulin, "Disease transmission in nineteenth-century Egypt," p. 46.
53. D. Kumar, *Science and the Raj*, p. 235.
54. Toby Huff, *The Rise of Early Modern Science: Islam, China and the West*, see Chapter 1.
55. Ali Ansari, *Modern Iran*, p. 12. However, attitude toward Western modernity and science was not the same at different periods and by different people. Under the constitution and the Reza Shah's rule, Western technology and science were looked at with great sympathy by the elite. We find a reverse attitude during the 1979 Revolution and at least during the first years of the clerical regime that ensued.

1. The State of Medical Theory and Practice in Nineteenth-Century Iran

1. Mirzâ Mussâ Sâvaji, *Dastur al-atebbâ fi ᶜalâj al-wabâ*, p. 54. James Morier, *Sargozasht-e hâji bâbâ esfahâni*, p. 179.
2. This form of education, according to George Makdisi, preceded the madrasa institution and began with the "Muslim education that was born with the Prophet's mission and the institution of *suhba* [fellowship] that served for the transmission of his *sunna*, and went on to serve other fields." George Makdisi, *The Rise of Islamic College*, pp. 128–129.
3. For the medieval period, see Gary Leiser, "Medical Education in Islamic Lands from the Seventh to the Fourteenth Century."

4. ʿAbdollâ Mostowfi, *Sharh-e zendeqâni man*, vol. 1, p. 714.
5. Gilmour, *Report on and Investigation into the Sanitary Conditions in Persia*, League of Nations, Geneva, 1925, pp. 31–32.
6. Archives of Sâzmân-e asnâd, no. 297031714.
7. See, for example, the license of practice for Dr. Esmâʿil Khân Eftekhâr al-hokamâ, whose internship at the American hospital of Hamadan was underlined. The license was dated Jamâdi II 1327/July 1909. Rustâyee, *Târikh-e teb va tebâbat*, vol. 2, p. 115.
8. Schneider, "La médecine persane. Les médecins français en Perse: leur influence," *Bulletin de l'Union Franco-Persane*, deuxième année, no. 5, décembre 1910–janvier 1911, p. 14.
9. Anonymous, MSS 505, p. 58; see Ebrahimnejad, *Medicine*, pp. 181, 224–225.
10. Mohammad-Taqi Mir, *Pezeshkân-e nâmi-ye fârs*, pp. 51–52.
11. Eʿtemâd al-Saltaneh, *Ruznâmeh ye khâterât*, pp. 28, 658.
12. Gilmour, ibid, p. 31.
13. See for example, *Dastur al-atebbâ fi ʿalâj al-wabâ* by Sâvaji; *Wabâiyya* (On cholera) and *Tâʿuniya* (On plague) by Shirâzi; *Dar wabâ* (On cholera) by Tehrâni; *Meftâh al-amân dar tebb* (The key of safety in medicine [during epidemics]) by Kani. Several books were titled *Hefz al-sehheh*, such as the *Tohfat al-sehheh-ye nâsseri*, by Rashti, the *Hefz al-sehheh* of Mirzâ Mohammad-Taqi Shirâzi, the *Hefz al-sehheh-ye nâsseri* of Mirzâ Nosrat-e Quchâni.
14. Although Arabic was part of the curriculum in the *madrasa*s, it was neither the official language, nor used in everyday conversations or correspondences. The statement of Mohammad-ʿAli Bâb that with his new religion Arabic grammar "*sarf va nahv* has been freed from the prison [of grammatical rules]", suggests that Bâb was not strong in Arabic. Although Bâb's deficiency in Arabic was partly due to the fact that he had not completed his madrassa curriculum when he was younger, it can provide us with an idea about the state of Arabic knowledge even amongst those who claimed to know Arabic, as Bâb wrote his Bayân in Arabic. For the statement of Bâb, see Sepehr, *Nâssekh al-Tavârikh*, vol. 2, p. 828.
15. J. Schacht & M. Meyerhof, *The Medico-Philosophical Controversy*, p. 77.
16. Schneider, "La médecine persane," ibid, p. 14.
17. On the different branches of knowledge necessary for a learned physician see: Gary Leiser, "Medical Education in Islam," p. 64.
18. Willem Floor, *Public Health in Qâjâr Iran*, pp. 141 ff. J. Morier, *Sargozasht-e haji bâbâ-ye esfahâni*, pp. 121–123.
19. See, for example, Sâvaji, *Dastur al-atebbâ*. For the section on faith healing and magic in this treatise, see pp.64–82. See also Nur al-Din Mohammad Hakim Shirâzi, *ʿAlâjât-e Dâr al-Shokuhi*.
20. Jacques Léonard, *La France médicale au XIXe siècle*, p. 42.
21. The association of magic and medicine, incantation and herbal remedy has a long history. In Egypt, Pharaoh Ramses III sent a priest for incantation and a physician for herbal treatment to the King of the Hittites, whose sister

was sterile. See Moulin, *Le médecin du prince*, p. 12. Likewise, according to the Vendidad, the sacred book of the Zoroastrians, medicine was threefold: medicine of knife, medicine of plant, and medicine of divine words. Abbas Naficy, *La médicine en Perse*, pp. 13–14.

22. *Iran-e Soltâni*, Year 57, no. 9 (July 19, 1904), rpnt, p. 241–242.
23. *Mehnat al-tâ'un* (Scourge of plague) Persian translation, 1247/1831–32, fol. 1.
24. On the precept *chon qazâ âyad tabib ablah shavad* (when Destiny arrives the physician becomes an idiot), see MSS 505, p. 35. Edited in H Ebrahimnejad, *Medicine*, p. 213. In this edition, I have suggested the two readings but have mistakenly preferred *âbeleh* and translated the word *ablah* (idiot) as *âbeleh* (smallpox). For the use of this precept see also Morier, Mirzâ Habib Esfahâni, *Hâji bâbâ*, p. 231.
25. Schneider, "La Medicine Persane," pp. 14–15.
26. *Gazette médicale d'orient*, VIIIème année, no. 2, Mai 1864, p. 19. This situation was a far cry from the received idea that genuine medicine was found in the "medical universities" or the "teaching hospitals" where medical education was well organized and monitored by competent physicians in Islamic lands. In fact if medical education and practice appeared ubiquitous it was because there was no organized and centralized system.
27. 'Aqili, *Kholâsat al-hekmat*, p. 6. Such a curriculum was supported by multi-disciplinary literature that developed in the aftermath of the Mongol invasion and the collapse of the Caliphate, such as the *Rashahât al-fonûn* (anonymous author), and Amoli, *Nafâyes al-fonûn fi 'arâyes al-'oyun*.
28. Seyf al-din Mohammad Ja'far Astarâbâdi, *Safineh-ye Nuh*, translated and edited in H. Ebrahimnejad, "Religion and Medicine in Qâjâr Iran," p. 421.
29. Emilie Savage-Smith, "Attitude towards dissection in Medieval Islam." Cyril Elgood, on the other hand, had an ethnical approach in the relationship between Greek medicine and Islamic medicine, believing that most physicians who had Greek medical knowledge and promoted it were either non-Muslims or Persians who were Shi'a and were not bound by Sunna. See C. Elgood, Tibb-ul-Nabbi or Medicine of the Prophet, p. 38. Elgood probably considered that Avicenna or Râzi, both of Sunni rite, were not orthodox Sunnis.
30. Ibn Qayyim al-Jawziyya, *Medicine of the Prophet*, translated by p. Johnstone, see preface by Seyyed Hossein Nasr and Introduction of Johnstone, pp. xvii, xxiii.
31. Mir, *Pezeshkân-e nâmi*, pp. 111, 137.
32. *Pezeshkân-e nâmi*, p. 149.
33. This led some contemporary physicians to blame those court physicians who ignored medical ethics for financial profit. 'Abdol-Hossein Zenuzi Tabrizi, *Matrah al-anzâr*, p. 142.
34. See E'temâd al-Saltaneh, *Ruznâmeh ye Khâterât* (Diary), pp. 24–25, 218, 831, et passim. *Nâsser al-Din Shah, Ruznâmeh-ye khâterât*, p. 37; Jean Baptist Feuvrier, *Trois ans en Asie*, pp. 70–71; Willem Floor, *Public Health in Qâjâr Iran*, p. 171.

35. ᶜEmâd al-Din Mahmud-e Shirâzi, court doctor to Shah Tahmasp Safavid, used his time for writing and experimenting in medicine. His secure financial situation allowed him to do this while also treating patients free of charge at the mausoleum of Imam Reza, or probably at the *dâr al-shafâ* that belonged to the mausoleum. ᶜEmâd al-Din Mahmud, *Resâleh-ye âtashak*, p. 2.
36. M. Bamdad, *Sharh-e hâl*, vol. 4, pp.192–3. Zenuzi, *Matrah al-anzâr*, p. 318.
37. Under Hâji Mirzâ Aqâssi, Mirzâ Bâbâ was apparently paid as any ordinary physician of the army. In June 1839, a cheque (*barât*), of unspecified amount, was signed for his six months' salaries. The Archives of Sâzmân-e Asnâd, no. 295005729, dated 7 RabiᶜII 1255/June 1839.
38. PRO, FO/60/90, Tehran, September 6, 1842.
39. *Iran-e Soltâni* No. 10, July 29, 1905, reprinted vol., p. 342. *Tiyul* (or *tuyul*), a Turkic (Chagatai) term, was a piece of land that the Qâjâr princes gave to a person in their service so that they could exploit it, usually to the end of their life, in place of a salary.
40. The Archives of Sâzmân-e Asnâd, no. 295001034, 1281/1865.
41. Rustâyee, vol. 2, p. 394).
42. Ibid, no. 9, July 5, 1905, reprinted volume, p. 338.
43. Reza Arasteh, *Educational and social Awakening of Iran*, p. 134.
44. See, for example, the books of Mirzâ Nosrat, *Hâfez al-sehheh-ye nâsseri*, pp. vvii–x, 3–4, and Mirzâ ᶜAbdolvahhâb Tafreshi, *Matlaᶜ al-tebb-e nâsseri*. Quchâni, a graduate of the Dâr al-Fonun, ornamented the preface of his book, *Hâfez al-sehheh-ye nâsseri*, with four pages of poetry in praise of the Shah and his minister, and in his introduction to the book acknowledged that the creation of the Dâr al-Fonun and other modern schools were made possible because of the Shah's order and will. Mirzâ Nosrat stressed that he was one of the creations (*parvardegân*) of the court that he had been serving for 28 years. Apparently he was one of the 42 students sent to Europe to study in 1859.
45. Sepehr, *Nâssekh al-Tavârikh* (NT), vols. 1–2, pp. 61, 309, 353. Mirzâ Ahmad Esfahâni, represented by the pejorative name of Mirzâ Ahmaq (dolt) to rhyme with his true name Mirzâ Ahmad, is described by James Morier in *Hâji Bâbâ*. According to Morier's tale, Mirzâ Ahmad hakimbâshi was a close ally to Malek al-shoᶜarâ Mirzâ Fathᶜ Ali who was also from Esfahan. Mirzâ Ahmad endeavored to secure the dismissal of the physician of Sir Harford Jones's Mission. See *Sargozasht-e hâji bâbâ esfahâni*, op. cit., pp. 163ff.
46. Sepehr, vol. II, p. 640.
47. Zenuzi, *Matrah al-anzâr*, p. 117. ᶜAbdol-Sabur-e Khoi Tabrizi, known as Hakim Qoboli (Turkish: with a wooden leg—one of his legs had been amputated and he used a wooden leg), translated the smallpox treatise of Dr. Cormick with the title: *Taᶜlim-e âbele-kubi*. He also wrote various books on modern medicine, including the *Anwâr-e nâsseriyeh*, which contained chapters on anatomy, modern medicine, and treatment and was lithographed in 1272/1856. *Majalleh-ye dâneshkadeh-ye adabiyât wa ᶜolum-e ensâni*, (*Journal of the Faculty of Literature and Human Science*), Tebriz, Spring 1351, no. 101.

48. Bâmdâd, II, pp. 327–328; Sepehr, II, p. 854. Elgood, *A Medical History*, pp. 488–489.
49. ᶜAbbas Eqbâl Ashtiyâni, *Mirzâ Taqi-Khân Amir-Kabir*, p. 100.
50. Tonekâboni authored only two books, *Matlab al-so'âl* and *Resâleh-ye eshâliyeh* and translated the *Bor' al-sâᶜa* of Râzi. Zenuzi, op. cit., p. 250.
51. ᶜAlinaqi Hakim al-mamâlek, *Ruznâmeh-ye safar-e Khorâssân*, Tehran.
52. Mirzâ ᶜAli, *Amrâz-e ᶜasabâni* (lit. *Nervous diseases*) p. 3. Dr. Mirzâ ᶜAli, by translating several medical books from French became the first major author who transmitted modern anatomy and pathology to Iran.
53. The author of MSS 506 attended this school and frequented the lectures of Polak. In the introduction to his book, *Amrâz-e ᶜasabâni*, a translation of Grisolle's *Traité de pathologie nerveuse*, Mirzâ ᶜAli Hamadâni highlights the fact that the Qâjâr government revived both traditional and modern sciences. Mirzâ ᶜAli Hamadâni, *Amrâz-e ᶜasabâni*, p. 4.
54. United States Department of State (USDS), online archive: http://digicoll.library.wisc.edu/cgi-bin/FRUS/FRUS-idx?type=article&did=FRUS.FRUS188788.i00 31&id=FRUS.FRUS188788&isize=M, no. 914, Pratt to Bayard, Legation of the US, Tehran, May 27, 1888; no. 912, Pratt to Bayard, Legation of the United States, Tehran, January 10, 1888.
55. USDS, Inclosure in no. 249, Legation of the US, May 28, 1896. There were also other Western physicians active in the capital, like Dr. J. G. Wishard, of the American mission in Shemran, the northern part of Tehran: no 45, Leg of the US, Tehran, December 21, 1898.
56. Anne Marie Moulin, *Le médecin du prince: voyage à travers les cultures*.
57. Sâvaji, *Dastur al-atebbâ*, pp. 3–4.
58. In fact, the state failed both to construct appropriate barracks for its troops in the capital and to organize and provide medical support for them. In 1864, during his mission in Iran, Carretto observed a soldier, allocated to guard their mission, suffering from typhoid under the shade of a tree. He recommended that the soldier be immediately transferred to a military building. However, a few days later Carretto found the same soldier in agony under the shade of another tree in a neighboring garden. As if Carretto answered directly to Sâvaji's remarks about the state's responsibilities for the health of its subject, he observed: "Such is the solicitude that the government has for the defenders of the throne." See *Gazette Médicale d'Orient*, Mai 1864, VIIIème année, no. 2, p. 18.
59. In the cholera epidemic of 1860, Nâsser al-Din-Shah ordered that physicians of note convene at the school of Dâr al-Fonun and address the question of the epidemic and set the necessary preventive or curative measures. See Kani Fakhr al-Atebbâ, *Meftâh al-amân*, pp. 1–2.
60. Tholozan, *Histoire de la peste bubonique en Perse*, p. 26.
61. Sepehr, II, p. 694.
62. Tholozan, *Histoire de la peste bubonique en Mésopotamie*, p. 51.
63. Xavier de Planhol, "Cholera in Persia," *Encyclopaedia Iranica*.

64. FO/284, vol. 325, copie d'une note de la Légation Impériale au ministre persan en date du juillet 20, 1877, no. 88.
65. Râzi Kani Fakhr al-Atebba, *Meftâh al-amân*, pp. 6–8.
66. John Freyer observed in 1680 that "Persia is the most temperate country in the whole of Asia." See *A New Account of East India and Persia in Eight Letters*, p. 328. Rousseau, the French consul in Baghdad stated in 1807 that "in Persia one does not find all these fatal maladies that afflict humanity elsewhere." See *Archives du ministère des affaires étrangères, Mémoires et documents*, vol. 6, fol. 71.
67. Tholozan, *Prophylaxie du choléra en Orient*, pp. 54–56.
68. Mohammad-Taqi Shirâzi, *Mofarraq al-heizeh v'al-wabâ*, p. 1; Mohammad-Pâdeshâh (Shâd), *Farhang-e Anendrâj*, p. 4471.
69. MSS 506, p. 90.
70. FO/60, vol. 20, no. October 14, 1821, fol. 108 and no. 22, December 16, 1821, fol. 117.
71. The medieval authors had observed the waning of the epidemic of cholera with the coming of the cold season. This might be because "when the cold intensifies, as Ibn Ridwan stated, the digestion becomes strong, the air stabilizes, the natural heat turns inward, the earth is covered with plants and its rottenness subsides. Then men's bodies become healthy." See Michael Dols, *Medieval Islamic Medicine*, p. 98. In the epidemiology of the nineteenth century the arrival of cold weather was known to stop the progress of cholera, which fits well into the climatic understanding of the disease. Later in the century, the emphasis was put on the individual's constitution or temperament. The nineteenth-century physicians in Iran knew from experience that the population was progressively immunized after the outbreak of the epidemic. According to modern bacteriology, it may have been that the choleric vibrios in potable water were killed by bacterial killers, the viruses called bacteriophages (meaning bacteria eater), described independently by Frederick W. Twort (in 1915) and Félix D'Herelle (in 1917). F. W. Twort, "The Discovery of the 'bacteriophage'," *The Lancet*, 1925, vol. 205, no. 5303: 803–852.
72. Sepehr, I, p. 338.
73. Sepehr, I, p. 340.
74. Idem, p. 382.
75. According to Tholozan, the plague of 1824 in Erzurum had left its germs in the Armenian Plateau and reappeared in 1825. *L'histoire de la peste bubonique en Mésopotamie*, p. 23.
76. Sepehr, I, pp. 430, 442.
77. Idem, I, p. 457.
78. *Histoire de la peste bubonique en Perse*, p. 41. According to Elgood, however, the Kermanshah plague was imported from Baghdad that year: *A Medical History*, p. 462.
79. Polak, *Safarnâmeh*, p. 502
80. Elgood, *A Medical History of Persia*, p. 460.
81. Due to the subsequent poor harvest, the Shah abated the annual taxes to be paid by the governors of Esfahan, Mazandaran, and Guilân. As a famine provoked

riot and disorder in Tehran, Fath-Ali-Shah ordered the opening of the grain stores and the feeding of the poor for seven months. Sepehr, I, p. 493.
82. Idem, vol. I, p. 499.
83. Public Record Office, FO/249, vol. 29, Tehran, October 10, 1833.
84. Sepehr, I, p. 502. Dr. Cormick's son remained in Azerbaijan at the service of the court of the heir apparent. In March 1853, he accompanied the British envoy to Trebizond on his way to England. *RVE*, no. 115 (5 Rajab 1269/April 14, 1853).
85. Sepehr, Ibid, pp. 502, 511.
86. Ibid, p. 515.
87. Sepehr, II, p. 649.
88. Shirâzi, *Wabâiyeh-ye kabireh* (On cholera) 1251/1835, p. 1.
89. Sepehr, II, 640.
90. Sepehr, II, 658.
91. Sepehr, II, pp. 640, 654.
92. Astarâbâdi, *Safina-ye Nuh*, text edited in Ebrahimnejad, "Religion and medicine in Qajar Iran," 2004, p. 417; Mirzâ Musâ Sâvaji, op. cit., pp. 6–7.
93. Elgood, *A Medical History*, pp. 496–497.
94. Sepehr, III, 1207–1208, 1222
95. Mohammad Hasan-Khân Malek al-hokamâ, *sargozasht (autobiography)*, reproduced in M. Rustâyee, vol. 2, pp. 503 ff.
96. Rustâyee, vol. 2, p. 506
97. Dr. Basil, *Resâleh-ye wabâiyyeh*, p. 5.
98. FO/248, vol. 325, Foreign Office, October 17, 1877, which refers to a dispatch of Churchill from the British consulate in Anzali, no. 14, July 31, 1877.
99. Rustâyee, vol. 2, p. 513–514.
100. Ibid, p. 516.
101. MS 506, pp. 28–31.
102. Esmâ-il Jorjâni uses the term *basâm* for *zât al-janb*, and *âmâs-e shosh* (inflammation of lung) for *zât al-riyeh*. According to Jorjâni, the first can transform to the second. See *Al-aghrâz al-tebbiyeh v'al-mabâhes al-'alâiyyeh*, vol. 1, pp. 602, 613.
103. Sâvaji, *Dastur*, pp. 14–15.
104. Shirâzi, *Wabâyeeyh kabireh*, p. 2.
105. Shirâzi, *Mofarraq al-hayza v'al wabâ*, p. 2; *Wabâyeeyh kabireh*, pp. 2–3.
106. J. Schlimmer, *Dictionnaire* . . . , p. 162.
107. The classification of epidemic fevers under the putrid fevers (*hummây-e khelt or 'afan*) was held by most nineteenth-century physicians. See, for example, Sâvaji, *Dastur*, p. 16.
108. Mirzâ Mohammad-Taqi Shirâzi, *Wabâiyyeh kabireh*, pp. 4–5.
109. Mirzâ Mohammad-Taqi Shirâzi, *Wabâiyyeh kabireh*, pp. 9–10.
110. Sâvaji, *Dastur al-atebbâ*, p. 12.
111. Sâvaji, *Dastur*, pp. 6–7.
112. MSS 506, pp. 32, 140.
113. Shirâzi, *Wabâeeyeh kabireh*, p. 3.
114. Anonymous, MSS 506, pp. 27–28.

115. Schlimmer, *Dictionnaire*, pp. 196–197. Dr. Basil, writing in 1898, observed that "typhus in Tehranis scarce, but I know that most local physicians disagree with me. I have been practising in Tehran for eight years and met more than fifty patients who, according to the Iranian doctors, were infected with *hasba* (حصبه), while I diagnosed typhus in only two of them, and by consulting other European doctors who were in Iran, they also confirmed that they saw only a few typhus patients which included probably the same whom I had visited." See his article in *Tarbiyat*, no. 87 (3 ZiHajj 1315/April 25, 1898), reprinted vol. by National Library, Tehran, 1376/1997, vol. 1, p. 348.
116. Polak, *Safarnâmeh-ye Polak, Iran va Iranian"*, p. 140.
117. Polak, *Resâleh dar mo'âlejâte va tadâbir-e amrâz*, pp. 30–31.
118. Sâvaji, p. 28.
119. Théophile Edouard Aubert, *Coup d'œil sur la médecine envisagée sous le point de vue philosophique*, pp. 77–78.
120. Polak, *Resâleh dar mo'âlejâte va tadâbir-e amrâz*, pp. 51–52.
121. Dr. Seref Etker, personal communication, June 2012.
122. ᶜAqili, *Kholâsat al-hekmat*, p. 10.
123. Grisolle, *Amrâz-e asabâni*, translated from French by Dr. Mirzâ ᶜAli, p. 2.
124. For Broussais and his ideas see *Examen des doctrines médicales et des systèmes de nosologie*. For an account on the ideas of physiologists see E. Ackerknesht, *A short History of Medicine*, pp. 148, 150; R. Porter, *The Greatest Benefit to Mankind*, pp. 313–314.
125. On Clot and his ideas and career, see Anne-Marie Moulin, "The construction of disease transmission in nineteenth-century Egypt," pp. 42–58.
126. On postmortem dissection in Egypt see Khaled Fahmy, "The police and the people in nineteenth-century Egypt," p. 359; idem, "The anatomy of justice: forensic medicine and criminal law in nineteenth-century Egypt."
127. ᶜAqili, *Kholâsat*, p. 10.
128. Polak, *Safarnâmeh*, p. 500.
129. *Hippocratic Writings*, see, for example, Epidemics Book I & II, pp. 78–138, and pp. 150–157
130. H. Scoutetten, *Histoire chronologique, topographique et étymologique du choléra*, p. 35. According to Scoutetten, Hippocrates also divided cholera into humid and dry, ibid, pp. 32–34. Michael Dols has a different reading of Hippocrates and believes that miasma was the most important cause of epidemics. See *Medieval Islamic Medicine*, p. 18.
131. Tholozan, *Histoire de la peste bubonique en Perse*, pp. 6–7.
132. Tholozan, *Histoire de la peste bubonique*, p. 7.
133. Shirâzi, *Moffaraq*, pp. 2–3.
134. Seyyed ᶜAli b. Mohammad Tabrizi, *Qânûn al-ᶜalâj*, p. 21.
135. Anonymous, *On Diseases Affecting Soldiers*, MSS 506, p. 23.
136. Tabrizi, *Qânûn al-ᶜalâj*, p. 56.
137. *Ruznâmeh-ye khâterât-e (the Diary of) E'temâd al-Saltaneh*, p. 368.
138. Shirâzi, *Wabâyeeh kabireh*, pp. 10–11.

139. Sâvaji, op. cit., p. 16.
140. Sâvaji indicates that "in order to understand the meaning of their order (opinion), *hekmat-e entezâm-e ishân*, while thinking to God I made utmost reflection and thanks to the divine lights thrown on me I found that." *Dastûr al-atebbâ*, p. 56.
141. In humoral medicine, the point of crisis in an illness is when the disease approaches the end either toward recovery or toward death.
142. Sâveji, *Dastûr al-atebbâ*, pp. 55–56.
143. Sâvaji, *dastur al-atebbâ*, pp. 9–10. According to Jean Chardin who visited Iran in the second part of the seventeenth century, summer and autumn were considered the unhealthy seasons while the occurrence of illnesses in the winter and spring were rare. *Voyages du chevalier Chardin en Perse et autres lieux de l'Orient*, vol. 2, pp. 259–260.
144. Sâvaji, pp. 32–33. As one can see the theory of unnatural weather as the cause of disease belongs to Hippocrates. See *Hippocratic Writings*, pp. 88 ff.
145. Clot Bey, *Relation de l'épidémie de choléra qui a régné en 1831, en Arabie et en Egypte*, pp. 45–46. Clot followed the theory and method of Broussais that became very popular in France. On Broussais and his method see Erwin H. Ackerknecht, "Broussais or a forgotten medical revolution"; Jean-François Braunstein, *Broussais et le matérialisme: médecine et philosophie au XIXe siècle*. On a critical study of Broussais' theory see Georges Canguilhem, *Le normal et le pathologique*, pp. 14 et 74. On the influence of Broussais on Clot, see Anne-Marie Moulin, "The construction of disease transmission in nineteenth-century Egypt," op. cit.
146. Shirâzi, *Wabâiyyeh kabireh*, p. 4.
147. Shirâzi, Mofarraq, pp. 4–5. This argument makes sense if we accept that Shirâzi considered the yellow bile was hot.
148. Sâvaji, pp. 34–35.
149. Sâvaji, p. 42.
150. Viral, bacterial and parasitic infections can cause splenomegaly.
151. Ibid, p. 42.
152. See above, quote of Avicenna, footnote 130.
153. Mohammad-Taqi Shirâzi, *Tᶜuniya*, pp. 1–2.
154. Henry Scoutetten believed that the most efficient preventive measure or cure for cholera was to transport a population to elevated areas. In the cholera of 1835 in Algiers, the French commander of the city, advised by Scoutetten who was sent there to fight against cholera, ordered the transfer of the affected Jewish population to a mountain close to the city. Cholera stopped almost instantaneously. A similar measure was taken during the Crimean War in Constantinople in 1854 with similar results. Henry Scoutetten, *Histoire chronologique*, pp. 4–5.
155. Seyyed ᶜAli b. Mohammad Tabrizi, *Qânûn al-ᶜalâj*, pp. 19–20
156. Tabrizi, pp. 18–19.
157. Ibid, p. 19.

158. As Stuart Hawthorne indicated, "By its eccentricity [cholera] has set all speculation as to the laws which regulate its course, at defiance." George Stuart Hawthorne, *The true pathological nature of cholera and an infallible method of treating it*, p. 8.
159. United States Department of State (USDS), No. 63, American Legation, Tehran, July 23, 1904.
160. Barthélémy Clot, in addressing the question of whether cholera was contagious or epidemic (infectious), and if contagious, transmitted by brief or prolonged contact, immediate or not, or whether the cause of cholera resided in miasma infecting the ambient atmosphere, took a very cautious stance. He did not side with any theory, nor did he pronounce any opinion, believing that medical science and physiology were not advanced enough to be able to answer those questions with certainty. Instead, he provided 12 cases or observations that according to him proved the noncontagious nature of cholera. Although Clot did not state that cholera was epidemic, or was caused by miasma, most of his observations implied that he was thinking so: (1) Cholera appeared across Egypt almost at the same time, with such inconceivable speed that it would not have been able to be transmitted through one-to-one contamination. (2) It penetrated areas that had no direct or indirect contact with the pilgrims affected by cholera. (3) In the port of Alexandria many ships were in the best position of isolation and had no communication with the sick or with the pilgrims. Nevertheless, the mariners on these vessels suffered from cholera more than the individuals in Cairo. Although Clot avoided any theoretical stance against contagion, he provided a selection of factual observations that went in favor of the noncontagious nature of cholera. The implication of this idea was that miasma rendered the creation of quarantine devised by the "contagionists" futile. François Delaporte in his book, *Le Savoir de la maladie*, argues that in 1832, the choice between contagiousness and epidemicity (miasma) could not be decided. See also J. K. Stearns, *Infectious Ideas: Contagion in Premodern Islamic and Christian Thought in the Western Mediterranean*.
161. B.W. Richardson & Wade Hampson Frost, *Snow on Cholera*.
162. On this question see L. Conrad and D. Wujastyk (eds), *Contagion: Perspectives from Pre-Modern Societies*.
163. Sâvaji, op. cit., pp. 20–21.
164. MSS 506, pp. 116–118.
165. Jean-Baptist Rousseau, *De la contagion du choléra-morbus de l'Inde*, pp. 6–7.
166. Rousseau, *De la contagion du choléra morbus de l'Inde*, pp. 7, 9.
167. Lois I. Magner, *A History of Medicine*, p. 495. For similar opinion see also: Christopher Hamlin, "Predisposing Causes and Public Health in Early Nineteenth-Century Medical Thought"; E.A. Heaman, "The Rise and Fall of Anticontagionism in France."
168. Vivian Nutton, "Did the Greeks Have a Word For It? Contagion and Contagion Theory in Classical Antiquity," p. 151. In the plague of 1831 in Baghdad and

Basra, Dr. Baigrie, the Residency Civil Surgeon in Baghdad smeared his fingers with camphorated oil and sprinkled concentrated vinegar on his clothes to protect himself against infection when moving among the victims. C. Elgood, *A Medical History*, pp. 462–463.

169. *Resâleh dar wabâ* (*Treaty on cholera*), translated by Mosio Gabriel and Mohammad Hossein Qâjâr, 1262, Tehran, National Library.
170. *Wabâiyyeh kabireh*, completed in Rabi' II 1251/August 1835, lithograph edition, Majles Library, Tehran, p. 2.
171. Sâvaji, *Dastur al-atebbâ;* Tehrâni (on *wabâ*) and Afshar (on *wabâ*).
172. In addition to the above-mentioned work, see Anonymous, *Majmu'eh-ye wabâiyeh* (*Collection of Essays on Wabâ*); 1291/1874, Majles Library; *Dastur-e* jelowgiri az wabâ va tâʿun (*Prescription for Prophylactic Measures against Cholera*), Dr. Babayef et al., undated, Library of Astân-e Qods; Anonymous, *Resâleh dar wabâ va tâʿun*, undated, national Library, Tehran; Mohammad-Razi Tabâtabâyee, *Tâʿun* (*Plague*), second part of the nineteenth century, Library of Gowharshâd, Mashhad; *Resâleh dar keyfiyat-e maraz-e tâʿun* (*Treatise on Plague*), under the supervision of Dr. Tholozan, 1293/1876, Library of Gowharshâd.
173. Mohammad Hossein Afshâr, *Resale-ye mo'âlejeh-ye maraz-e wabâ*, pp. 4 ff.
174. Mohammad-Hossein Afshâr, p. 6–8.
175. Shirâzi, *Mofarraq al-heyzeh v'al wabâ*, p. 13.
176. Mohammad-Hossein Afshâr, p. 10.
177. Quoted by Tholozan, *Observations sur le choléra*, pp. 25, 30.
178. Shirâzi, *Mofarraq al-heyzeh v'al-wabâ*, pp. 12–13. See also H. Ebrahimnejad, "Un traité d'épidémiologie de la médecine traditionnelle persane."
179. Although the term "suppression" in this clinical case can be ambiguous insofar as it conveys at the same time the elimination and retention of urine. What is significant is that the theoretical transition sometimes exploited these semantic ambiguities in order to give a new meaning or understanding to a term or a text without entirely breaking with the old concepts. Whatever the truth of the matter in the nineteenth century, in modern medicine the physiology of "suppression of urine" is hypoperfusion of the kidneys: insufficient blood to kidneys, due to dehydration.
180. On the influence of rhetoric, culture, and discourse on medical practice and healing, see David Harley, "Rhetoric and the Social Construction of Sickness and Healing." See also Byron Good & Mary-Jo DelVecchio Good, "The Comparative Study of Greco-Islamic Medicine: The Integration of Medical Knowledge into Local Symbolic Contexts."
181. Parviz Badiʿi (ed.), *Yâddâshthây-e ruzâneh-ye Nâsser al-Din Shâh*, p. 223.
182. *Yâddâshthâ-ye ruzâneh*, p. 377.
183. D. Schneider, *La Medicine Persane*, p. 15.
184. Dr. Basil advised quinine for the treatment of cholera. *Resâleh-ye wabâiyyeh*, p. 6.
185. Polak, *Resâleh dar moʿâlejât*, MSS 6164, p. 34; MSS 506, pp. 78, 83. Quinine was also used as preventive medicine. On his first trip to Europe in 1873,

Dr. Dickson, who accompanied the Shah, gave him a syrup of quinine. It seems that this European product was preferred by the royal patients or the wealthy classes. Naser al-Din Shah, *Ruznâmeh-ye khâterât dar safari avval-e farangestân*, p. 14.
186. These physicians were photographed by Nâsser al-Din Shah, who beneath the photo wrote: "This was taken on the day we used European purgative." See H. Ebrahimnejad, *Medicine*, pp. 38–40.

2. The Physicians and Their Encounter with Western Medicine

1. For Bahâ'ud-dowleh and his *Kholâsat al-Tajârob* (*Quintessence of Experience*), see C. Elgood, *Safavid Surgery*, pp. xvii–xix. ʿEmâd al-Din Mahmud-e Shirâzi was known for his work on *Bikh-e Chini* (China Root), and *Atashak* (*Syphilis*).
2. When Mohammad Shah (r. 1719–1748) ascended the throne in India, "he weighed Hakim ʿAlavi Khan in silver and appointed him to a high position with six thousand horsemen." See Nayyar Wasti, "Iranian physicians in the Indian Sub-continent," p. 280.
3. On the encounter of Persian physicians in India with Ayurveda see Seema Alavi, *Islam and Healing*, pp. 28ff.
4. Nayyar Wasti, op. cit., p. 280. ʿAlavi-Khân was probably the father of the famous physician Mohammad-Hossein ibn Mohammad-Hâdi-ye ʿAqili-ye ʿAlavi-ye Khorâssâni-ye Shirâzi, referred to in this book as ʿAqili, the author of *Makhzan al-Adwiyeh*.
5. Rostam al-Hokamâ, *Rostam al-Tavârikh*, p. 419.
6. ʿAqili's books were produced in lithograph editions in India and Iran, including the *Qarâbâdin-e kabir* (*Greater Pharmacopea*) in 1276 and 1277.
7. Jalâlu'd-Din Abd'ur-Rahman As-Suyuti, *Tibb-ul-Nabbi*, p. 3; See also *Medicine of the Prophet*, published by the Ta-Ha Publishers, London, 1994, Introduction.
8. ʿAqili, *Kholâsat al hekmat*, p. 8.
9. ʿAqili, *Kholâsat al-hekmat*, pp. 62–63.
10. ʿAqili, *Kholâsat al-hekmat*, pp. 11, 28.
11. Idem, *Moʿâlejât-e amrâze mokhtasseh az sar ta qadam*, p. 370.
12. *Kholâsat al-hekmat*, p. 6.
13. Porman, Savage-Smith, *Medieval Islamic Medicine*, p. 44.
14. Mirzâ Mohammad Taqi Shirâzi, *Wabâiyya Kabira*, p. 1; idem, *Tâʿunia*, p. 1; Mirzâ Musâ Sâvaji, *Dastur al Atebbâ*, pp. 6–7.
15. Astarâbâdi, *Safineh-ye Nuh*, edited and translated in H. Ebrahimnejad, "Religion and medicine in Qâjâr Iran," p. 420.
16. ʿAqili, *Kholâsat al-hekmat*, op. cit., fol. 31 (pp. 62–63).
17. Ebn-e Sinâ, *Qânun dar tebb* (Canon of medicine), volume 1, translated by ʿAbdol-rahmân Sharafkandi, edited by A. Pakdaman, M.-R. Ghaffâri, H. ʿErfâni, Tehran University Press, Tehran, 1357/1978, p. 4.
18. See A. Z. Iskandar, *Arabic Manuscripts On Medicine and Science*, p. 243.

19. Hakim Mohammad wrote his book during the reign of Shah Safi I (1629–1642 AD), to whom he dedicated it. He was a medical officer of one of the Turkish divisions of the Safavids. See C. Elgood, *Safavid Surgery*, pp. xxi–xxii.
20. Hakim Mohammad, *Zakhirah-ye Kâmela*, pp. 3–4, 13–14, 19, 22, 34, 36, 46, et passim.
21. Abu Nasr As'ad b. Ilyâs ibn Matrân, *Bustân al Atibbâ wa rawzat al alibbâ*, pp. v, iic (nvado hasht).
22. Both Aflâtun (Plato) and Massih al zaman (Jesus of the time) might refer to Christian physicians. Massih was sometime used to indicate healer. See Hakim Mohammad, *Zakhira-ye Kâmela* (*The Treasury of Perfection*), pp. 14, 93, et passim. For a short presentation of Hakim Mohammad and his book, see C. Elgood, *Safavid Surgery*, pp. xxi–xxii. On Western medical influence in Islam during the sixteenth century, see Haskell Isaacs, "European Influences in Islamic medicine."
23. On the role of Europeans in the introduction of new "drugs" see Nur al-Din Mohammad b. Hakim 'Abdol-Malek-e Shirâzi, *Alfâz al-adwiya*, fols. 122–123, 132–134. See also 'Atâollâh b. Mohammad al-Hosseini, *Anis al-Majâles*, fols. 231–235.
24. W. Floor & M. H. Faghfoory, *The First Dutch-Persian Commercial Conflict*, pp. 1–3. On the introduction of tobacco and its use in Persia see Willem Floor, "The art of smoking in Iran and other uses of tobacco," pp. 47–85. Willem Floor, *The Persian Gulf*, pp. 237, 240.
25. Qâzi Ibn-e Kâshef-e Mohammad-e Yazdi, *Resâleh-ye chub-e chini va chai va qahveh*, fols. 128–129.
26. Denis Wright, *The English Amongst the Persians*, p. 1.
27. For a recent study on the fall of the Safavids, see Rudolph Matthee, *Persia in Crisis: Safavid Decline and the Fall of Ispahan*.
28. C. Issawi, *The Economic History of Iran 1800–1914*, p. 16.
29. C. Elgood, *A Medical History*, p. 440. Dr. George Briggs returned to Persia with John Malcolm's second mission in 1808. See Major Evans Bell, *Memoir of General John Briggs*, p. 32.
30. Iradj Amini, *Napoleon and Persia*, p. 83.
31. AMAE, Paris, Mémoires et Documents, Perse, vol. 7, pp. 113–122 (document dated February 20, 1808).
32. James Morier, *A Journey Through Persia, Armenia, and Asia Minor, to Constantinople in the years 1808 and 1809*, p. 79.
33. C. Elgood, *A Medical History of Persia*, pp. 440–445.
34. On Jukes, see Shirin Mahdavi, "Andrew Jukes, British East India Company surgeon and political agent (1774–1821)."
35. Denis Wright, *The Persians Amongst the English*, p. 123.
36. For the intrigues Mirzâ Ahmad tried to initiate against British doctors, see J. Morier, *The Adventures of Hajji Baba of Ispahan*, pp. 115 ff.
37. AMAE, Paris, Mémoires et Documents, Perse, vol. 7, p. 116.
38. Sâleh b. Nasrollâh al-Halabi, *Tebb-e jaded-e kimiyâyee ta'â berklesus*.

39. AMAE, Paris, Mémoires et Documents, Perse, vol. 7, p. 116.
40. ᶜAbdol-Sabur-e Khoi, *Taᶜlim-nâmeh yâ resaleh-ye âbeleh kubi* (*Instructions on inoculation of smallpox*).
41. M. Najmabadi, "Les relations médicales entre la Grand Bretagne et l'Iran et les médecins anglais serviteurs de la médecine contemporaine de l'Iran," p. 705. We have not found any copy of this document in English.
42. ᶜAbdol-Sabur-e Khoi, "*Taᶜlim-nâmeh yâ resaleh-ye âbeleh kubi,*" op. cit., fols. 4–7.
43. ᶜAbdol Sabur Khoi, *Morakkabât-e jowhariyyeh* (*Chemical drugs*), fols. 1–3. Mohammad b. ᶜAbdol-Sabur-e Khoi-ye Tabrizi (called Hakim Qoboli) studied modern medicine with Dr. Cormick, physician of prince ᶜAbbas-Mirzâ (1789–1833). In addition to the above books on vaccination and Paracelsus's medicine, ᶜAbdol-Sabur wrote *Majmaᶜal-hekmatain wa jâmeᶜal-tebbain* by order of Mohammad Shah, and *Jâmeᶜal-anwâr-e nâsseriyyeh*, in three volumes: on anatomy, on principles of modern medicine, and on pathology and treatment. See *Anwâr-e nâsseriyyeh*, p. 4.
44. According to Pasha Khân, a general in ᶜAbbas Mirzâ's army, the latter took his Western and Iranian doctors with his army during the Caucasus war, in order to ensure treatment of injured officers or soldiers. Hâfez Farmânfarmâyân (ed), *Khâterât-e siyâsi-ye Mirzâ ᶜAli-Khân-e Amin al-Dowleh*, pp. 12–13.
45. In addition to Mirzâ Ahmad-e Tonekâboni, FathᶜAli Shah was served by other traditional doctors, including Mirzâ Mohammad-Taqi Shirâzi and Mirzâ Ahmad-e Qâjâr, the author of *Tohfat al-khâqân*.
46. Elgood, *A Medical History*, pp. 455–456. Most Qâjâr sovereigns, with the exception of FathᶜAli Shah and Nâsser al-Din Shah, were bedridden and died relatively young. Their medical needs might be a major incentive for their predilection to consult European doctors. On the illness of the Qâjâr Shahs, see Amir Arsalan Afkhami, "The Sick Men of Persia."
47. E. Treacher Collins, *In the Kingdom of the Shah*, pp. 107, 159–160, 276.
48. E. Treacher Collins, op. cit., p. 276.
49. Iradj Amini, *Napoleon and Persia*, p. 15.
50. G.-A. Olivier, *Voyage dans l'empire ottoman, l'Egypte et la Perse*.
51. J.M. Tancoigne, *Lettres sur la Perse et la Turquie d'Asie*, vol. 1, Preface; vol. 2, pp. 163–166. Two major documents surviving from Gardanne's mission include a geographical and zoological study of the province of Guilan by Trezel, "Notice sur le Guilan et le Mazanderan, par Mr Trezel Ingénieur," *Géographe*, novembre 26, 1808, no. 8, Ministère de la guerre, Archives historiques, Vincennes (Paris); and a travel account by Tancoigne, *Lettres sur la Perse* . . . . Op. cit.
52. Maurice Spronk, "La question Persane," *Bulletin de l'Union Franco-Persane*, 2e année, no. 5, décembre 1910–janvier 1911, pp. 9–10.
53. The award of concessions by the Qâjâr state in the second part of the nineteenth century was aimed at modernization, like the Reuter Concession, and sometimes, or for the purpose of receiving money in exchange, like the tobacco concession. Tobacco concession was proposed by Amin al-Soltân, the prime

minister during the third trip of Nâsser al-Din Shah to Europe in 1889. The concession, officially awarded to Major Gerard Talbot, an advisor to Lord Salisbury, the British prime minister, was for a duration of 50 years and the Qâjâr state would receive an annual sum of £15,000, a quarter of the yearly profits and a dividend of 5 percent on the capital. See J. Savaqeb, "Abʿâd-e farhangi-ye emtiyâzenâmeh-ye tanbâku," *Meshkât*, nos. 62–65 (1978/1999): 339–360, pp. 339–340.
54. James Morier, *The Adventures of Hajji Baba*, pp. 555–556.
55. About ʿAbbas Mirzâ's attempts of modernization see, Abdol-Sabur-e Khoi, *Morakkabât-e Jowhartiyyeh*, op. cit. fols. 1–3.
56. Denis Wright, *The Persians Amongst the English*, pp. 70–74.
57. Wright, *The Persians*, op. cit., p. 73.
58. Wright, *The Persians*, p. 74.
59. Wright, *The Persians*, p. 77.
60. Wright, *The Persians*, p. 80.
61. Dr. Riach, for example, in 1841, was sent from Tehran to Herat to negotiate the evacuation of Ghourian by the Qâjâr troops. Elgood, *A Medical History*, pp. 486–489. In Iran, since Antiquity, physicians were often part of the court. Borzoe, first minister to Khosrow Anushirvân (r. 531–79), was also his physician. See Djalal Khaleghi-Motlagh, "Borzuya," *Encyclopedia Iranica*, online access: October 11, 2010. Rashid al-Din Fazlollâh was also the prime minister and physician-cum-advisor to Ilkhân.
62. Wright, *The English Amongst the Persians*, p. 124.
63. Wright, *The English*, p. 124.
64. Denis Wright, *The Persians Amongst the English*, p. 76.
65. *Bulletin de l'académie royale de médecine, tom 2 (1837–38)*, novembre 21, 1837, p. 170, quoted in Silvia Shiffoleau, "Genèse de la coopération sanitaire internationale," vol. 2, p. 21. The Ottoman empire was opposed to any close relationship between Europe and Iran, particularly since the Safavid time. When in March 1714, Mohammad-Reza Beg, the mayor of Erevan, was sent as ambassador to France, he had to cross Constantinople in the guise of a pilgrim to Mecca and was still imprisoned (before being released thanks to Comte des Alleurs, the French ambassador in Constantinople, who helped him to reach Marseilles on October 23, 1714) (Iradj Amini, *Napoleon and Persia*, pp. 20–21). Similarly, Amédé Jaubert, the envoy of Napoleon to Iran in 1805, had to endure eight months of confinement in the Ottoman empire before finally leaving the country and reaching Tehran (Amédé Jaubert, *Voyage en Arménie et en Perse*).
66. Letter of Cl. Sheil, Tehran, September 6, 1842, PRO, FO/60/90. Elgood, *A Medical History*, p. 495. Dr. Labat's departure might have been caused by his differences with Comte de Sartiges, the French ambassador in Tehran, who lambasted Labat for having tried to "undermine his (Sartiges's) situation and promote his own interests." Dr. Labat had published articles in *La Revue de Paris* (November 1844 and April 1845) against Sartiges. See Archives du Ministère

des Affaires Etrangères (AMAE), Correspondances Politiques Perse (CPP), vol. 21 (September 30, 1845, no. 65), fols. 107–108.
67. Âghâssi had fixed 500 toman (6,000 French francs) for the travel fees and 3,000 tomans (36,000 FF) as the doctor's annual salary. CPP, vol. 21, October 06, 1845; fols. 117–118 & October 08, 1845, fols. 121–123, & correspondence of October 2, 1845, fol. 126. Âghâssi confided to Comte de Sartiges his opinion on Labat as "an untrustworthy doctor" and described his departure "as if he fled Iran," ibid.
68. J. Théodoridès, "Tholozan et la Perse," p. 287.
69. I. Amini, op. cit., pp. 165–166.
70. I Amini, op. cit., p. 173.
71. AMAE, CPP, vol. 21, no. 24, Paris, January 20, 1845, fols. 7–8; Schneider, "La médecine en Perse," pp. 3–4.
72. Ibid., fols. 57–58.
73. Abbas Eqbal Ashtiyâni, *Mirzâ Taqi-Khân-e Amir-Kabir*, pp. 272–275. A book on the use of cartridge as opposed to canon, and its advantage in mobility was translated by the order of Âghâssi. See Persian Manuscript, dated 1264/1848, anonymous, National Library, Tehran.
74. Abbas Eqbal Ashtiyani, op. cit., p. 157. Adamiyat, *Amir-Kabir va Iran*, p. 290.
75. Ashtiyani, op. cit., p. 158.
76. Although the Dâr al-Fonun of Iran was established before the first short-lived Ottoman *Darülfünun*, which was inaugurated in 1863, the idea was taken from the *Tanzimat* reforms. It is, however, interesting that the first institution addressing the establishment of a university in the Ottoman empire was the *Encümen-i Daniž* (literally Science Society), which convened in 1851, the year when the Dâr al-Fonun in Terhan was inaugurated. See E. Dölen, *Türkiye Üniversite Tarihi, 1: Osmanlı Döneminde Darülfünun (1863–1922)*, p. 7–21. On the influence of Istanbul's intellectual environment on Qâjâr reformists, see Mohammad Fazlhashemi, "Istanbul's Intellectual Environment and Iranian Scholars of the Early Modern Period."
77. It was Mirzâ MohammadᶜAli Shirâzi, who introduced the newly arrived teachers to the Shah and inaugurated the Dâr al-Fonun on December 29, 1851, as by that time Amir-Kabir was already dismissed; Adamiyyat, *Amir-Kabir*, pp. 362–363; Ashtiyâni, p. 160.
78. Ashtiyani, op. cit., p. 161. Borowski was the son of a Polish officer whom ᶜAbbas Mirzâ had employed in his army. General Borowski was killed in the first war in Herat in 1838. But a village was assigned to his family as tuyul, so they could live on its income. Adamiyyat, *Amir-Kabir va Iran*, p. 288; M. Bamdad, *Sharh-e hâl*, vol. 2, p. 129.
79. Elgood, *A Medical History*, op. cit., pp. 497–500.
80. Another Englishman, Mr Taylor, was the teacher of English at the school in 1882. See J. Gurney and G. Nabavi, "Dâr al-Fonun" (online access, August 24, 2011), *Dânesh*, no. 2, June 24, 1882, reprinted version, p. 7.

81. *Dânesh*, no. 7, September 14, 1882, p. 1/reprinted, p. 25, and no. 10 (October 13, 1882), p. 1/37.
82. Elgood, *A Medical History*, op. cit., p. 498.
83. Jean Théodoridès, "Un grand épidémiologiste franco-mauricien": Joseph Désiré Tholozan (1820–1897), p. 105; Jean-Louis Plessis & Jean Théodoridès, "Tholozan: médecin militaire à compétence étendue," p. 280.
84. Nâsser al-Din Shâh, *Ruznâmeh-ye Khâterât dar safar-e avval-e farangestân*, op. cit. pp. 23–24.
85. Elgood, *A Medical History*, op. cit. 509.
86. Dr. Dickson went to London in 1865, conveying the contract that was prepared in Tehran, and returned with the approval of his government in 1867. See RDEI, no. 606 (July 24, 1867), II, p. 1011. See also *RDEI*, nos. 566 & 597 (January 11, 1865, Jan.23, 1867), II, pp. 741, 954–955.
87. Kazollani arrived in Tehran via Erevan, accompanied by ᶜAziz Khân Ajudan Bâshi on March 12, 1851. *RVE*, no. 7 (17 jamad I 1267/March 20, 1851).
88. It is significant that despite his close relationship with the United States, Mohammad-Reza Shah called upon French doctors Jean Bernard and Georges Flandrin to treat his leukemia in the 1970s. See Farah Pahlavi, *Mémoires*.
89. C. Elgood, *A Medical History of Persia*, op. cit., p. 439.
90. Ackerknecht, *A Short History of Medicine*, p. 151.
91. H. Nategh, *Iran dar râhyâbi-ye farhangi 1834–1848* (*Iran and cultural enlightenment*), pp. 181ff. To my knowledge, it is the pioneering work of Homa Nategh that first revealed the reformist dimension of the government of Hâji Mirzâ Âghâssi.
92. See the letter of Tholozan to Nâsser al-Din Shah, reproduced in *Ruznâmeh-ye khâterât-e Nâsser al-Din Shâh dar safar-e avvale farangestan*, pp. 555–557.
93. Homa Nategh, *Kârnâmeh-ye farhangi-ye farangi dar Iran*, p. 52.
94. H. Nategh, *Iran dar râhyâbi-ye farhangi*, op. cit., p. 104.
95. The original work of Cormick in English could not be found, but its translation into Persian by ᶜAbdol-Sabur Khoi (see footnote 41) was lithographed. For the treatises on chemical medicine, see *Tebb-e jadid-e kimiyâyee*, translated from original Arabic version of Sâleh b. Nasrollâh Halabi.
96. He refuted the essences (*jowhariyât*) in his *Jowhariya*, 1269/1853, in Oeuvres (collection) of Shirâzi's treatises, lithographed edition, Library of Majles, Tehran. One of Shirâzi's pupils, Astarâbâdi, argued that the "essences" of drugs introduced by the Europeans were harmful because they were altered through their transformation from their natural state to their essence, and because they were produced in a climate different from that of Iran and did not fit the temperament of the local population. *Safineh-ye Nuh*, fol. 3, in H. Ebrahimnejad, "Religion and Medicine in Qâjâr Iran," op. cit., p. 411.
97. ᶜAbdol-Hossein Zenuzi Tabrizi, *Matrah al-anzâr*, pp. 319–320.
98. *RDEI*, no. 536 (January 15, 1863), rpnt, vol. 1, p. 512.
99. *Matrah al-anzâr*, op. cit., p. 319.

100. *"namak-e mos'hel ke az Englis mi'âvarand,"* Shirâzi, *Wabâiyyeh kabireh*, op. cit., p. 5.
101. Schlimmer, *Terminologie médico-pharmaceutique*, pp. 219–220, 301, 528.
102. Asnâd (Archives), Sâzmân-e asnâd-e ketâbkhâneh-ye melli, Tehran, no. 295000922, undated. See fig. 2.2.
103. Jacob Polak, *Safarnâmeh*, op. cit., p. 415.
104. Sajjadi, Dentistry, Encyclopaedia Iranica, 1994.
105. *Ruz-nâma-ye Irân*, no. 649, 26 Jomâdâ I 1305/Feburary 9, 1888, repr., IV, p. 2615.
106. Rashti, *Hefz al-Sehheh*, Tehran, National Library, manuscript no. 439, or 5–10439, pp. 24, 316; Bâmdâd, II, pp. 305, 328, 437
107. *RDEI* (no. 500 (October 6, 1861), repnt, vol. I, p. 240. Ibid., no. 616 (March 25, 1868), II, p. 1,066.
108. *Ruznâmeh-ye ᶜElmi*, nos. 3, January 29, 1877, 4 & 11, Tehran, Majles Library, no. 452. Fakhr al-Atebba, *Meftah al-amân*, 1861, p. 3. About the Majles-e Hefz al-Sehha and Ryznâmeh-ye ᶜElmi, see Chapter 3.
109. Mirzâ Mohammad Kâzem Rašti, *Hefz al-sehheh-ye nâsseri*, p. 22.
110. Fereidun Adamiyyat, *Amir-Kabir va Iran*, p. 335. In 1851 Schlimmer was still residing in Rasht. *RVE*, no. 21 (26 Shaᶜbân 1267/June 26, 1851).
111. Malek al-atebbâ, *Hefz al-sehha nâsseri*, pp. 24–25.
112. Astarâbâdi was a cleric who had studied medicine under physicians like Shirâzi, Mirzâ Musâ Sâvuji, and Mirzâ Khalil. See Astarâbâdi, *Safina-ye Nuh*, edited and translated in H. Ebrahimnejad, "Religion and Medicine," p. 421. Mohammad-Taqi Mir, *Pezeshkân-e nâmi-ye fârs*, p. 50.
113. *Ruz-nâma-ye ᶜelmi*, nos. 1 (January 8, 1877) and 2 (January 15, 1877). On this subject see H Ebrahimnejad, KâZem Rašti, *Encyclopaedia Iranica*, vol. 15.
114. *Ruznâmeh-ye ᶜelmi* (*Journal of Sciences*), nos. 4 (February 5, 1877) & 5 (February 12, 1877).
115. Mirzâ Mohammad al-Râzi al-Kani Fakhr al-Atebbâ, *Meftâh al-Amân*, p. 65. Mehdi-Aoli Hedâyat Mokhber al-Saltaneh, *Khâterât wa Khatarât*, p. 2.
116. Kani, *Meftâh al-Amân*, pp. 7–8.
117. Kani, *Meftâh al-Amân*, p. 37.
118. Kani, *Meftâh al-Amân;* pp. 2–3.
119. Eᶜtemâd al-Saltaneh, *Târikh-e montazame nâsseri*, vol. 3, p. 1,728.
120. As the remains of a dead Muslim must rest in peace, dissection was tantamount to punishment after death. Therefore no dissection was permitted on Muslim corpses. During the Babis uprising, Dr. Polak was asked to dissect dead Babis as a mark of disrespect to their cadavers, a request that Polak rejected, "lest he became an object of hatred to the Babis." Polak, *Safarnâmeh*, p. 211.
121. On the August 15, 1852, Nâsser al-Din Shah was shot by three Babais, adepts of the new religion. While rumors of the death of the Shah spread, Dr. Cloquet extracted the bullets without difficulty. Elgood, *A Medical History*, op. cit., p. 504; Sepehr, *Nâssekh al-Tavârikh*, vol. 3, p. 1,184. After his lectures that

took two hours every day, Dr. Polak demonstrated surgical operations on the patients of hospitals. *Safarnâmeh*, op. cit., p. 211.
122. Lutz Richter-Bernburg, *Persian Medical Manuscripts at the University of California* (Los Angeles, Malibu, 1978), p. 185.
123. ʿAbd al-Razzâq, *Kholâsat al-Tashrih*, fol. 3.
124. The twofold meaning is well illustrated in the poem of Mawlavi Jalâl al-Din Mohammad Balkhi (1207–1273): (*sineh dâram sharha sharha az farâq—tâ beguyam sharh-e dard-e eshtiyâq*: I have a heart in pieces (*sharha sharha*) from separation, to express (*sharh*) the pangs of my yearning (for my love). See *Masnawi-ye Maʿnawi*, Book I. p. 1, electronic source.
125. *Ruznâmeh-e ʿElmi*, no. 5, February 12, 1877.
126. *Aksar-e joruh va qruhi ke behâlat-e solh beham miresad hame az amrâz-e mazâjiyyeh mahsub mishavad ke bâyad be eslâh-e mazaj moʿâlejeh shavad* Anonymous MSS 505, "On the establishment of hospitals" in Ebrahimnejad, *Medicine*, p. 228.
127. Ernest Cloquet was the nephew of the eminent anatomist Dr. Jules Cloquet, Dean of the Faculty of Medicine in Paris, who sent him to Iran at the request of Hâji Mirzâ Âghâssi in c.1845. Schneider, "La médecine Persane," p. 10.
128. Polak, *Tashrih-e badan al-ensân* (Anatomy of the human body), lithographed, Tehran, 1270/1854, p. 4.
129. Polak, *Safarnâmeh*, p. 210.
130. Persian MSS, written (or copied) in 6 RabiʿI 1277/September 1860, no. 4603, Ketâbkhâne-ye markazi, University of Tehran, pp. 1–2.
131. Anonymous, MSS 506 (Commentary on Polak), National Library, Tehran. The document bears no title or date, but it appears it was written around 1858. The author says "it is now nearly seven years since Polak entered Iran" MSS 506, p. 85. In another place, providing an account of the epidemic of *mothbeqa* in 1271/1855, and *wabâ* in 1856, the author explains the development of *nowbeh* 'this year = *emsâl* [1858]) among the regiment of Khalkhâl, pp. 61–62. The author of this critical commentary on Polak seems to be the author of another treatise on the establishment of state hospitals (MSS 505, edited in Ebrahimnejad, *Medicine*,) that is also anonymous and undated. The similarities in their style and information suggest that both were authored by one and the same person, who introduces himself as the chief army physician. According to Eʿtemâd al-Saltaneh (*Târikhe montazam*, III, pp. 1,730–1,731), following the death of Kazullani in March 1852 (Jamadi I 1268), who was the '*hakimbâshi-ye nezâm*, the chief physician of the army, Mirzâ Mohammad-Vali replaced him. Was Mirzâ Mohammad Vali the author of our documents? We cannot confirm.
132. Polak, *Dastur al ʿamal wa taʿlimât-e moʿâlejât wa tadâbir-e amrâz*. This treatise is undated and bears no title but in the introduction Polak explains what is about: *dastur al-ʿamal va taʿlimât-e moʿâlejât va tadâbir-e amrâz-e chandi keh aksar-e owqât be afvâj-e qâhereh mostowli va bâʿes-e halâk-e ishân migardad az qabil-e nowbeh va eshâl al-dam va mothbeqa va taʾsir-e borudat* (Treatise on

intermittent fevers, dysentery, typhoid, and cold). It is undated but seems to be written a year of two after Polak's arrival, ca.1853.
133. MSS 506, p. 124.
134. Mohit-e Tabâtabâ'i, "Dâr al-Fonun wa amir-Kabir," p. 192; Mahbubi Ardakâni, *Târikh-e mo'assessât*, pp. 189–195; Ringer, *Education, Religion and Discourse of Cultural Reform*, p. 48.
135. It is likely that he knew Cloquet personally; he mentions him when talking about his son's illness and that through Cloquet or Focceti he studied French medical texts. MSS 506, p. 5.
136. MSS 506, pp. 1–2.
137. Morier, *The Adventures of Hâjji Baba*, op. cit., p. 116.
138. Polak, MSS 6164, pp. 4, 8.
139. Commentary on authoritative texts in Islamic medicine constituted a major form of medical literature from the medieval period, such as the commentary written by Ibn Nafis on the Canon of Avicenna in which he refuted the opinion of Galen and Avicenna that the blood from the right ventricle of the heart passed to the left ventricle through seeps or invisible pores between the two. The *Shokuk alâ Jâlinus* (*Doubts on Galen's ideas*) written by Razi is another example. On this format of Islamic medical literature, see H. Ebrahimnejad, "Medicine in Islam or Islamic Medicine."
140. Ibid., pp. 58–59.
141. MSS 506, p. 18; the book in question, by August François Chomel, which the author calls *Osul-e mabâhes-e amrâz*, is the "*Elements de pathologie générale.*"
142. MSS 506, pp. 91–92, 100, 103, 131, et passim.
143. Arthur de Gobineau, *Trois and en Asie, Œuvres*, II, p. 223.
144. *Safarnâmeh*, p. 214.
145. See *On the Establishment of State Hospitals*, edited and translated in H. Ebrahimnejad, *Medicine*, pp. 211–212.
146. H. Ebrahimnejad, *Medicine*, pp. 213–214.
147. Ibid., p. 212.
148. The commentaries "*sharh*" presented opportunities for physicians to oppose or to correct the ideas of their predecessors or contemporaries. Râzi's *Shokuk alâ Jâlinus* and Ibn Nafis's *Sharh-e Qânun* illustrate this genre of medical literature. Criticism also came in other forms. Mirzâ Qâzi in his treatise on *Chub-e Chini* criticized ʿEmâd al-Din Mahmud-e Shirâzi's idea of cold and hot, proposing that matters are *morakkab al-qowâ* (composed of various qualities). See Qâzi b. Kâshef al-Din Mohammad Yazdi, *Resâleh-ye Mirzâ Qâzi*, Wellcome Trust Library, WMS. Per. 293 (B).
149. See Claude Bernard, *Principes de médecine expérimentale*, pp. XIII, 2–5.
150. Mirzâ ʿAbdol-Karim Tehrâni, *Ketâb-e ʿAlâyem al-Amrâz* (*Treatise on the symptoms of diseases*), p. 4.
151. *Ketâb-e ʿAlâyem*, op. cit., pp. 6–8.
152. For Paracelsus (*Tebb-e jadid-e Kimiyâyee*), see above, footnotes 36 & 93.
153. *Ketâb-e jarrâhi moshtamel bar do jeld va yek resale dar kahhâli az tasnifât-e ʿalijâh doctor Polak-e namsavi* . . . "the book on surgery in two volumes and one treatise

on ophthalmology authored by His Excellency Dr. Polak from Austria . . . ," undated, p. 2.
154. Polak, *Ketâb-e jarrâhi*, p. 2.
155. Polak, *Ketâb-e jarrâhi*, pp. 3–8.
156. Ibid., p. 8; Shirâzi, *Mofarraq al-heyzeh w'al-wabâ*, pp. 2–3.
157. Polak, *Ketâb-e jarrâhi*, pp. 2–3.
158. In traditional medicine bloodletting was also called *estefrâq-e kolli* (general evacuation), because blood contain all the four humors, and by opening a vein all humors were evacuated. See Mirzâ Mohammad-Kâzem Rashti Malek al-Atebbâ, *Hefz al-sehheh-ye nâsseri*, op. cit., p. 37.
159. Dr. Albo, *Jarrâhi* (Surgery), pp. 2–3. A note on p. 3 of the text explains that "in the traditional medicine the term *enserâf* (diversion) meant that one of the four humours (*akhlât*) did not move in its natural route and by bloodletting, the surgeons tried to change its trajectory."
160. Albo, ibid., pp. 3–4.
161. *Ruznâmeh-ye ᶜelmi*, nos. 1 (January 8, 1877) & 9 (March 12, 1877).
162. J. Schlimmer, Mirzâ Mohammad-Taqi Kâshâni, *Meftâh al-Khawâss-e nâsseri*, p. 2.
163. J. Schlimmer, *Meftâh al-Khawâss*, pp. 4–5.
164. J. Schlimmer, *Meftâh al-Khawâss*, pp. 16–17. The term *akalat* (or *akalit*) that Schlimmer used seems to be a drug that dried out ulcers in syphilis or chancre. See his *Terminologie Medico-Pharmaceutique*, pp. 122, 535.
165. *Meftâh al-Khawâss*, pp. 15–18.
166. Rashti, *Hefz al-Sehheh*, pp. 31–181.
167. Pormann & Savage-Smith, *Medieval Islamic Medicine*, p. 44.
168. Mirzâ Nosrat-e Tabib-e Quchâni, *Hâfez al-sehheh-ye nâsseri*, pp. 8–12.
169. Anonymous, *Ketab dar elme tebb*, Persian MSS, no. 179, pp. 305–306. Mirzâ Mussâ Sâvaji, *Dastur al Atebbâ*, op. cit., pp. 13–14. See also Chapter 1, diagram no. 1.
170. *Ruznâmeh-ye Dowlat-e ᶜElliyeh-ye Iran*, no. 502 (Oct. 17, 1861), reprinted by the National Library, Tehran, 1370/1991, vol. 1, p. 253.
171. Mirzâ Rezâ, *Pâtologi*, pp. 2, 30–31. Based on the translation of Augustin Grisolle's *Traité élémentaire et pratique de pathologie interne*, 2 vols. The manuscript we use has no mention of the translator, but Dr. Mirzâ ᶜAli in his *amrâz-e ᶜasabâni* (Grisolle's *Traité des maladies nerveuses*), p. 5, indicates that this book was translated by "the late Mirzâ Rezâ-ye doktor."
172. Grisolle/Mirzâ Rezâ, MSS 1519, p. 8.
173. Grisolle/Mirzâ Rezâ, MSS 1519, pp. 2–3. Etymologically, *mothbeqa*, from *mothbeq* (covering) means continuous. *Mothbeqa* fever was a fever that did not interrupt during days and nights. It was often referred to as *hasba*, a disease that is now known as typhus. See ᶜAli Akbar Dehkhoda, *Loghatnâmeh*, vol. 13, pp. 21,054–21,055.
174. *Iran*, nos. 1 (11 Moharram 1288/April 2, 1871), 11 (17 Safar 1288/May 8, 1871) & 18 (21 Rabi I 1288/June 10, 1871).

175. *Hommâ-ye yawm* was identified by Schlimmer to muscle sore. See *Terminologie Medico Pharmaceutique*, p. 162. See also Mirzâ Abol-Hassan-Khân b. ᶜAbdol-Wahhâb-e Tafreshi, *Matla ᶜal-tebb-e nâsseri*, pp. 7–8.
176. Mirzâ Rezâ, *Pâtologi*, pp. 4–5.
177. *Matla ᶜal-tebb-e nâsseri*, p. 6.. On Gabriel Andral (1791–1876) and his theory of blood chemistry that dismantled Broussais's theory see E. Ackerknecht, *A Short History of Medicine*, p. 151. See also Gabriel Andral, *A Treatise on Pathological Anatomy*, 1929.
178. Augustin Grisolle, *Traité élémentaire et pratique de pathologie interne*, vol. 1, p. 4.
179. *Ruznâmeh-ye ᶜElmi*, no. 54 (February 9, 1879).
180. On this question, see H. Ebrahimnejad, *Medicine*, pp. 6–7.
181. *Resâleh-ye wabâiyyeh*, translated by ᶜAbdol-Majid Eᶜtezâd al-Atebbâ, pp. 2–3.
182. *Resaleh-ye wabâiyyeh*, pp. 18–20.
183. J. Snow, *Snow on Cholera*, p. 15.
184. Snow, idem, pp. 18–23.
185. Report of a meeting of 23 Rajab 1307/March 15, 1890, on prophylactic measures against cholera, in 20 pages, National Library, Tehran. On Abdol ᶜAli Seif al-Atebbâ, see Roustayee, vol. 2, p. 300.
186. MSS 506, pp. 124, 128.
187. We have seen how this question of identity influenced the formation of Islamic sciences. See H. Ebrahimnejad, "Islamic medicine and medicine in Islam," op. cit. The appropriation of sciences in order to constitute an "imperial ideology" under Sasanid Iran represents another experience of this kind. See D. Gutas, *Greek Thought, Arabic Culture*.
188. The letter reproduced in F. Qâzihâ (ed.), *Ruznâmeh-ye khâterât-e Nâsser al-Din Shah dar safar-e dovvome farangestân*, pp. 291–293. In this letter Nâsser al-Din Shah avows that his trip to Europe was initiated and organized by his prime minister, Mirzâ Hossein Khân Sepahsâlâr.
189. J. Schlimer and Mirzâ Mohammad-Taqi Kâshâni, *Meftâh al-khavâss*, p. 4.

3. The Reform Movement and Medical Institutionalization

1. While the Rousseau family praised the potential of commerce in Persia, reports by Brouguière and Olivier, who were sent by the Directorate to explore relationships with Persia, were quite negative about Persian possibilities. See Irène Natchkebia, "Jean-Baptist-Louis-Jacques Rousseau's Manuscript": "Tableau general de la Perse modern (1807): Military possibilities and Trades of Persia in the Context of the Indian Expedition," (unpublished manuscript). I am grateful to Irène Nachkebia for giving me a copy of this article.
2. For more details about this intellectual movement see Homa Nategh, *Irân dar râhyâbi-ye farhangi 1834–1848*, chapters 2 and 3; Fereidun-e Adamiyyat and Homa Nategh, *afkâr-e ejtemâᶜi wa siyâsi eqtesâdi dar âsâr-e montasher nashodeh-ye dowreh-ye Qâjâr*.

3. According to Homa Nategh, the rise of Babism in 1844 represented a social movement and not the invention of a new religion. "It stands to reason that the masses in villages and cities never turn against their religion overnight." *Iran dar râhyâbi-ye farhangi*, p. 62. For an account of Babist movement both as doctrinal dispute against the Usuli Shiite establishment and as sociopolitical protest against the economic context of mid-nineteenth-century Iran see Abbas Amanat, *Resurrection and Renewal: The Making of the Babi Movement in Iran*. For the influence of the Babism, and the anti-Babi movement that it triggered, on the formation of political Islam and in general on the political awakening in Iran see M. Tavakoli-Targhi, *Bahâ'i-setizi va eslâmgarâ'i dar Iran* (Anti-Babism and Islamism in Iran).
4. Baron de Bode, *Travels in Luristan and Arabistan*, vol. 1, p. 74.
5. See, for example, Mollâ Ardeshir, *Vaqâye^c-e pârsiyân*; Anonymous, *Fehrest-e ketâb-e majma^c al-moluk*.
6. AMAE, Mémoire et Documents, Perse, vol. 9, 1851.
7. This question is discussed in B. Lewis, *What Went Wrong?*, pp. 41ff.
8. James Morier, *The Adventures of Hajji Baba*, pp. 482–489.
9. *Tarbiyat*, no. 178, Ziqa^cda 6, 1317/March 8, 1900, reprinted version, vol. 1, pp. 709–711. Reprinted by the National Library, Tehran, 1376/1997.
10. For the major works of the Dâr al-Fonun, see Qodratollâh Rowshani Za^cferânlu (ed), *Amir-Kabir wa Dâr al-Fonun*. See also J. Gurney; G. Nabavi, "Dâr al-Fonun."
11. *RVE*, no. 364 (Jamadi II 5, 1274/January 21, 1858).
12. E^ctemâd al-Saltaneh, *Târikhe Montazam*, vol. 3, p. 2042. On the relationship between religious endowments and education in Qâjâr Iran, see H. Ebrahimnejad, "The waqf, the state and medical education in the nineteenth century."
13. Johannes L. Schlimmer, *Terminologie Médico-Pharmaceutique*, pp. 225–226.
14. *Ruznâmeh-ye Vaqâye^c-e Ettefâqiyeh (RVE)*, no. 89 (29 Zihajja 1862/October 14, 1852).
15. *RVE*, no. 377, September 7, 1274 (April 21, 1858), reprinted by National Library, Tehran, 1374/1995, vol. 4, p. 2530.
16. *RVE*, no. 415 (January 13, 1859), rpnt, vol. 4, pp. 2,798–2,799. The Governor of Kashan executed this order by posting the royal order at the (traditional) school Soltâni and appointing teachers to test the progress of the students every month. Ibid., no. 420 (February 16, 1859), rpnt, vol. 4, pp. 2,830–2,831.
17. *Dânesh*, no. 1, p. 2 & no. 3, p. 2, reprinted, pp. 2, 10.
18. *Iran*, no. 590 (January 03, 1886), III, p. 2,377. According to *Iran*, the founder of this school was (Ja^cfarqoli Khân-e) Nayyer al-Molk, the cadet brother of ^cAliqoli Khân-e Mokhber al-Dowleh, who was, at the time, the Minister of Education (*vazir-e ^colum*), while ^cAliqoli Khân was the Head of the Dâr al-Fonun. In 1882, when the order to create this school was issued, ^cAliqoli Khân was both Director of the Dâr al-Fonun and Minister of Education. See also E^ctemâd al-Saltaneh, *Ruznâmeh*, p. 343.

19. *RVE*, no. 456 (Jamâdi I 16, 1276/December 14, 1859), rpnt, p. 2991.
20. Ibid.
21. Isâ Sadiq, *Modern Persia and Her Educational System*, p. 18.
22. *RDEI*, no. 544 (June 5, 1863), rpnt, vol. 1, p. 572.
23. *RDEI*, no. 589 (July 7, 1866), II, pp. 907–910.
24. Ibid.
25. *RVE*, no. 433 (Shawwâl 15, 1275/May 18, 1859), IV, rpnt, p. 2,896; idem, no. 456, ibid, p. 2,991.
26. For an account of the reopening of the school in 1875 and 1893, see Samad Sardâriniâ, *Dâr al-Fonun-e Tabriz*, pp. 13–14. The journal *Akhtar* reports that six years after its inauguration, the school was closed down in 1881. Cited in Sardâriniâ, *Dâr al-Fonun-e Tabriz*, p. 17.
27. Sardâriniâ, *idem*, pp. 75–77; See also annex to journal *Dânesh*, the four monthly journals of *Madrasa-ye mobârakeh –ye Dâr al-Fonun-e Tabriz*, October, and November 1893, January and February 1894; M. Ringer, *Education, Religion*, pp. 87ff.
28. *RDEI*, no. 502 (October 24, 1861), rprnt vol. 1, pp. 252–254; *RDEI*, no. 582 (January 14, 1866), vol. 2, pp. 859–61.
29. Eᶜtemâd al-Saltaneh, *Tărikh-e montazam-e nâsseri*, vol. 3, p. 1726; Gurney, Nabawi, op. cit.
30. *Dânesh*, no. 2 (June 24, 1882), p. 3.
31. Asnâd (archives) of Ketâbkhâneh-ye Melli (National Library), Tehran, no. F/438. According to the *RVE* (nos. 394 & 395, August 19 & 26, 1858), the Dâr al-Fonun had about 91 students.
32. ᶜAli Khân, *Hekmat-e tabiᶜi. Osul-e ᶜelm-e fizik*.
33. *RVE*, no. 102 (Rabi II 3, 1269/January 14, 1853). This Mirzâ Seyyed ᶜAli might be the same as Seyyed ᶜAli b. Mohammad Tabrizi, the author of the *Qânun al-ᶜalâj*, 1853. See Chapter 1, footnote 153.
34. *RVE*, no. 98 (Rabi I 5, 1269/December 17, 1852).
35. *RVE*, nos. 98.
36. *RVE*, no. 367 (Jamâdi II 26,1274/February 11, 1858).
37. *RVE*, no. 456 (December 14, 1859).
38. *RDEI*, no. 501 (October 17, 1861), rpnt, vol. 1, p. 252.
39. *Iran*, no. 55 (Ramazan 17, 1288/November 30, 1871).
40. *Merrikh* [*The Journal of the Military*], no. 18 (May 26, 1880), lithograph edition, Tehran.
41. *Iran*, no. 55.
42. Focceti from Italy who taught physics and chemistry for several years at the Dâr al-Fonun terminated his contract in 1862 and left the country (*RDEI*, no. 523, June 13, 1862, rpnt vol. I, p. 419).
43. *RDEI*, no. 535 (January 8, 1863), rpnt, I, p. 504.
44. *RE*, no. 1 (ZiHajja 22, 1293/January 8, 1877).
45. *RDEI*, no. 541 (April 17, 1863), rpnt, I, p. 550.

46. Schneider, "La Médecine Persane, les médecins français en Perse et leur influence," p. 14.
47. These diplomas are reproduced in M. Rustâyee, *Târikh-e tebb wa tebâbat dar iran*, op. cit., vol. 1, Introduction, pp. *sado panjâho shesh- sado panjâho hasht*.
48. *RDEI*, no. 629 (March 18, 1869), rpnt, II, p. 1137.
49. *RDEI*, no. 629 (March 18, 1869), rpnt, II, p. 1137. Mirzâ Seyyed Razi authored a book, called *Qarâbâdin-e Sâlehi*, on compound drugs, in 1868. *RDEI*, no. 616, 25 March 1868, II, p. 1068. The lithographed book that we have found by this army physician is titled *Tohfat al-Sâlehin*.
50. *RDEI*, no. 634 (November 9, 1869), rpnt, II, p. 1168.
51. ʿAli-Qoli-Mirzâ Eʿtezâd al-Saltaneh was a staunch opponent of traditional astrology and wrote a book refuting it: *Falak al-Saʿâda*, lithograph edition, Tehran, 1278/1861. Eʿtezâd al-Saltaneh was among the group that translated Descartes' *Discours sur la method*.
52. Mirzâ ʿAli Doktor, *Amrâz-e Asabâni*, p. 4; ʿAli Khân, *Osul-e ʿelm-e fizik*, p. 2.
53. See, for example, Mirzâ ʿAbdol-Karim Tabib Tehrâni, *ʿAlâyem al-Amrâz*, pp. 2–3. Also see Mirzâ ʿAli Doktor, who informs us that he was a teacher of traditional medicine (*tebb-e ʿatiq*) and modern anatomy (*tashrih-e jadid*). *Amrâz-e Asabâni*, op. cit., preface, and p. 3.
54. Mohamma-Taqi Mir, *Pezeshkân-e nâmi-ye Fars*, pp. 94–95.
55. *RVE*, no. 102 (January 14, 1853); *Dânesh*, no. 6 (August 31, 1882), p. 2. Mirzâ Hossein Afshâr, a student of Dr. Polak, was the son of Mirzâ Ahmad Hakimbâshi, a traditional court doctor. See *Resâleh-ye moʿâlejeh-ye Maraz-e wabâ*, p. 3.
56. *RVE*, no. 45 (Safar 17, 1268/December 12, 1851).
57. Polak, *Safarnâmeh*, pp. 205–206.
58. *RVE*, no. 22 (Ramazan 20, 1267/July 19, 1851).
59. *RVE*, no. 25 (Ramazan 24, 1267/July 23, 1851).
60. *RVE*, no. 3 (Rabiʿ II 19, 1267/February 21, 1851).
61. F. Adamiyyat, *Amir Kabir va Iran*, p. 334.
62. *RVE*, no. 55 (Rabi II 28, 1862/February 20, 1852). This number was not high but it was through the elites that information was to be spread among the population at large who were unable to read or understand the guidelines.
63. Âqâ Mirzâ Mohammad-e Tehrâni, *Resâleh*, dated Shaʿbân 1, 1269/May 10, 1853, pp. 13–14; Mirzâ Mohammad Râzi Kani Fakhr al-Atebbâ, *Meftâh al-Amân*, p. 6.
64. *RVE*, no. 397 (September 9, 1858), rpnt, IV, pp. 2,677–2,680.
65. *RVE*, no. 439 (June 29, 1859), rpnt, IV, p. 2,920.
66. *RVE*, nos. 450 (October 26, 1859) & 452 (November 17, 1859), rpnt, IV, pp. 2,966, 2,973–2,974.
67. "It has always been a custom that the servitors (*châkerân*, i.e. ministers or high ranking officers) of the state, in order to obtain the government of a city or province, gave gifts (*pishkesh*) to His Majesty. It is evident that they could not have given gifts had they not been able to levy twice as much tax from

the subjects of that city or province, who were consequently ruined. . . . His Majesty has now decided that for the comfort and benefit of his subjects this custom be abolished and that he, himself, would refrain from sending officials to the provinces to receive gifts and will not dismiss any governor without justifiable reason either . . . ," *RVE*, nos. 429 (April 20, 1859) & 463 (March 28, 1860), rpnt, IV, pp. 2,879, 3,032.
68. *RDEI*, no. 472 (September 6, 1860), rpnt, vol. 1, p. 6.
69. Two decades later in 1883, during Nasser al-Din Shah's journey to Mashhad, inhabitants gathered along the way to complain about the collection of *suyursât* by force. Mirzâ Qahramân Amin-Lashkar, who accompanied the Shah, observed that despite the collection of *suyursât* for the army of the Shah, fodder for transport animals was scarce. Mirzâ Qahramân Amin-Lashkar, *Ruznâmeh-ye Safar-e Khorâssân*, pp. 57–60.
70. Fakhr al-Atebbâ, *Meftâh al-Amân*, pp. 1–3. According to Fakhr al-Atebbâ, this council met twice a week but the *RDEI* (no. 481, January 4, 1861) tells us that it took place once a week. Iraj Mirzâ was an ophthalmologist. M. Bamdad, *Sharh-e hâl-e rejâl-e Iran*, vol. 6, p. 267.
71. Sepehr, *Nâssekh al-Tavârikh*, p. 1,449.
72. Polak, *Safarnâmeh*, pp. 214–215; Sepehr, *Nâssekh al Tavârikh*, vol. 3, pp. 1,438–1,448.
73. *RVE*, no. 421 (February 23, 1859), IV, p. 2835; idem, no. 427 (April 6, 1859), IV, p. 2872. These students were all related to notable people and ministers and therefore they were not all of the same age. Some of them were too young, such as Yahyâ Khân, the son of Hassan-ᶜAli Khân, the ambassador to Paris, whose (childish behavior caused problems in his education, or Qahramân Khân who was the youngest and studied with children. After five years of study he received his baccalauréat. They studied at different schools before doing practical and specialized training in other schools such as engineering, military, and carpentry. The duration of their studies also varied. *RDEI*, nos. 500, 548, 555 (October 6, 1861; September 30, 1863; February 18, 1864), I, pp. 242, 610, II, pp. 664–665.
74. On the relationship between the political influences of Britain, Russia, and France and the reform movement in Qâjâr Iran, see F. Adamiyyat, *Andisheh-ye Taraqqi*, Chapter 6.
75. *Comptes-Rendus des séances de la Société d'Ethnographie*, Paris, Challamel Ainé, Editeur, Librairie de la Société d'Ethnographie, 1860, Tom premier, pp. 19–20, 69.
76. Literally, House of Forgetfulness, Farâmushkhâneh sounded similar to Farâmâsunery (Freemasonry) and its members meant to "forget" the discussions held during their meeting beyond the society (in other words, to keep them secret).
77. Hamid Algar, "Freemasonry ii In the Qâjâr Period," *Encyclopaedia Iranica*, op. cit.
78. F. Adamiyyat, *Andishe-ye Taraqqi*, p. 125.

79. F. Adamiyyat, *Andisheh-ye Taraqqi*, p. 82.
80. Comte de Gobineau, *Les Religions et les Philosophies dans l'Asie centrale*, in *Œuvres*, II, p. 476. This book was translated into Persian and lithographed in Tehran in 1863 under the title *Osul-e hekmat-e Diâkart (Principles of the philosophy of Descartes)*, but there is no mention in this book that Gobineau wrote or translated it. See the notes of the editors of Gobineau's *Oeuvres*, II, p. 1,113. It was translated by the cooperation of several scholars including ʿAliqoli Khân-e Eʿtezâd al-Saltaneh. See footnote 51
81. On Tholozan, see Jean Théodoridès, "Tholozan et la Perse," pp. 287–296.
82. Théodoridès, "Tholozan et la Perse," p. 289. This was the work of Laennec that revolutionized medicine in Europe. For a list of Tholozan's works, see J. Schlimmer, *Terminologie*, p. 227.
83. Fereidun Adamiyyat, *Andisheh-ye Taraqqi wa Hokumat-e Qânun*, p. 77.
84. *RDEI*, no. 552 (December 23, 1863), II, p. 639.
85. The state announcement said: "Some people who leave the country for pilgrimage to Mecca, go to Europe for tourism, as soon as they cross the border.... From now on those pilgrims who intend to travel to Europe must acquire the permission of the government." *RDEI*, no. 566 (January 11, 1865), II, pp. 740–741. Emigration to Europe in the guise of pilgrimage sometimes also happened when a person was dismissed from his duties or dispossessed of his properties, as in the case of Dust-Mohammad Khân-e Moʿayyer al-Mamâlek, the son-in-law of the Shah, who, under the pretext of going on pilgrimage, fled to Paris in 1883. Mirzâ Qahramân-e Amin-Lashkar, *Ruznâmeh*, pp. 53–54.
86. As A.-R. Sheikholeslami observed, "The most characteristic aspect of Nâsser al-Din Shah's reign [was] the absence of any consistent policy," although the patrimonial structure of his authority guaranteed the stability of his state. A.-R. Sheikholeslami, *The Structure of Central Authority in Qâjâr Iran*, pp. 209–213.
87. We find no report of this council's activity in the state journal (which announced its creation in 1861) yet throughout the 1860s it published news of epidemic diseases, such as intermittent fevers, cholera, and smallpox. See, for instance, *RDEI*, no. 554 (February 4, 1864), II, p. 658.
88. See, for example, *RVE*, nos. 94 (Safar 6, 1269/November 19, 1852), which reports that physicians had not yet found any treatment for cholera, but had been able to find out that the source of the epidemic was in the filthy and unclean environment and recommended that the inhabitants of cities and villages keep their houses and streets clean. In issue 117 (Rajab 19, 1269/April 28, 1853), it published a short notice by Dr. Cloquet that recommends treating diarrhea (which was considered a precursor to cholera in the patient) by adding six to ten drops of white petrol to a cup of spirit of wine, or wine or tea and drinking it. If diarrhea and vomiting did not stop, this drink with a higher dose of white naphtha (*naft-e sefid*) should be repeated. Two weeks later this journal reported that the remedy proved to be efficient. *RVE*, no. 119 (May 12, 1853).

89. *RVE*, no. 123 (Ramazan 1, 1269/June 8, 1853).
90. *RVE*, no. 120 (Saʿbân 10, 1269/May 19, 1853).
91. This period of reform began when Mirzâ Hossein Khân (at the time Moshir al-Dowleh) was the Plenipotentiary Minister of the Shah in Istanbul before being promoted to Minister of Justice, Pensions, and Endowments in December 1870, then Minister of War in September 1871, and finally prime minister in November 1871. Adamiyyat, *Andisheh*, p. 126.
92. Eʿtemâd al-Saltaneh, *Târikh-e montazam-e nâsseri*, vol. 3, pp. 1933, 1935.
93. *Târikh-e Montazam*, ibid, p. 1934. Mirzâ Hossein Khân-e Sepahsâlâr was dismissed from the post of prime minister in 1873, but remained active as Minister of War until 1881, when he was dismissed from this post also, in the guise of an appointment to Administrator of the Imam Rezâ Endowment in Mashhad. However, shortly after he died under suspicious circumstances.
94. Similar to Greek and Roman legends where Mars is god of war, the Merrikh planet, called also Bahrâm in Persian legend, is a symbol of war, hostility, and bloodshed. It is said that the name Mars derives from *merrikh*. Loghatnâmeh Dehkhodâ, vol. 13, pp. 20,728–20,729.
95. *Ruznâmeh-ye Mellati*, Rajab 13, 1285/October 30, 1868.
96. *Iran*, no. 1 (Moharram 11, 1288/April 2, 1871).
97. *RE*, no. 3 (January 29, 1877).
98. *Iran*, no. 303 (Ziqaʿdeh 18, 1293/December 5, 1876).
99. *RE*, nos. 3 and 4 (Moharram 21, 1294/February 5, 1877).
100. *Iran*, no. 446 (Rabi I 14, 1298/February 14, 1881).
101. *RDEI*, no. 604 (June 20, 1867); vol. 2, pp. 996–7.
102. *RE*, no. 5 (Moharram 28, 1294/February 12, 1877).
103. *RE*, no. 10 (Rai I 3, 1294/March 19, 1877).
104. Schneider, "La médecine Persane," op. cit., p. 15.
105. *RDEI*, nos. 481 & 482 (January 04 & 11, 1861), rpnt, I, pp. 97, 107.
106. Sâvaji, *Dastur al-Atebbâ*, p. 6. Paradoxically, the title emphasized "treatment" rather than the "preservation of health." But the treatise indicated first the preventive and then the curative measures.
107. Sâvaji, op. cit., see, for example, pp. 43, 49.
108. *Ruznâmeh-ye Mellati*, nos. 22 & 25, op. cit.
109. *Iran*, no. 445 (Rabi I 6, 1298/April 6, 1881).
110. *Iran*, no. 646 (January 10, 1888), rpnt, vol. 4, p. 2602.
111. Tholozan, "Resâlehye Mokhtasar dar Bâb-e Vaccin," in Tholozan, *Badâyeʿ al-hekmat Nâseri*. Long before Tholozan, the practice of variolization as a preventive measure against smallpox had been recognized by Western visitors in the East. See Anne Marie Moulin and Pierre Chuvin, *Lady May Montagu L'Islam au péril des femmes*.
112. Mohammad-Râzi Kani Fakhr al-Atebbâ, *Meftâh al-Amân*, Op.cit., pp. 1–2.
113. *Dânesh*, no. 9 (October 15, 1882).
114. Document of Sâzmân-e Asnâd (National Archives), no. 118, quoted in Rustâyee, vol. 1, p. dliii. According to Mohammad-Hassan Khân-e Eʿtemâd

al-Saltaneh (Sani⁽al-Dowleh), one of the purposes of the MHS was to send examiner physicians to all cities and provinces. *Al-Ma'âser w'al-âsâr*, op. cit., p. 115.
115. *Dânesh*, no. 1 (June 10, 1882).
116. *Dânesh*, no. 3 (July 17, 1882) and 9, op. cit.
117. Rustâyee, vol. 1, pp. dlii–dliii.
118. As in the case of the health officers of Brujerd, the two brothers Mirzâ Mohammad-Sâdeq and Mirzâ Mohammad-Hossein. Rustâyee, vol. 2, pp. 191–199.
119. Pierre-Jean-Georges Cabanis, *Sketch of the Revolutions of Medical Sciences*, p. 324.
120. Cited in C. Elgood, *Safavid Surgery*, p. 21. See also Francis Richard, *Raphaël du Mans missionnaire en Perse au XVIIe s.* 2 vols.
121. Elgood, *Safavid Surgery*, Ibid.
122. The *Madresseh-ye Dâr al-Shafâ* (literally, hospital school) has sometimes been wrongly represented as a *dâr al-shafâ* (see H. Tâjbakhsh, *Târikh-e bimârestânhâ-ye Iran*, pp. 216–217.) This was not a "philanthropic facility attached to a school or mosque" (see W. Floor, *Public Health*, p. 190), but rather a religious school constructed by Fath⁽Ali Shah, probably on the model of the madresseh-ye dâr al-shafâ of Qom, in which religion and philosophy were taught. See *Iran*, nos. 649 (February 9, 1888) & 980 (March 19, 1889), rpnt, vol. 4, pp. 2,613, 2740.
123. In documents dated 1835, there are references to "modern surgeons" (*jarrâhân-e jadid*) whose names were given by Mirzâ Bâbâ Hakimbâshi to the army accountant for their salary. Sâzmân-e Asnâd-e Ketâbkhâneh-ye Melli (Archives of the National Library), hereafter Asnâd, Tehran, no. 295000455, dated Jamâdi II 16, 1252/September 28, 1836,
124. *RVE*, no. 27 (Shawwâl 9, 1267August 7, 1851). Dr. Polak refers to different cases where divisions of the army, accustomed to cool areas, were sent on missions to warmer climates and were, as a result, affected with *nowbeh* (intermittent fever). See Polak, *Resâleh dar dastur al-⁽amal wa ta⁽limât-e mo⁽âlejât-e amrâz-e nowbeh, eshâl al-dam, mothbeqa va brudat ke aksar-e owqât be afwâj-e qâhereh mostowli wa bâ⁽es-e halâkat-e ishân migardad* (*Guidance for the treatment of intermittent fever, bloody flu, typhus, and hypothermia that often affect the victorious divisions and cause mortality*), pp. 9–10.
125. Polak, *Safarnâmeh*, pp. 408–409.
126. Contrary to Polak's commentary, there were always surgeons in the army even in the previous centuries and at least before the 1850s, for example, in 1835. See footnote 123. This tradition continued with modernization of the army. In 1883, in the list of the medical staffs of the royal artillery and the regiments (*afwâj*) stationed in the capital, we find physicians and surgeons attached to each regiment. *Iran*, no. 510 (March 3, 1883), rpnt, III, p. 2,054.
127. *RDEI*, no. 562 (August 4, 1864), II, p. 712.
128. *RDEI*, no. 559 (June 2, 1864).
129. *RVE*, no. 41 (Moharram 19, 1268/November 14, 1851).

130. *RVE*, nos. 102 & 103 (Rabi II 3 & 10, 1269/January 14 & 21, 1853). In no. 103 (January 21, 1853), the number of patients treated at the hospital is given as 2,238. The journal added that "during the last week only 30 individuals were treated there." This hospital was built outside of the city, between the two Gates of Dowlat and Qazvin (see Figure 3.2). *RVE*, nos. 67 (Rajab 23, 1268/May 13, 1852) and 99 (12 Rabi I 1269/December 24, 1852).
131. *RVE*, nos. 98 (5 Rabi I 1269/December 17, 1852) and 99 (12 Rabi I 1269/December 24, 1852).
132. *Rowzat al-safâ-ye nâsseri*, vol. 10, p. 813.
133. For this discussion see H Ebrahimnejad, *Medicine*, Chapter 3.
134. The economic crisis had awakened even the Shah to the inefficiency of his state. But the introduction of modern ideas and the impact of reforms undertaken in the Ottoman Empire during the Tanzimât, in addition to the impact of Western ideas through intellectuals, like Malkam Khân, who studied there, also played a crucial role in this change. It was in 1859 that Malkam Khân wrote his *Ketâbcheh-ye Gheibi Daftar-e Tanzimât* and presented it to Nâsser al-Din Shah via Mirzâ Jaʿfar Khân-e Moshir al-Dowleh, who was the Head of the newly established State Council (*showrâ-ye dowlati*). Mehdi Bamdad, *Sharh-e hâl-e rejâl-e Iran*, vol. 1, pp. 243–244.
135. *RDEI*, nos. 538, 539, 541 (March 11 & 18 & April 17, 1863,), I, pp. 525, 533, 550.
136. Mirzâ Mohammad Khân was from the Davalu clan of the Qâjârs that had long been a political ally of the reigning Qavânlu clan. Initially, the Sarkeshikchi-bâshi (Head of the Royal Guard), he became Sepahsâlâr (Head of the Army) and then a member of the State Council in 1859. M. Bamdad, *Sharh-e hâl*, vol. 3, p. 231.
137. In May 1865, he ordered the police (*mohtaseb*) of Tehran to ask inhabitants to remove their rubbish to the outside of the city and to keep alleys and streets clean. When meat became scarce in Tehran, he ordered that butchers be finan-cially helped throughout that period, but committed them to provide the city with meat of affordable price in other seasons. *RDEI*, nos. 571, 574 (May 25 & August 10, 1865), II, pp. 777, 798.
138. *RDEI*, no. 576 (October 19, 1865), II, p. 811. See also Mehdi Bamdad, *Sharh-e hâl*, vol. 3, pp. 229–232.
139. *RDEI*, nos. 592 & 606 (September 6, 1866, July 24, 1867), II, pp. 928, 1007.
140. For the edition of this document, see H. Ebrahimnejad, *Medicine*, part 2.
141. Tehran was extended about one kilometer in each direction. Eʿtemâd al-Saltaneh *Al-Mʾâser wʾal-âsâr*, pp. 72–73.
142. *Iran*, no. 59 (Shawwâl 9, 1288/December 22, 1871). This hospital might be a new one, but it is also possible that Mokhber al-Dowled meant the 1867 hospital, which would have been closed for a while.
143. N. Najmi, *Terhrân-e ʿahd-e nâsseri* (Tehran, ʿAttâr Publishers, 1364/1985), p. 421.
144. *Merrikh*, no. 8 (Jamad II 8, 1296/May 30, 1879).

145. *Iran*, no. 798 (April 27, 1893), rpnt, vol. 4, pp. 3209–3210.
146. See R. Cooter, M. Harrison, & S. Sturdy (eds.), *War, Medicine and Modernity*.
147. Anonymous, *Resâleh dar favâyed-e feshang-e jangi (On the benefits of military cartridge)*, pp. 6–14, et passim. The book is undated but it was written on the order of Hâji Mirzâ Âghâssi. It must have been written before 1847.
148. *Ruznâmeh-ye ᶜelmiyyeh wa adabiyye*, no. 7 (12/Moharram 1294/Janurary 27, 1877).
149. Boqrât al-Molk presented a short treatise on Teryak (opium) as a proof of his knowledge and skill, in support of his request for license. Asnâd, no. 297028920, date of the application letter Ramazan 15, 1330/August 28, 1912; date of the recommendation letters 1907.
150. *ᶜElm bar mardomân vâjeb ast khosusan ᶜelm-e badan ke lâzem ast va baᶜd az dânestan-e ᶜelm-e tebb, ᶜelm-e jarrâhist*. See: *Zakhirah-ye Kâmela*,op. cit., p. 3.
151. MSS 505, quoted in Ebrahimnejad, *Medicine*, pp. 227–228.
152. *Qawâᶜed-e kolliyeh-ye nezâmiyeh (the general rules in the army)*,1270/1854, National Library, Mss. No. 3,223.
153. ᶜOnsor al-Maᶜâli Kaykâvus b. Eskandar, a contemporary of Avicenna, advised physicians to work at hospitals to acquire medical knowledge and skills. See *Gozideh-ye Qâbus nâma*, p. 215.
154. See MSS 505, in Ebrahimnejad, *Medicine*, pp. 224–235.
155. *RDEI*, no. 532 (13 Jamadi I 1279/November 6, 1862), I, p. 485.
156. *RDEI*, no. 499 (September 15, 1861), rpnt, I, pp. 231–232.
157. *RDEI*, no. 554 (February 4, 1864), II, p. 657.
158. *RDEI*, no. 500 (October 6, 1861), I, p. 240. See also Chapter 2, p.?
159. *RDEI*, no. 644 (August 25, 1870), II, pp. 1231, 1234.
160. On Rashti, see H. Ebrahimnejad, "KÂẒEM RAŠTI," *Encyclopaedia Iranica*, op. cit.
161. See, for example, Mirzâ Mohammad-Taqi Shirâzi, *Tâᶜuniya (On Plague)*, op. cit.
162. On the origins of quarantine see Pierre-Louis Laget, 'Les Lazarets et l'émergence de nouvelles maladies pestilentielles au XIXe et au début du XXe siècle, online document (accessed on October 31, 2007).
163. Quarantines were set up in Iran after the International Sanitary Council in Constantinople (1866), under French influence, advocated and enforced them in other countries. At the beginning of his career in Iran, Tholozan was in favor of quarantines, but as he pursued his research on cholera and plague, he opposed them, arguing that they were founded more on political and commercial interests than based on scientific evidence. Tholozan, 1873a, pp. 29–34. For political and economic interests that might have been behind his opposition, alongside scientific reasons, see Hormoz Ebrahimnejad, "L'introduction de la médecine Européenne en Iran," *Sciences sociales et santé*, vol. 16, No.4, 1998, 69–96, pp. 86–89. Dr. Basil believed that quarantines not only did not prevent epidemics of cholera but could also reinforce their spread, even though one might speculate that this position could have been due to his pro-British leanings. Basil, *Resâleh-ye wabâiyyeh*, p. 15.

164. Homa Katouzian, *State and Society in Iran. The Eclipse of the Qâjârs and the Emergence of the Pahlavis*, London, New York, I.B. Tauris, 2006, pp. 2–5.
165. A. -R. Sheikholeslami, *The Structure of Central Authority in Qâjâr Iran*, pp. 8 ff.
166. Another case is the truncated and impaired authority of the Shah within the Qâjâr tribal structure. On this question, see H. Ebrahimnejad, *Pouvoir et succession sous les premiers Qâjârs*.
167. *Iran*, no. 1 (Moharram 11, 1288/April 2, 1871). A journal of the military (*Ruznâmeh-ye nezâmi-ye dowlat-e ᶜelliyeh-ye Iran*) announced that "this journal is not a state and official paper. It is like many European journals in which all kinds of questions, provided that they are not against the custom and law of the kingdom, can be published." *Ruznâmeh-ye nezâmi-ye dowlat-e ᶜelliyeh-ye Iran*, nos. 1–6 (from 29 Ziqaᶜdeh 1293/December 16, 1876, to 5 Moharram 1294/January 20, 1877).
168. For an opposite view see C. Schayegh, *Who is Knowledgeable is Strong*, op. cit., pp. 64–66.
169. *RDEI*, no. 502 (October 24, 1861), rpnt, vol. 1, p. 253.
170. Saᶜid Nafisi, "Doktor ᶜAli-Akbar Khân-e Nafisi Nâzem al-Atebbâ (1263–1342/1847–1924)," *Majalleh-ye Yâdegâr*, no. 4, pp. 52–53.
171. Rustâyee, vol. 2, pp. 556–557.
172. Sepehr, *Nâssekh al-Tavârikh*, vol. 3, p.1438; Polak (*Safarnâmeh*, pp. 214–215. Mirzâ ᶜAlinaqi, who later received the court title *Hakima al-Mamâlek* and then *Wâli* (as he was also appointed Governor of Qom), was one of the personal physicians of Nâsser al-Din Shah and always present during his trips. He died in Tehran in April 1903. *Irân-e Soltâni*, no. 3 (May 5, 1903), rpnt, p. 21. *RDEI*, no. 523 (June 13, 1862), rpnt, vol. 1, pp. 420–421.
173. *Ruznâmeh-ye Dowlat-e ᶜElliyeh –ye Iran* (continuation of *RVE*), no. 479, December 13, 1860, reprinted by the National Library, Tehran, 1370/1991, vol. 1, pp. 79–80.
174. *RDEI*, no. 482 (January 24, 1860), rpnt, vol. 1, pp. 106–107.
175. *RDEI*, no. 502 (October 24, 1861), I, p. 252.
176. *RDEI*, no. 523 (June 13, 1862), I, p. 419.
177. *RDEI*, no. 536 (January 15, 1863), I, p. 515.
178. *RDEI*, no. 583 (February 22, 1866), II, pp. 868–9.
179. The witness of Dr. Schneider quoted above particularly illustrates this. See footnote 46.
180. *RVE*, no. 456 (December 14, 1859), rpnt, vol. 4, p. 2990. About arithmetic (*hesâb*), for instance, the *RVE* reported: "Previously, the students studied only the Digest (*Kholâsat al-Hesâb*) of Sheikh-e Bahâyee and once they had learned some of its chapters, they exaggerated the complexity of the matter and avoided teaching it to others, as it was explicitly advised by Sheikh-e Bahâyee himself at the end of his book. However, nowadays, the students of second and third class can write treatises far better [i.e., clearer] than the *Digest* of Sheikh-e Bahâyee and even refute the *Digest*'s principles. Mirzâ ᶜAbdol-Ghaffâr, one of the maths

students has recently resolved one of the unresolvable questions of Sheikh-e Bahâyee." *RVE*, nos 456 (Ibid), and 464 (April 18, 1860).
181. *RE*, no. 40 (February 25, 1878).
182. This idea is also examined by Schayegh, *Who is Knowledgeable is Strong*.

4. Medical Transition under the Constitution

1. *Be dânesh shavad kâr-e 'âlam besâz; ze bidâneshi kâr-e 'âlam derâz*. Anjoman-e *moqaddas-e melli-ye esfahân*, year 1, no. 1 (January 6, 1906). This poem is a modified version of the poem of ʿAbdol-Rahmân-e Jâmi (1414–1492) of Herat: *ze dânesh shavad kâr-e giti besâz; ze bidâneshi kâr gardad derâz. Agar dar jahân nabvad âmuzegâr; shavad tireh az bikherad ruzegâr. Cho nâdân ze dânâ konad sarkashi; nabinad ze dowrân-e giti khashi*. For Jâmi, *dânesh* (knowledge) had the broader meaning of "wisdom" and education. Clearly our constitutionalists took it to mean science in modern terms, or "knowledge of modern science."
2. Fereydoun-e Adamiyat, *Andisheh-ye taraqqi va hokumat-e qânun*, p. 7.
3. The themes (modern) science and/or progress were a leitmotif in most constitutional periodicals. See, for example, *Omid*, no. 2 (November 9, 1906); *Anjoman-e moqaddas-e melli-ye esfahân*, year 1, no. 2, January 13, 1906, and no. 4, January 27, 1906.
4. The list of Muslim intellectuals who, as a result of the influence of modern science have reinterpreted Islam and the tenet of the Koran or the Shariʿa, is long. For an analysis of the view of some Muslim intellectuals on Christianity and modern science and their reaction to Western influence see Hugh Goddard, *Muslim Perceptions of Christianity*, see for example, pp. 40 ff. and 83ff.
5. On this question see Roy MacLeod and M. Lewis (eds.), *Disease, Medicine and Empire* (London: Routledge, 1988).
6. The pro-*Babist* tendency of Edward G. Browne, to give just one example, illustrates this. Henry Corbin's analysis of *Shi'ism* is a far cry from the positivist studies of most Western scholars on Islam. His approach in studying *Shi'ism* and Islamic philosophers was fundamentally influenced by these philosophers' ideas and visions.
7. Jean Calmard, "Feuvrier," *Encyclopaedia Iranica*.
8. *Affaires Diverses Politiques*, vol. 4, dossier 23, letter of the Ministry of War, January 20, 1890; letter of Tholozan, April 1st, 1893; letter of Feuvrier, Avril 17, 1893. Feuvrier, with the support of de Balloy, endeavored to expel Tholozan and take his place, but, as the latter had the sympathy and protection of Nâsser al-Din Shah, he did not succeed. What is, however, essential in this rivalry is that for the French government it was important to maintain the French medical presence at the Qâjâr court, regardless of who fulfilled this function.
9. *Archives du Ministère des Affaires Etrangères, Affaires Diverses Politiques (ADP)*, vol. 4, dossier 22 (Le Colombier, le Janvier 16, 1890); and dossier 26 (Téhéran, le Mai 5, 1890). In fact, the French chargé d'affaires, de Balloy, worked against the presence of Dr. Tholozan in Persia to the benefit of Dr. Feuvrier (Principal

Military Physician, Second Class), as Dr. Tholozan revealed in his letter of April 1st, 1893.
10. J. Calmard, ibid; J.-B. Feuvrier, *Trois ans à la cour de Perse*, pp. 20–22.
11. See Tholozan's letter to Marcel Dieulafoy, November 23, 1892, Bibliothèque de l'Institut, Ms. 2691, fol. 246, reproduced by Jean Théodoridès, "Tholozan et la Perse," *Histoire des sciences médicales*, Tome XXXII, no. 3, 1998, p. 294.
12. J. Calmard, "Feuvrier," op. cit.
13. Tholozan's Letter to Mr Flury, April 1, 1893, ADP, op. cit.
14. *ADP*, op. cit. According to J. Calmard his mission in Iran was ended in March 20, 1893, by the order to the Ministry of War (idem, ibid).
15. *Iran*, no. 480 (March 5, 1882), rpnt, III, p. 1928.
16. Eʿtemâd al-Saltaneh, *Al-Maʾâser w'al-âsâr*, op. cit. p. 116.
17. AMAE, ADP, dossier 93, Dépêche Télégraphique, Téhéran, le Juillet 29, 1893. Schneider's salary in France was reduced by about 12,000 francs, equivalent to the salary that the prince regent had pledged to pay. In addition to reducing his salary, the government did not pay his travel and installation expenses. See ibid, letter dated Juillet 1893.
18. AMAE, ADP, dossier 93, letters of de Balloy, Paris, June 10, 1893, Tehran, July 31, 1893, and Tehran, February 15, 1894. According to the contract signed between the Shah's ambassador in Paris, Nazar-Aqâ, and Dr. Schneider, the latter was uniquely attached to Nâyeb al-Saltaneh, but the way the French government saw it, he was attached to the French Legation in Tehran. See ibid, contract dated September 15, 1893, and the letter of "Ministre de la Guerre," Paris, June 19, 1895.
19. *Iran*, no. 821 (Ramazan 7, 1311/March 11, 1894), rpnt, vol. 4, p. 3,302.
20. AMAE, ADP, dossier 93, letter dated May 29, 1894.
21. AMAE, ADP, dossier 93, Tehran August 26, 1894, letter no. 27.
22. *Ruznâme-ye khâterât*, op. cit., p. 961.
23. *Ruznâmeh-ye khâterât*, p. 916.
24. *Ruznâmeh-ye khâterât*, p. 1,015.
25. *Iran*, nos. 788 (January 2, 1893) & 821 (Ramazan 7, 1311/March 11, 1894), rpnt, vol. 4, pp. 3, 169–3,170, 3,302–3,303.
26. *Iran*, no. 789 (January 16, 1893), vol. 4, p. 3,173.
27. شلمه the original spelling is unclear as it can be read Chalmet, Chelmet, or Cholmet. *Iran*, nos. 575 (June 16, 1885) rpnt, vol. 3, p. 2,317, & 656 (May 19, 1888), rpnt, vol. 4, pp. 2,641 & 821, op. cit, p. 3,303.
28. AMAE, ADP, Dossier Schneider, 93, letter of the Minister of War to the Minister of Foreign Affairs, July 17, 1894.
29. *The Illustrated London News*, June 14, 1873.
30. For such a differentiation made between the attitudes and opinions of different members of the British Legation in Iran toward the Constitutional Revolution see Mansour Bonakdarian, *Britain and the Iranian Constitutional Revolution of 1906-11: Foreign Policy, Imperialism and Dissent*.

31. Nicolas de Khanikof, a Russian-born French scholar and diplomat, was the author of several ethnographic and historical works, including *Mémoire sur la partie méridionale de l'Asie Centrale*, Paris, de L. Martinet, 1861. On C. Elgood, see Bagley, F. R. C., 'Elgood, Cyril Lloyd, *Encyclopaedia Iranica*, vol. 13.
32. *Iran*, no. 649 (February 9, 1888), IV, p. 2614.
33. Tholozan's letter to Dieulafoy, November 1893, in Jean Théodoridès, op. cit.
34. Eᶜtemâd al-Saltaneh, *al-Ma'âser v'al-âsâr*, op. cit., p. 82.
35. *AMAE*, dossier 68 (3) for Mirzâ Mohammad-Khân: letter of Nazar Aqâ, April 12, 1889 and response of the Minister, May 8, 1889. Dossier 68 (5) for Moᶜin al-Atebbâ: letter of Nazar Aqâ, Mai 14, 1891, and letter of the Minister, June 22, 1891.
36. The list of these publications is given in his *Tashrih (Anatomy)*, Tehran, lithographed, 1312/1895. Mirzâ ᶜAli Hamadâni's books were also all translations from French books.
37. He was indeed persuaded that "lacking common borders with Persia, whereby France could secure a political influence, propagation of language and creation of a French school constituted France's major means to achieve industrial and commercial goals [in Persia]." Schneider, *La médecine persane*, op. cit., p. 13.
38. Cited in H. Nategh, "*Les Iraniens à Lyon*," online source, op. cit.
39. H. Nategh, "*Les Iraniens à Lyon*," op. cit.
40. United States Department of State (USDS), Nos: 548–550, January 1887 to August 1889. See also W. Floor, *Public Health in Qâjâr Iran*, p. 198. The USDS dates the donation by the prime minister in 1885, but this is a mistake as Amin al-Soltan died in 1883.
41. USDS, no. 546, Legation of the United States, Tehran, November 29, 1886.
42. Ibid, no. 912, Tehran, January 10, 1888.
43. . See, "Probing History of Medicine and Public Health in India," *Indian Historical Review*, vol. 37, No. 2 (2010): 259–273, pp. 238–9.
44. Talcott Parsons, "The Place of Ultimate Values in Sociological Theory," *International Journal of Ethics*, vol. 45, No. 3 (April 1935), pp. 282–316.
45. Dr. Schneider, *La Médecine Persane: Les Médecins Français en Perse*, p. 5.
46. Talcott Parsons, "Ultimate Values in Sociological Theory," p. 293. See also Parsons, *Structure of Social Action*, New York, McGraw-Hill, 1937, pp. 641–42.
47. Robert E. Speer, *The Hakim Sahib, The Foreign Doctor, A Biography of Joseph Plumb Cochran*, New York, London, etc., Flemming H. Revell Company, 1911, pp. 64–65.
48. Speer, *The Hakim Sahib*, op. cit., p. 323.
49. Parsons, "Ultimate," p. 294.
50. De Balloy, *MAE* (CPC/NS), Perse (1897–1917), vol. 1, Mai 1896.
51. De Balloy, MAE (CPC/NS), Perse (1897–1917), vol. 1, Mai 1896.
52. Ibid., April 9, 1897.
53. Using the terms of Alan Lester, "Imperial Circuits and Networks: Geographies of the British Empire," *History Compass* vol. 4, No. 1 (2006), pp. 131–2.
54. E. Malekzâdeh, *Negâhi be omur-e kheiriyeh*, pp. 53–54.

55. *Iran-e Soltâni*, Year 57, no. 17 (November 22, 1904), rpnt, p. 275.
56. For example, the mamluks in Egypt converted their conquered lands into *waqf*. A. Ragab, "Le Bimâristân al-Mansuri: L'institution dans le contexte politique et social mamelouk (1285–1390)."
57. On this, see Adam Sabra, *Poverty and charity in medieval Islam: Mamluk Egypt, 1250–1517*.
58. Mohammad Moradi, "Ganj-anduzi va habs-e mâl," *Faslnâmeh-ye waqf mirâs-e jâvidân*, no. 71.
59. Seyyed Mostafa Matbaᶜeh-chi Esfehâni, "Negâhi be waqf-shenâsi-ye faqihân (2)" *Faslnâmeh-ye waqf mirâs-e jâvidân*, no. 72.
60. In his *waqf* deed for the Pasteur Institute, ᶜAbdol-Hossein Mirzâ Farmânfarmâ indicates that the ultimate goal of the creation of the *waqf* for the Pasteur Institute was to "seek God's satisfaction" (*lemorzât-allâh*) by providing treatment for the inhabitants of the capital, so that this charitable foundation remains as his legacy, which is a "treasure for the world beyond" (*zakhireh-ye okhraviyyeh*). See the *waqfnâmeh* of the Pasteur Institute, quoted in E. Malekzâdeh, op. cit., pp. 142–143.
61. For a case study of sociopolitical aspects of *waqf* and its evolution, see Christoph Werner, *An Iranian Town in Transition*, pp. 97 ff.
62. See the long list of mosques and madrasas given by Mirzâ Hassan Hosseini Fassâyee, *Fârsnâmeh-ye nâsseri* vol. 2, pp. 1205 ff.
63. This cannot be generalized to other Islamic countries. In the Ottoman Empire under the Seljuks, for instance, there were more *dâr al-shafâs* or hospices based on *waqf* than in Iran. By the same token, the use of *waqf* for public benefit began there in the nineteenth century, earlier than in Iran, as in the case of the *waqf*-endowed modern hospital (Zeynep-Kamil) that was established in 1875. See F. Günergun and S. Etker, "*Waqf* endowments and the emergence of modern charitable hospitals in the Ottoman Empire."
64. Among hundreds of *waqf* deeds preserved in the *Waqf* organization of the city of Neishabur that I have examined, there are only a few destined for public welfare. Most of them, even those created in the twentieth century, were to fund religious ceremonies (*rowzeh-khâni*) that were deemed to gain God's grace for the *waqf* founder.
65. Mohammad-Esmâᶜil Rezwâni (ed), *Gozideh-ye farmânhâ* (Royal decrees), pp. 17–21.
66. Hospital Najmiyeh was built by Malek-Tâj, the daughter of Firuz-Mirzâ Nosrat al-Dowleh and the mother of Mohammad-e Mossadegh, in Khordad, 1305/June 1926, and the income of various properties were endowed for the expenses of the hospital and its patients. See Omid-e Rezâyee, "Bimârestân-e Najmiyeh," *Mirâs-e Jâvidân*, year 7, no. 4.
67. *al-mâlo v'al-banuna zinat ul-hayât –el-donyâ v'al-bâqiyât-ul-sâlehâto khayron ᶜenda rabbaka sawâban va khayron amalan.* "Wealth and children are the adornment of the present life; but good works, which are lasting (*albâqiyât al-sâlehât*) are better in the sight of thy Lord as to recompense and result." See *The*

Koran, Sura al-kahf, âya 46. Translated from Arabic by J. M. Rodwell, Phoenix, London, 2009 (first published 1909), p. 192.
68. Hassan Tajbakhsh, *Târikh-e bimârestânhâ-ye Iran*, pp. 91–93.
69. S. Hossein Omidiyâni, "Gâhanbâr dar Iran-e bâstân," *Waqf Mirâs-e Jâvidân*, no. 3, Tehran Autumn 1372/1993, p. 46, quoted in Elhâm Malekzâdeh, *Negâhi be omur-e kheiriyeh dar dowreh-ye Qâjâr*, p. 15.
70. It is significant that the reason for the leading cleric in Tabriz to occupy the Roshdiyyeh School in July 1895 was that he had been the administrator (*motevalli*) of the *waqf* that funded the *maktab* that Roshdiyyeh had used for his school. In fact this *maktab*, had been abandoned and no one claimed its "*towliyat*" (administration). Once the leading cleric found that it was to be used for modern education he claimed that the said *maktab* was under his administration. See *Akhtar*, Year 23, no. 4, Moharram 23, 1313/July 16, 1895.
71. *Maᶜâref*, no. 1 (Shaᶜbân 1, 1316/December 15, 1898).
72. These donations were both punctual and annual. In 1899, the Shah donated 5,000 tomans, the prime minister 500 tomans, the Minister of War 500 tomans, Beglarbegi, the former Minister of Customs, 500 tomans, Nâzem al-Dowleh, 50 tomans, and so on. Mozaffar al-Din Shah committed 2,000 tomans annually and Beglarbegi, 100 tomans, and so on. The Embassy of Iran in St Petersburg collected money from the members of the embassy as well as from Russian persons of note. *Maᶜâref*, no. 10 (Zihajj 15, 1316/April 26, 1899). Some of the Zoroastrian Parsis in India also, in solidarity, contributed to the creation of the schools. *Maâref*, no. 8 (Ziqadeh 11, 1316/March 23, 1899).
73. According to the journal *Maᶜâref*, most of the students at Madreseh-ye Eftetâhiyeh in Tehran were children of noblemen (*aᶜâzem va ashrâf*), *Maᶜâref*, no. 72 (Shwwâl 24, 1318/February 14, 1901). For example, among the students of the madreseh-ye Eftetâhiyyeh in the Sanglaj quarter we find Mirzâ Es'hâq, son of the late Mirzâ Taqi Khân-e Lashkarnevis; Mirzâ ᶜAbbas Khân, son of Adib al-Saltaneh; Mirzâ Mehdi Khân, son of Mojir al-Dowleh. *Maᶜâref*, no. 8 (Shavval 15, 1316/February 26, 1899).
74. *Maᶜâref*, no. 10 (Ziqadeh 1, 1316/March 13, 1899)
75. *Maᶜâref*, no. 33 (Shawwâl 20, 1317/February 21, 1900).
76. ᶜAbdollâh Mostowfi, *Sharh-e zendegâni-ye man*, vol. 2, p. 18.
77. In 1911, for example, ᶜAref Qazvini sang in a concert organized by the Democrat Party in order to collect money for the construction of a school. Hâmed Jalilvad, "*Bozorgân-e musiqi-ye Iran: Aref-e Mashruteh,*" online source, www.persianpersia.com, access: December 26, 2011.
78. *Maᶜâref*, no. 11 (Moharram 1, 1317/May 12, 1899).
79. *Maᶜâref*, no. 12 (Moharram 15, 1317/May 26, 1899).
80. *Nedâ-ye Vatan*, no. I (December 27, 1906), p. 6.
81. E. Malekzâdeh, *Negâhi be omur-e kheiriyeh*, p. 54.
82. E. Malekzâdeh, op. cit., pp. 150–151.
83. A. Kasravi, *Târikh-e Mashrutiyat*, p. 429.
84. *Omid*, nos. 6 (December 8, 1906), and 8 (December 21, 1906).

85. This movement was in fact germinated under Mirzâ Taqi-Khân-e Amir-Kabir who wanted to establish a "modern state" administered along modern lines, a state based on a "regulation" or as he once confided to Malkom-Khân "constitution." Adamiyyat, *Amir Kabir va Iran*, pp. 223–224.
86. On this question see the outstanding work of Ali Mirsepassi, *Democracy in Modern Iran: Islam, Culture, and Political Change*, particularly chapter 4.
87. *Majles*, Shawwâl 8, 1324/November 25, 1906, no. 1, p. 3.
88. *Irân-e Soltâni*, no. 3 (May 5, 1903), rpnt, pp. 21–22.
89. *pâsebâni-ye vojud-e moqaddas-e shakhs-e saltanat va esbât-e hoquq-e dolat bar raᶜiyat.Merrikh*, no. 1 (Moharram 5, 1296/December 30, 1878).
90. *Maᶜâref*, Year 3, no. 68, Ramazan 22, 1318/January 13, 1901).
91. Malek al Motekallemin believed that once 5,000,000 individuals (nearly 50 percent of the population) in Iran acquired literacy, Iran could claim to be a country ranked among the civilized nations. *Suresrâfil*, no. 1 (17 Rabiᶜ II 1325/May 30, 1907).
92. *Omid*, no. 1 (November 2, 1906).
93. *Omid*, no. 5 (December 1, 1906).
94. Verses such as *hal yastavi-allazina yaᶜlamuna v'allazina lâ-yaᶜlamun?* "Are those who learn and those who do not equal?" or *Tatab-al 'alm farizaton 'alâ kolle muslim va muslima* "The acquisition of science is a duty for all Muslims," were referred to by the constitutionalist periodicals. The latter verse was the epigraph printed in all issues of the journal *Omid*. See, no. 2 (November 9, 1906).
95. *Omid*, no. 2 (November 9, 1906).
96. *Akhtar*, year 20, no. 22 (Jamâd II 4, 1311/December 13, 1893).
97. *Irân-e Soltâni*, Year 57, no. 17 (November 22, 1904), rpnt, pp. 275–276.
98. *Anjomanhâye asre mashruteh*, pp. 277–278.
99. *Tamaddon*, Year 1, no. 54 (Moharram 26, 1326/June 8, 1908). The article is signed by Mohammad-Hossein b. ᶜAbdolwahhâb-e Tehrâni. Management of water was as a major consequence of modern public health in the West. On this subject see the pioneering work of Jean-Pierre Goubert, *La conquête de l'eau*.
100. *Tarbiyat*, no. 17 (April 8, 1897), rprnt, vol. 1, pp. 65–66.
101. *Tarbiyat*, no. 15 (March 25, 1897), vol. 1, pp. 59–60, & no. 18 (April 15, 1987), vol. 1, pp. 69–70.
102. *Omid*, no. 2 (Ramazan 21, 1324/November 9, 1906). For the constitutionalists, nationalism was not only a revival of the past, but also, and more importantly, the sovereignty of the nation, which was the source of state power. *Omid* compared the might of Nâder Shah and Shâh ᶜAbbas the Great with Mozaffar al-Din Shah, stating that no doubt the first two were powerful and did good work for Iran, but neither accepted limits to (*mashrut*) their sovereignty, and sharing it with their subjects as Mozaffar al-Din Shah had done.' As far as *Omid*, was concerned, the authority and strength of the past Shah only benefited him and his entourage, while the new state, based on the sovereignty of the subjects (nation) (*ekhtiyâr-e tâmmeh be raᶜyat*), strengthened the whole country. *Omid*, no. 1 (Ramazan 15, 1324/November 2, 1906).

103. *Omid*, no. 7 (December 15, 1906).
104. *Omid*, no. 13 (January 25, 1907).
105. A. Valentine Williams Jackson, *From Constantinople to the Home of Omar Khayyam*, p. 113.
106. *Kashkul*, Year 1, no. 25 (Shawwâl 9, 1325/November 15, 1907).
107. Ibid.
108. Archives of the Sâzmân-e Asnâd, dossier 89, no. 297/215593, dated July 23, 1311 (October 1932). Such practice was commonplace in Iran for years to come. Still around 1948, the "health officers" who were trained for primary care, such as injections, or even those who had acquired some experience by working with a friend or relative, were considered skilled enough to practice medicine in their villages, where they did not face control. Ebrahim-e Khorrami, a doctor of my native village, Mashhad Zolfâbâd in Farâhân, was one of these "doctors." "Dr" Khorrami once injected a drug, probably penicillin, into one of his own relatives, who was apparently suffering from a fever. At that time penicillin was considered to have a miraculous effect. The hand of the poor patient swelled as a result and he died shortly thereafter. An inspector was sent to investigate the cause of the death, but family relationships saved "Dr" Khorrami from being prosecuted, as the elders of the man's family withheld their complaint. For this information, I am grateful to Dâriush-e Hemmati (interviewed in the summer 2000). Mirzâ Hedâyat, Dâriush's father, especially skilled in bloodletting, was one of the traditional practitioners in the same village, but judging from a lithograph book that he used it appears he had also some knowledge of modern medicine. This book, entitled *Kholâsat al-hekmat* see figure 4.1, handed on to me by Dâriush, is a translation of a French medical text, based on modern medicine of the late nineteenth century. It was written by Zeyn al-'Abedin Khân-e Doktor, son of Shokrollâh Khân-e Qâjâr-e Qovânlu. The publication date of the book is not known as the first and last pages are missing, but it was dedicated to Mozaffar al-Din Shah. There is a prescription left in this book, dated February 20, 1332 (February 1953), that follows the way drugs were composed and prescribed according to the description given on p. 3 of this book. See fig. 4.2.
109. Asnâd, dossier 93, no. 297/21597, dated June 7, 1312 H. Sh.
110. Asnad, ibid, dated July 8, 1312 H. Sh.
111. See Mohsen Nâsseri, "Dr Ahmadiyeh and clinical traditional medicine in Iran," in *Yâdnâmeh ye Hakim Doktor Abdollâh-Khân-e Ahmadiyeh*, published by the National Museum of the history of medical sciences, 1381/2002, pp. 28–29.
112. *Irân-e Soltâni*, no. 9 (July 28, 1903), rpnt, p. 70. We find advertisements for his business in successive issues of the journal. rpnt journal, pp. 79, 87, . . .
113. *Omid*, no. 5 (December 1, 1906).
114. *Irân-e Soltâni*, no. 7 (June 16, 1903), rpnt, p. 54.
115. *Irân-e Soltâni*, no. 20 (December 15, 1903), rpnt, p. 160.
116. *Omid*, no. 13 (January 25, 1907).
117. Asnâd, Dossier 148, no. 297/35886, Esfand 1305/February 1926.
118. Asnâd, dossier 151, no. 297/35886, dated July 24, 1314 (October 1935).

119. Asnad, dossier 91, no. 297/21595, dated January 51313 (March 26, 1934).
120. Asnad, dossier 158, no. 251, dated November 27, 1305 (February 15, 1926).
121. Asnad, dossier 156, no. 251, dated October 1, 1308 (December 12, 1929).
122. Philippe Rochard & H. E. Chehabi, "The identities of the Iranian Zurkhanah," p. 318; idem, "Le sport antique des zurkhâne de Téhéran: formes et significations d'une pratique contemporaine," pp. 153–159. On pre-Islamic sport in Iran see Mohsen Zakeri, *Sâsânid Soldiers in Early Muslim Society: the origins of Ayyârân and Futuwwa*.
123. This is illustrated in the poetry and songs of the Zurkâneh. In one of these songs, Morshed Morâdi says: "Iran the birthplace of the Khosroes, Jamsjhid, and Kiyân, is our heart, our house and our hope; Iran is our body, and our life. We do not sell out the smallest part of Iran, non! as this soil is dear; we the people of Iran do not know except Iran . . . It is our duty to sacrifice our life for the *vatan*." The term *vatan* did not bear the same meaning across history in Iran. The generic term in the case of Iran rendering *vatan* is "homeland" that in terms of geographical extent has always been perceived differently by different observers, going from native village to empire "Iranshahr" and "Iranzamin" (the land of Iran). Metaphorically, *vatan* can be imagined as land of father or mother, but this has also changed throughout history. According to Tavakoli-Targhi, under the Constitutional Revolution the concept *vatan* "as a home headed by crowned father" was contested by *matriotic* discourse that imagined *vatan* as a dying 6,000-year-old mother. *Refashioning Iran*, Ch. 7, pp. 113, 123, et passim. See also F. Kashani-Sabet, *Conceiving Citizens*, pp. 32–32; idem, *Frontier fictions, Shaping the Iranian Nation, 1804–1946*, p. 216. We should not, however, ignore the influence of Western "patriotism," and particularly the French concept of "*patrie*" and its intrinsic link with the "nation," the central state and a well-defined territory. For an account of "*nation*" and "*patrie*," "national independence" and "national interest," when threatened by "clericalism" in France, see René Rémond, *L'anticléricalisme en France*, pp. 24–25.
124. Qahreman-Mirzâ Ein al-Saltaneh (edited by Salur, Afshar, vol. 3, pp. 2049–2050, cited in Fâruq Khârâbi, *Anjomanhâye asre Mashruteh*, p. 275.
125. Dânesh, no. 1 (Ramazan 10, 1328/September 15, 1910).
126. *Shekufeh*, Year 1, no. 2 (Moharram 23, 1331/January 2, 1913) See fig. 4.3.
127. *Shekufeh*, Year 1, no. 2 (Moharram 23, 1331/January 2, 1913)
128. In Polak's view the benefit conferred on the state by medicine was a matter of economics: the state spent money training soldiers and it was thanks to the art of medicine, which could save their lives, that this investment was not wasted. In addition, such soldiers could go on to contribute to the economy of their village once their military service was over. Polak, *Resâleh dar moʿâlejât va tadâbir-e amrâz (diseases that mostly affect soldiers in the barracks)*, National Library, MSS 6164, op. cit., p. 2. Mirzâ Mussâ Sâvaji, a contemporary of Polak, also emphasized, in his panegyric style, the role of doctors in the prosperity of the population. In providing a medical service, they kept the population

healthy so that it could serve the King "especially if, or when, the King supports physicians." *Dastur al-Atebbâ fi ʿAlâj al-Wabâ*, op. cit., pp. 3–4.
129. Mirzâ Seyyed Razi, who died in 1304/1887, was the chief physician of the army under Nâsser al-Din Shah.
130. The association of education of women and children, public health, and population increase is a leitmotif of journal *Shekufeh*. See, for example, no. 17, 9 Zi Hajjeh 1331 (November 9, 1913). In another article, no. 8 (Jamadi I 17, 1331/April 24, 1913), the journal argued that lack of public health caused population decline and as a result a decline in commerce and industry.
131. In his book on the first visit of Nâsser al-Din Shah to France, Comte de Croizier indicated that "if the Shah cannot revive the bygone glorious days of Persia and is not as strong as the Achaemenid Shahs, who removed the yoke of Assyrian domination from Iran, to take back the lands conquered by Peter the Great and to drive the Turks beyond its Western borders, he at least has the glory of having prepared the prosperity of his country that is likely to become, as in the pre-Islamic past, the arbiter of the Asiatic equilibrium." Cte de Croizier, *La Perse et les Persans*, op. cit., p. 63. We have indicated the closer relationship between Iran and France in the second part of the nineteenth century, in the aftermath of the Crimean War in 1856 and conflicts between Iran and England over Afghanistan. In 1875, the spirit of the post-Crimean War still informed French diplomacy.
132. Mirzâ Aqâ-Khân Kermâni, *Ey Jalâl al-Dowleh*, p. 131–132. According to Kermâni, "The Iranians have become mentally and morally degenerate because they have been alienated by Islam. They have lost any zeal and sense of honour. Otherwise, instead of mourning the death of Horr ibn Yazid-e Riyâhi every year, or his parents, whom they don't even know, they would mourn the death of their ancestors killed by the Arabs who looted their homeland. The Iranians do not cry for the daughter of their king [Yazdgerd] who was sold like a slave, but cry for the false story of Zeinab and Kolsum taken as slaves in the court of Yazid." *Ey Jalâl al-Dowleh*, pp. 142–143. Yazin ibn Muʿâwiya ibn Abi Syfyân (645–683) was the second caliph of the Umayyad who killed Hossein, the third imam of the shiʿas.
133. *Ey Jalâl al-Dowleh*, pp. 166, 169.
134. *Ey Jalâl al-Dowleh*, pp. 181–2.
135. C. Shâyegh has examined the influence neo-Lamarkian genetics had on Iranian intellectuals but more in relation to the twentieth-century developments. Cyrus Shâyegh, *Who is Knowledgeable is Strong*, see, for example, pp. 116, 139, and 143.
136. These conditions included vast spaces or gardens, where the air is clean and healthy. The surroundings of the school must be clear of any rubbish or animal and vegetable waste, as in a great number these "create an air saturated with *hidrojen surfureh* [Hydrogen Sulfide H_2S], which in warm and humid seasons cause head and eye aches. They can also produce ammoniac, which affects the mucus tissues of the nose, eyes, throat and mouth and causes pain in the throat

and eyes." *Adab*, no. 19 (Moharram 28, 1320/May 7, 1902). Other requirements for the school children included gymnastics and sport, learning how to use instruments in carpentry and weaving, which were necessary for the development of the country and the use of weapons. It said if children between the age of eight and nine begin to be drilled and wear uniform, they would be able to carry rifles and weapons and perform military gymnastic at the age of 10 and 11 and at 13 or 14, they would be ready to resist (the invasion of) more powerful states. *Adab*, Year 2, no. 29, Rabi II 27, 1321/July 23, 1903.
137. *Suresrâfil*, no. 24 (Moharram 24, 1326/February 27, 1908).
138. *Anjoman-e Moqaddas-e Melli-ye Esfahân*, no. 1 (January 6, 1907).
139. *Majles*, Shawwâl 8, 1324/November 25, 1906, no. 1, p. 3. It is highly significant that under the constitution, the anthropomorphist image of the country was almost reversed, even compared to the reform era of Mirzâ Hossein Khân-e Sepahsâlâr (see footnotes 89 and 90). During the second wars with Russia (1826–1826), in his *Resâleh-ye Jahâdiyyeh* and in order to mobilize population and state against the Russians, Mollâ Rezâ Hamedâni said the Shah was the brain or the heart and the army was the body (hands and legs). "Resâleh-ye jahâdiyyeh-e Hâji Mollâ Rezâ-ye Hamedâni (died 1247/1831)," see http://www.hawzah.net/en/ArticleView.html?ArticleID=81882 (access date, April 19, 2012).
140. *Asnâd*, no. 297031826 (۱۲۴ ک ۴اپ), Salakh Moharram 1323/April 6, 1905).
141. *Asnâd*, no. 240008825 (۵۱۰ص۱۳ ۱ ۱) July 15, 1907.
142. *Asnâd*, no. 297000166 (۱۱ ۷ ۱ ۱۱ب ۱), see fig 4.5.
143. In 1908, Mirzâ Seyyed ᶜAli Boqrât al-Molk was praised for his knowledge of traditional medicine alongside modern medicine in a certificate for medical practice issued by the National Society for Health Council (*anjoman-e hefz al-Sehheh-ye Melli*). *Asnâd*, no. 297028926, Zul Hajja 1325/January 1908.
144. See, for example, the cases of Fathᶜ Ali Tabib, a medical officer (*Hâfez al-Sehheh*) of Khorâssân, who was sent to Sistan by Dr. Schneider during a plague epidemic. To support an application for Order of the Ministry of Foreign Affairs, grade three, he presented recommendations written by physicians of the British and Russian consulates, and a certificate from Dr. Kelley. See *Asnâd*, no. 297029371 (۱ب۱۱ ۱ک۲۵۸), Safar 29, 1325/April 13, 1907. Similarly, requesting a Golden Scientific Order (*neshân-e talây-e ᶜelmi*) from the government in 1909, Mirzâ Abolqâsem Asᶜadol-Hokamâ mentioned that he had studied at the Dâr al-Fonun under Dr. Albo. *Asnâd*, no. 297029457 (۱ب۱ق۵۰۷), Shaᶜbân 19 & 23, 1327/5 & September 9, 1909.
145. *Majmu 'eh-ye qavânin* (Record of Parliamentary Legislation], 1906-11, quoted in Amin Banani, *The Modernization of Iran*, p. 62.
146. *Asnâd*, Ministry of Finance (Vezârat-e Mâliyeh), Jamâd I 19, 1327/June 8, 1919.
147. See *Asnâd*, no. 240008825 (۵۱۰ص۱۳ ۱ ۱), translation of the contract, Ministry of Foreign Affairs, May 1910.
148. *Asnâd*, see fig. 4.7.
149. *Irân-e Soltâni*, no. 1, March 31, 1906, rpnt, pp. 1, 2.

150. *Iran*, Year 57, no. 11 (August 30, 1904) & year 58, no. 11 (August 12, 1905), rpnt, pp. 253, 346.
151. *Iran*, Year 57, no. 9 (July 19, 1904), rpnt, pp. 241–242.
152. On this question see also H. Ebrahimnejad, "The waqf, the state and the medical education in nineteenth century Iran," in *Development of Modern Medicine in Non-Western Countries*, op. cit., pp. 59–81.
153. *Asnâd*, no. 240000334 (١١٢١ 725), dated Shawwâl 21, 1325/November 27, 1907.
154. *Asnâd*, no. 2970018235 (ش اب320), letter of the Ministry of the Interior to the Ministry of *Owqâf and Maʿâref va fawâyed-e ʿAmmeh* (Endowment, Education and Public Welfare), Jumâda II 8, 1329/June 6, 1911). The hospital, however, burned (down) in May 1911, and due to lack of funds and furniture, the madrassa was not operational. The *Anjoman-e Maʿâref-e Guilân* tried to collect funds for the reconstruction of the hospital. But, the Minister of the Interior warned that the use of such donations must be closely supervised, as "nowadays, some people may collect money under the pretence of charity, but use it for their own benefit." Ibid., latter of Jowzâ 16, 1329 (June 1911).
155. E. Malekzâdeh, op. cit., pp. 194–195.
156. M. Nikpour & A. Ghaffârinejad, *Pishineh-ye Pezeshki-ye Kerman dar sadeh-ye akhir*, pp. 88–91.
157. J. Gilmour, *Report on and Investigation into the Sanitary Conditions in Persia*, pp. 26–27.
158. Gilmour, *Report*, p. 26.
159. Gilmour, *Report*, p. 27. See also A. Afkhami, "Institut Pasteur," in *Encyclopaedia Iranica*, op. cit.
160. Mehrmâh Farmânfarmâiyân, *Zendeginâmeh*, vol. 1, p. 197.
161. *Asnâd*, no. 297029340 (١پ١١ق327), Safar 12, 1326/March 16, 1908.
162. In addition to his political undertakings, Farmânfarmâ was a proactive entrepreneur, who increased his wealth by playing the market, mostly by buying and selling properties but also through agricultural, commercial, and industrial activities. *Irân-e Soltâni*, no. 5 (Rabi I 5, 1321/June 1, 1903), rpnt, p. 39. While he was in Fars, he purchased a large amount of land and even during his exile in Iraq he continued to manage his wealth in Iran. On his return from exile and during his governorship of Kermânshâh, he obtained a government concession to construct a road to the border with Iraq. *Irân-e Soltâni*, no. 14 (October 6, 1903), rpnt, p. 112. Farmânfarmâ was the Governor of Kerman when the constitution was granted and when the preparations for parliamentary elections began, he gave his support to the election. One of the initiatives of the constitutionalists was the creation of a national bank, which issued shares. Farmânfarmâ was one of the first Qâjâr princes to give 10,000 tomans to buy shares, pledging more later. See *Nedâ-ye Vatan*, no. 2 (January 3, 1907), p. 7. This contribution was important, considering that many of the wealthy who pledged to purchase shares never honored their words. In early December 1906, pledges were made by several notable people to contribute about 27,000 toman and Zell al-Soltân, the Governor of Esfahan, pledged 150,000 toman; but with the exception of

1,500 toman from two merchants, 1,000 toman from Hâji Seyyed Ali Qazvini and 500 toman from Hâji Mohammad Jaᶜfar-e Esfahâni, no one else had paid by Shawwâl 22, 1324/December 9, 1906. See the letter of Hâj Mohammad Ebrahim-e Esfahâni to Hâj Mohammad Hassan-e Amin al-Zarb, Shawwâl 22, 1324; edited in *Majalleh-ye barrasihâ-ye târikhi* (*A journal of History and Iranian Studies*), edited by Jahângir Qâem-maqâmi, Tehran, nos. 1–2, Spring–Summer 1345/1966, p. 134.
163. According to T. Parsons, the conformity to the norm may be inherent in the moral authority of the norm, but it may also be a means to the realization of the actor's private ends, apart from the common value system. Conformity to the "common value system" may also be due to the positive advantages attached to it (such as social esteem) and to the desire to avoid the unpleasant consequences of nonconformity. See "The place of ultimate value in social theory," op. cit., p. 299.
164. W. Floor, *Public Health*, op. cit., p. 216.
165. C. Elgood, *A Medical History of Persia*, pp. 531–532, 536.
166. Dr. Schneider had abstained from the Council following his resignation the previous month, protesting that the local authorities in Sistan had not executed recommendations made by the Health Council. FO/371/114 (file 36259), October 29, 1906, fol. 500.
167. FO/371, vol. 105 (file 5758), February 19, 1906.
168. On the invitation of Dr. Tholozan, Edward Browne attended one of these meetings in 1887 and saw that "a majority of the physicians present knew no medicine but that of Avicenna." See Edward G. Browne, *Arabian Medicine*, Cambridge, Cambridge University Press, 1926, p. 93.
169. In the treatment of Hâjeb al-Dowleh sometime before the 1890s, several physicians, including Dr. Tholozan, Mirzâ Kâzem (Rashti), and Mirzâ Nazar ᶜAli Hakimbâshi, were involved. *Asnâd*, no. 295000711 (ر ١ ٣ ١ 390), undated. In another document from approximately the same period, a nobleman responding to Nâsser al-Din-Shah's inquiry about his state of health complained about the contradictory diagnosis and prescriptions of physicians. *Asnâd*, no. 295001012 (١ ر ١ ٣ ١ 694) undated.
170. Elgood, *A Medical History*, op. cit., pp. 535–536.
171. John Gilmour, *Report on and investigation into the sanitary conditions in Persia*, p. 32.
172. Gilmour, *Report*, pp. 32–33.
173. Document of Sâzmân-e Asnâd, dated Jamâd II 10, 1331/May 17, 1913, reproduced in Rustayee, vol. 2, p. 850.
174. Gilmour, Ibid., p. 31.
175. Gilmour, Ibid., p. 34.
176. Gilmour, Ibid., p. 36.
177. Ibid., p. 36.
178. Elgood, *A Medical History*, pp. 536–537.
179. FO/371, vol. 107 (file 11106) Mach 31, 1906, fol. 264.

180. FO/371, vol. 106 (file 8712), Foreign Office, March 19, 1906, fol. 237.
181. FO/369, vol. 32 (file 26543), August 3, 1906 (Ispahan, June 25, 1906), fol. 416.
182. According to M.-Q. Majd, Iran lost nearly 40% of its population during the famine of 1917-1919. *The Great Famine and Genocide in Iran: 1917-1919*. Although mortality due to famine and diseases might have attained several millions, the figure of 9 million given by Majd seems overestimated.
183. Ali M. Ansari, *Modern Iran*, pp. 3, 66. See also idem, *The Politics of Nationalisation*, pp. 110-111.
184. Ansari, *Modern Iran*, pp. 53, 70.
185. Cited in F. Kashani-Sabet, "The politics of reproduction: Maternalism and Women's Hygiene in Iran," p. 11.
186. F. Kashani-Sabet, "The politics of reproduction."
187. *Ettelâ῾ât*, no. 385, 24 Aban 1306/1927, cited in F. Kashani-Sabet, ibid., p. 12.
188. *Shekufeh*, no. 17 (Zihajja 9, 1331/November 9, 1913), rpnt, p. 65.
189. Gilmour, op. cit. p. 53., in 1924, when Gilmour wrote his report, the Municipality was only set up in Tehran, Anzali, Tabriz, and Hamadan. Ibid., p. 27. This system, known as *Assistance Public-Hôpitaux de Paris, de Bordeaux, de Marseille, and so on*, is still alive in France.
190. Gilmour, *Report*, pp. 53–54.
191. J. Gilmour, *Report*, p. 18.
192. When in 1300/1883 ῾Ali-Qoli Khân-e Mokhber al-Dowleh, Minister of Education, returned from Berlin, he took with him Dr. Isidor Albo to teach medicine at the Dâr al-Fonun. M. Rustâyee, "A῾lam al-Dowleh, Khalil Khân-e Saqafi, tabib va nâji-ye mashrutiyyat," p. 58.
193. Elgood, *A Medical History*, pp. 545–546.
194. This could not be reimbursed before Dr. Ilberg's departure with the German Legation in 1915 and the debt was still outstanding at the end of World War I, when the government of Wosuq al-Dowleh pledged to reimburse it with added annual interest of 6 percent. See *Asnâd*, 'ministère des finances (*vezarat-e maliyeh*), cabinet du ministère, no. 20909, dated September 29, 1915; *Asnâd*, no. 240030278, dated June 26, 1297 (September 1918).
195. Amin Banani, *The Modernization of Iran*, pp. 36–37.
196. In one of the documents it was called the German hospital but this term has been crossed out and corrected by adding *dowlati* (state). *Asnâd*, no. 240030278, dated June 26, 1297 (September 1918).
197. Elgood, *A Medical History*, op. cit., p. 550.
198. Elgood, *A Medical History*, op. cit., pp. 545–456.
199. See a series of documents in *Asnâd*, no. 2500003001, dated January 29, 1312 & August 12, 1312, etc. . . . See also E. Malekzâdeh, *Negâhi be omur*, pp. 163–168.
200. The hospital Ahmadiyeh was built in Rajab 21, 1332/June 15, 1914 by Amir-Khân-e A῾lam in Tehran near the Artillery Square. *Behdâsht-e Jahân (World Health Magazine)*, year 17, nos. 2–3, autumn 1383/2004, p. 63.
201. *Asnâd*, no. 240027454, dated November 8, 1913. A house was rented from a certain Mostafâ Khân in order to set up a military hospital in Fars. The rent and

the cost of some of the repairs for 1335 H (October 28, 1916–November 15, 1917) were 78 tomans and for 1336 H (November 16, 1917–October 6, 1918), 66 tomans. See *Asnâd*, no. 240027454, dated Rabi' II 15, 1335/February 1917.
202. Gilmour, op. cit., pp. 24–28.
203. *Asnâd*, no. 297024200, dated dalv 9, 1337 (1919).
204. *Asnâd*, no. 240030278, Qows 11, 1336/December 1, 1818.; *Asnâd*, no. 240030278, dated *Thowr* 1336/ca. April 1918.
205. 200 tomans of the hospital's budget was for the salary of employees, but the Ministry of Finance did not approve this because there were no physicians or surgeons. *Asnâd*, no. 240030278, dated Dalv 27, 1336/ca. January 1919.
206. *Asnâd*, no. 240030278, dated *Thowr* 1336/ca. April 1918.
207. *Asnâd*, no. 240030278, letter dated Jowza 6, 1336/ca. May 1918.
208. *Asnâd*, no. 250000755, letter to the Office of *Maᶜâref and Owqâf* (Education and Endowment), dated 10/9/1308 SH. (November 30, 1929).
209. E. Malekzadeh, op. cit., pp. 138–139.
210. Mohammad Hossein Azizi & Farzâneh Azizi, "Government-sponsored Iranian Medical Students Abroad (1811–1935)," p. 356.
211. H. Katouzian, *State and Society in Iran*, p. 83.
212. This question, given its extent, cannot be examined in this chapter and needs to be explored separately.
213. Ansari, *Modern Iran*, pp. 66–79; Keddie, *Modern Iran, Roots and Results of Revolution*, p. 99.
214. Avery, "The Pahlavi Autocracy: Rizâ Shâh, 1921–41," *CHI*, VII, p. 241; Keddie, *Modern Iran*, p. 98.
215. Avery, "The Pahlavi Autocracy: Rizâ Shâh, 1921–41," *CHI*, VII, p. 240.
216. For example, the British government wanted the dispensary of the British Legation and the doctor attending it to remain within the Legation precincts so that "His Majesty's Government receive full credit for the benefits it confers." FO/371, vol. 106 (file 8712), Foreign Office, March 19, 1906, fol. 237.
217. A.-A Afkhami, "Public health in Persia iii," pp. 157–163.
218. The Rezâ-Shahi Hospital in Mashhad, for example, was established and operated by German doctors. Ansari, *Modern Iran*, p. 66.
219. Charles Oberlin, *Barnâmeh-ye behdâsht barâ-ye Iran* (*A program for health preservation [public health] in Iran*), pp. 3–4.
220. See, for example, Dr. Hossein ᶜAli Qezel Ayâq, *Hefz al-sehheh yâ kelid-e sehhat (Preservation of health or key to health)*.

Conclusion

1. On the Iranian enlightenment in the nineteenth century and its complex development and roots, see Ali Ansari, *The Politics of Nationalism*, pp. 40ff.
2. Thomas Kuhn, *The Structure of Scientific Revolutions*, Chicago, London, The University of Chicago Press, 1996, p. 175.
3. On this issue see, Toby Huff, *The Rise of Early Modern Science*, pp. 30–32.

4. Soraya Tremayne, "The dilemma of assisted reproduction in Iran," FVV in ObGyn, 2012, Monograph pp. 70–74, p. 72; Mohammad Jalal Abbasi-Shavazi *et al*, "The 'Iranian Art Revolution': Infertility, Assisted Reproductive Technology, and Third-Party Donation in the Islamic Republic of Iran," pp. 4–5.

Bibliography

1. See, for example, M. Rustâyee, *Târikh-e tebb va tebâbate dar irân*; M. Nikpour & A.-R. Ghaffârinejad, *Pishineh-ye pezeshki-ye Kerman dar sadeh-ye ahkir*.

Bibliography

With the development of interdisciplinary approaches in history since the second half of the twentieth century, the history of medicine in Iran flourished as one of the specialized fields in social history. At the same time, many archival sources and manuscripts about medicine that had remained unstudied have come to the fore. The most important data on this subject are found in the Qâjâr journals and periodicals that thrived during the second half of the nineteenth century. Several of these journals have been published in offset form, but many remain unpublished. The Persian archival sources of the National Library (*Sâzmân-e Asnâd*) in Tehran also contain invaluable information on medicine and the medical profession. So far, the archives of the *Sâzmân-e Asnâd* have been more the objects of reproduction and edition, but they have rarely been examined.[1] The present book draws largely on journals and periodicals of the Qâjâr period as well as on the above-said archival sources. The traditional names in Iran are indicated according to the old system as they did not have family name. They usually precede or succeed by titles such as Mirzâ, Khân, Aqâ.

Persian sources

MSS & Lithographed

ᶜAbd al-Majid-e Eᶜtezâd al-Atebbâ (translated by), *Resâleh-ye wabâiyyeh*, National Library, Tehran, 1309/1892.

ᶜAbd al-Razzâq, *Kholâsat al-Tashrih*, Persian MSS, St Petersburg, Oriental Institute Library, no. 154.

Anonymous, *Fehrest-e ketâb-e majmaᶜal-moluk*, 1257/1841, National Library, St Petersburg, no. 88.

Anonymous, *Ketâb dar ᶜelm-e tebb*, Persian MSS, no. 179.

Anonymous, *Majmuʿeh-ye wabâiyeh*(*Collection of essays on wabâ*); 1291/1874, Majles Library.

Anonymous, *Mehnat al-tâʿun* (*The Scourge of Plague*), translated from Elyâs b. Ebrâhim Yahudi's *Mehnat al- tâʿun w'al-wabâ* in Arabic, ca. fourteenth–fifteenth century), 1247/1831–32, Terhan, Malek Library.

Anonymous, *Rashahât al-fonun*, Persian MSS, National Library, St Petersburg, no. 529.

Anonymous, *Resâleh dar favâyed-e feshang-e jangi* (*On the benefits of military cartridge*), Persian MSS, no. 11055, National Library, Tehran.

Anonymous, *Resâleh dar wabâ va tâʿun*, undated, national Library, Tehran.

Anonymous, untitled, written (or copied) in 6 Rabiʿ I 1277/September 1860, Persian MSS, no. 4603, Ketâbkhâne-ye markazi, University of Tehran.

Âqâ Mirzâ Mohammad-e Tehrâni, *Resâleh*, dated 1 Shaʿbân 1269/May 10, 1853.

ʿAtâollâh b. Mohammad al-Hosseini, *Anis al-majâles*, Persian MSS, Oriental Institute, St Petersburg, undated, no. D 420.

Dr [Isidor] Albo, *Jarrâhi* (Surgery), translated by Khalil b. Hâji Mirzâ ʿAbdol-Bâqi-ye Eʿtezâd al-Atebbâ, 1306/1891.

Dr Basil, *Resâleh-ye wabâiyyeh* (edited/translated) by Abdolmajid Eʿtezâd al-atebbâ, Tehran, Dâral-Fonun, 1309/1891.

Dr. Babayef et al., *Dastur-e jelowgiri az wabâ va tâʿun* (*Prescription for prophylactic measures against cholera*), undated, Library of Astân-e Qods.

ʿEmâd al-Din Mahmud-e Shirâzi, *Atashak (Syphilis)*, 997/1589, Persian MSS, Majles Library, Tehran, no. 6307.

ʿEmâd al-Din Mahmud-e Shirâzi, *Bikh-e chini* (China Root), 993/1585, published in the journal of Hamdard Society, India, *Studies in History of medicine and science*, vol. XII, no. 1–2(2001).

Hakim Mohammad, *Zakhira-ye kâmela* (*The Treasury of Perfection*), Persian MSS, no. 8825, National Library, Tehran.

Mirzâ ʿAbdol-Karim-e Tehrâni, *Ketâb-e ʿalâyem al-amrâz* (*Treatise on the symptoms of diseases*), 1267/1851, Persian MSS, no. 821, Tehran, Sepahsâlâr Library.

Mirzâ Abol-Hassan-Khân ebn-e ʿAbdol-Wahhâb-e Tafreshi, *Matla ʿal-tebb-e nâsseri*, Persian MSS, 1299/1881, Tehran, Majles Library, no. 18585.

Mirzâ Ahmad-e Qâjâr, *Tohfat al-khâqân*, an abbreviated form of Râzi's *Bar'al-sâʿa* for the attention of the Shah. Copy dated 1264/1848, National Library, Tehran, Persian MSS, no. 12369/1).

Mirzâ ʿAli [-Akbar] Hamadani, *Amrâz-e ʿasabâni* (literally *Nervous diseases*), a translation of Grisolle's, *Traité de pathologie nerveuse* (Persian MSS, 1297/1880, The Central Library of Tehran University, no. 1519).

Mirzâ ʿAli [-Akbar-e] Hamadani, *Tashrih* (*Anatomy*), Tehran, lithographed, 1312/1895.

Mirzâ Mohammad al-Râzi al-Kani-ye Fakhr al-Atebbâ, *Meftâh al-amân*, Persian MSS, Safar 1279/August 1862, no. 2522, National Library, Tehran).

Mirzâ Mohammad Kâzem-e Rashti, *Hefz al-Sehheh-ye Nâsseri*, Persian MSS, Tehran, National Library, manuscript no. 439, or 5–10439 (for the electronic copy); lithograph ed., Tehran, 1887–88).

Mirza Mohammad-Taqi-ye Shirâzi, *Hefz al-sehheh*, undated, lithographed, second part of the nineteenth century, Majles Library.

Mirza Mohammad-Taqi-ye Shirâzi, *Mofarraq al-heizeh v'al-wabâ*, ca. 1861, Theran, lithographed, Majles Library.

Mirza Mohammad-Taqi-ye Shirâzi, *Tǎᶜunia*, dated 1831, in Arabic, Lithograph, Tehran, Library of Majles.

Mirza Mohammad-Taqi-ye Shirâzi, *Wabâiyeh-ye kabireh* (On cholera) 1251/1835, Theran, lithographed, Majles Library.

Mirzâ Mohammad-e Tehrâni al-asl-e Najafi Maskan [born in Tehran, resident in Najaf], *Dar wabâ* (on cholera), Library of Majles, lithographed, Tehran, 1853.

Mirzâ Musâ Sâvaji Fakhr al-Hokamâ, *Dastur al-atebbâ fi ᶜalâj al-wabâ* (*Prescription of physicians on the treatment of cholera*), dated May 11, 1853, unknown date of lithograph edition, Tehran, Majles Library.

Mirzâ Nosrat-e Tabib-e Quchâni, *Hâfez al-sehheh-ye nâsseri*, 1309/1891, Persian manuscript, MSS no. 2265, microfilm no. 2835,Tehran, Library of Majles.

Mirzâ Reza, *Pâtology*, Persian manuscript, Tehran University Central Library, no. 1519.

Mohammad al-Râzi al- Kani, *Meftâh al-amân* (Key to the safety [during epidemics]), 1862, Tehran, National Library, MSS 2522, 1279/1862.

Mohammad ebn-e ᶜAbdol-Sabur-e Khoi-ye Tabrizi [known as Hakim Qoboli], *Morakkabât-e jowhariyyeh* (*Chemical drugs*), Persian MSS, Library of UCLA, AR. 96I.

Mohammad ebn-e ᶜAbdol-Sabur-e Khoi-ye Tabrizi, *Anwâr-e nâsseriyyeh*, 1272/1856, lithographed, no. 52066, Majles Library, Tehran.

Mohammad ebn-e ᶜAbdol-Sabur-e Khoi-ye Tabrizi, *Taᶜlim-nâmeh yâ resaleh-ye âbeleh kubi* (*Instructions on inoculation of smallpox*), dated Shaᶜbân 28, 1245/February 22, 1830.

Mohammad Hossein-e Afshâr, *Resale-ye moᶜâlejeh-ye maraz-e wabâ*, Persian manuscript, Tehran, National Library, no. 2479.

Mohammad-Hossein ebn-e Hâdi-ye ᶜAqili, *Moᶜâlejât-e amrâz-e mokhtasseh az sar tâ qadam*, Lithograph edition (3B/17–16), 1276/1859.

Mohammad-Hossein ebn-e Hâdi-ye ᶜAqili, *Kholâsat al-hekmat*, Bombay, 1261/1845.

Mohammad-Hossein ebn-e Hâdi-ye ᶜAqili, *Qarâbâdin-e kabir* (*Greater Pharmacopea*) in 1276 and 1277.

Mohammad-Razi Tabâtabâyee, *Tǎᶜun* (*Plague*), second part of the nineteenth century, Library of Gowharshâd, Mashhad.

Mollâ Ardeshir, *Vaqâyeᶜ – e pârsiyân*, Persian MSS, no. 342, National Library, St Petersburg.

Mosio Gabriel and Mohammad-Hossein-e Qâjâr (translated by), *Resâleh dar wabâ* (*Treaty on cholera*), 1262, Tehran, National Library.

Nasrollah Mirzâ Qâjâr, *Chahar maqâleh–ye nâsseri dar tebb* [1871] National Library, Tehran, MSS 722.

Nur al-Din Mohammad ebn-e Hakim ᶜAbdol-Malek-e Shirâzi, *Alfâz al-adwiya*, Persian MSS, composed in 1083/1628–29, the copy used made in

Rajab 1279/December 1862–January 1863), Library of Oriental Institute, St Petersburg.

Nur al-Din Mohammad ebn-e Hakim ᶜAbdol-Malek-e Shirâzi, ᶜAlâjât-e dâr al-shokuhi (Treatments of Dâr al-Shokuhi), 1778, Persian MSS, supplément persan, no. 342, Bibliothèque National de France, Paris.

Polak, Dastur al ᶜamal va taᶜlimât-e moᶜâlejât va tadâbir-e amrâz..., Persian MSS, National Library, Tehran, no. 6164.

Polak, Tashrih-e badan-e ensân (Anatomy of the human body), written with the help of his pupil, Mirzâ Mohammad-Hossein-e Afshâr, lithographed, 1270/1854, Tehran.

Polak, Ketâb-e jarrâhi moshtamel bar do jeld va yek resaleh dar kahhâli az tasnifât-e ᶜalijâh doctor Polak-e namsâvi... 'Book on surgery in two volumes and one treatise on ophthalmology authored by His Excellency Dr Polak from Austria...', undated.

Polak, Resâleh dar dastur al-ᶜamal va taᶜlimât-e moᶜâlejât-e amrâz-e nowbeh, eshâl al-dam, mothbeqa va brudat ke aksar-e owqât be afwâj-e qâhereh mostowli va bâᶜes-e halâkat-e ishân migardad (Guidance for the treatment of intermittent fever, bloody flu, typhus and hypothermia that often affect the victorious divisions and cause mortality), Persian manuscript, 1269/1853, National Library, Tehran, no. S6164.

Qarageuzlu, Qawâᶜed-e kolliyeh-ye nezâmiyeh (the general rules in the army), 1270/1854, National Library, Mss. No. 3223.

Qâzi ebn-e Kâshef al-Din Mohammad-e Yazdi, Resâleh-ye Mirzâ Qâzi, copy dated 1241/1833, Persian MSS, Wellcome Trust Library, WMS. Per. 293 (B).

Qâzi ebn-e Kâshef-e Mohammad-e Yazdi, Resâleh-ye chub-e chini va chai va qahveh, Persian MSS, British Library, Add. 16919 A.

Report of a meeting of Rajab 23, 1307/March 15, 1890, on prophylactic measures against cholera, in 20 pages, National Library, Tehran.

Sâleh ibn Nasrollâh al-Halabi, copy dated 1225 (ca. 1810, St Petersburg, Library of Oriental Institute, manuscript no. C 1615).

Sâleh ibn Nasrollâh al-Halabi, Tebb-e jadid-e kimiyâyee t'â Bereklus (Modern chemical medicine of Paracelsus) [originally written in 1669], anonymous translator, dated 1225/1810, St Petersburg, Library of Oriental Institute, manuscript no. C 1612, fols. 1b–69a.

Schlimmer, J., and Mirzâ Mohammad-Taqi Kâshâni, Meftâh al-Khawâss-e Nâsseri, Persian MSS, 1277/1861, no. 7379, Tehran, Library of Sepahsâlâr.

Schlimmer, Johannes, Terminologie Médico-Pharmaceutique et Anthropologique Française-Persane avec Traductions Anglaise et allemande des termes... Lithographed, 1874, Tehran.

Seyyed ᶜAli ebn-e Mohammad-e Tabrizi, Qânun al-ᶜalâj, Persian manuscript, no. 4115, 1269/1853, the Central Library of Tehran University.

Tholozan (under the supervision of), Resâleh dar keyfiyât-e maraz tâᶜun (Treatise on plague), 1293/1876, Mashhad, Library of Gowharshâd.

Tholozan, Badâyeᶜal-hekmat-e Nâseri, Tehran, National Library, MS, no. 6193.

FathʿAli, *Wasyyatnâmcheh, Irâdâti ke Mirzâ Fathʿ Ali be târikh-e Rezâ Qoli-Khân gerefteh ast* (criticism of the History of Rezâ Qoli-Khân (Târikh-e Rozat al-Safâ-ye Nâsseri)), copy dated 1301 (1884), Tehran, National Library, MSS no. 3111.

Zein al-Din abu Ebrahim Esmâʿil Jorjâni, *Zakhireh-ye Khârazmshâhi*, WMS. Per. 281, London, Library of the Wellcome Trust.

Persian Periodical

Adab
Anjoman-e Moqaddas-e Melli-ye Esfahân
Behdâsht-e Jahân (World Health Magazine), year 17, nos. 2–3, autumn 1383/2004
Dânesh
Nashrieh-ye madreseh-ye mobârakeh-ye dâr al-fonun-e Tabriz (*Journal of the Tabriz Dâr al-Fonun*)
Iran-e Soltâni
Kashkul
Maʿâref
Majalleh-ye barrasihâ-ye târikhi (*A Journal of History and Iranian Studies*)
Majles
Merrikh (*The Journal of the Military*)
Nedâ-ye Vatan
Omid
Ruznâmeh-ye dowlat-e ʿelliyeh-ye Iran (RDEI)
Ruz-nâmeh-ye ʿelmi (RE)
Ruz-nâmeh-ye Irân (Iran)
Ruznâmeh-ye ʿelmiyyeh vava adabiyye
Ruznâmeh-ye Mellati
Ruznâmeh-ye Vaqâyeʿ-e Ettefâqiyeh (RVE)
Majalleh-ye dâneshkadeh-ye adabiyât va ʿolum-e ensâni (*Journal of the Faculty of Literature and Human Science*)
Shekufeh
Suresrâfil
Tamaddon
Tarbiyat, Nakhostin nashriyeh – ye gheir-e dowlati-ye Iran (The first nongovernmental journal of Iran), published from December 1896 to March 1905, offset publication (Tehran, National Library, 1376/1997).

Archives

Abbreviations:

Asnâd = Archives of the Sâzmân-e Asnâd of the National Library (*Sâzmân-e Asnâd-e Ketâbkhâneh-ye Melli*), Tehran
PRO = Public Record Office, National Archives, London

FO = Foreign Office, National Archives, London
AMAE = Archive du Ministère des Affaires Etrangères, Paris
CPP = Correspondances Politiques, Perse (Archives du Ministère des Affaires Etrangères)
ADP = Affaires Diverses Politiques (Archives du Ministère des Affaires Etrangères)
CPC = Correspondances Politiques et Consulaires (Archives du Ministère des Affaires Etrangères)
USDS = United States Department of State
AMAE, ADP, dossier 68 (3) for Mirzâ Mohammad-Khân: letter of Nazar Aqâ, April 12, 1889 and response of the Minister, May 8, 1889. Dossier 68 (5) for Moʿin al-Atebbâ: letter of Nazar Aqâ, Mai 14, 1891 and letter of the Minister, June 22, 1891.
AMAE, ADP, dossier 93, Dépêche Télégraphique, Téhéran, le 29 Juillet 1893.
AMAE, ADP, dossier 93, letter dated May 29, 1894.
AMAE, ADP, dossier 93, letters of de Balloy, Paris, June 10, 1893, Tehran, July 31, 1893, and Tehran, February 15, 1894).
AMAE, ADP, dossier 93, Tehran August 26, 1894, letter no. 27.
AMAE, ADP, Dossier Schneider, 93, letter of the Minister of War to the Minister of Foreign Affairs, July 17, 1894.
AMAE, ADP, vol. 4, dossier 22 (Le Colombier, le 16 Janvier 1890); and dossier 26 (Téhéran, le 5 Mai 1890).
AMAE, ADP, vol. 4, dossier 23, letter of the Ministry of War, January 20, 1890; letter of Tholozan (to Mr Flurry), April 1, 1893; letter of Feuvrier, Avril 17, 1893.
AMAE, CPC, Perse, de Balloy's letter (1897–1917), vol. 1, April 9, 1897.
AMAE, CPC, Perse, de Balloy's letter (1897–1917), vol. 1, Mai 1896.
AMAE, CPC, Perse, de Balloy's letter (1897–1917), vol. 1, Mai 1896.
AMAE, CPP, vol. 21 (September 30, 1845, no. 65), fols. 107–108.
AMAE, CPP, vol. 21, October 06, 1845; fols. 117–118 & October 08, 1845, fols. 121–123, & correspondence of October 2, 1845, fol. 126.
AMAE, CPP, vol. 21, no. 24, Paris January 20, 1845, fols. 7–8.
AMAE, Mémoire et Documents, Perse, vol. 9, 1851.
AMAE, Mémoires et Documents, Perse, vol. 7, p. 116.
AMAE, Mémoires et Documents, Perse, vol. 7, p. 116.
AMAE, Mémoires et Documents, Perse, vol. 7, pp. 113–122 (document dated February 20, 1808).
AMAE, Mémoires et documents, vol. 6, fol. 71.
Asnâd, 'ministère des finances [*vezârat-e mâliyeh*], cabinet du ministère, no. 20909, dated September 29, 1915.
Asnâd, dossier 148, no. 297/35886, Esfand 1305/February 1926.
Asnâd, dossier 151, no. 297/35886, dated July 24, 1314 (October 1935).
Asnad, dossier 156, no. 251, dated October 1, 1308 (October 1929).
Asnad, dossier 158, no. 251, dated November 11, 1305 (February 15, 1926).
Asnâd, dossier 89, no. 297/215593, July 23, 1311 (October 1932).
Asnad, dossier 91, no. 297/21595, dated January 5, 1313 (March 26, 1934).

Asnad, dossier 93, no. 297/21597, date: July 8, 1312 H. Sh.
Asnâd, dossier 93, no. 297/21597, dated June 7, 1312 H. Sh.
Asnâd, Ministry of Finance (Vezârat-e Mâliyeh), Jamâdi I 19, 1327/June 8, 1919.
Asnâd, no. 240000334 (١١٢٥٧٢٥), dated Shawwâl 21, 1325/November 27, 1907.
Asnâd, no. 240008825 (١١٣ص٥١٠) July 15, 1907.
Asnâd, no. 240008825 (١١٣ص٥١٠), translation of the contract, Ministry of Foreign Affairs, May 1910.
Asnâd, no. 240027454, dated Rabiᶜ II 15, 1335/February 1917.
Asnâd, no. 240027454, dated November 8, 1913.
Asnâd, no. 240030278, Qows 11, 1336/December 1, 1818.
Asnad, no. 240030278, dated June 26, 1297 (September 1918).
Asnâd, no. 240030278, dated June 26, 1297 (September 1918).
Asnâd, no. 240030278, dated Dalv 27, 1336/ca. January 1919.
Asnâd, no. 240030278, dated *Thowr* 1336/ca. April 1918.
Asnâd, no. 240030278, letter dated Jowzâ 6, 1336/ca. May 1918.
Asnâd, no. 2500003001, dated January 29, 1312 & August 12, 1312.
Asnâd, no. 250000755, letter to the Office of Maᶜâref and Owqâf (Education and Endowment), dated 10/9/1308 SH. (November 30, 1929).
Asnâd, no. 295000711 (ر١٣١٣٩٠), undated.
Asnâd, no. 295001012 (١ر١٣ ٦٩٤) undated.
Asnâd, no. 295001034, 1281/1865.
Asnâd, no. 295005729, dated Rabiᶜ II 7, 1255/June 1839.
Asnâd, no. 297000166 (١١١٧ اب).
Asnâd, no. 2970018235 (٣٢٠ ش ا اب), latter of Jowza 16, 1329 (June 1911).
Asnâd, no. 2970018235 (٣٢٠ ش ا اب), letter of the Ministry of the Interior to the Ministry of *Owqâf and Maᶜâref va fawâyed-e ᶜAmmeh* (Endowment, Education and Public Welfare), Jumâda II 8, 1329/June 6, 1911).
Asnâd, no. 297024200, dated dalv 9, 1337 (1919).
Asnâd, no. 297029340 (٣٢٧ ق ١١ اب), Safar 12, 1326/March 16, 1908.
Asnâd, no. 297029371 (٢٥٨ ق ١١ اب), Safar 29, 1325/13/4/1907.
Asnâd, no. 297029457 (٥٠٧ ق ١١ اب), Shaᶜbân 19 & 23, 1327/September 5 & 9, 1909.
Asnâd, no. 297031826 (١٢٤ ک فاب), Salkh Moharram 1323/April 6, 1905).
Asnâd, no. F/438.
Notice sur le Guilan et le Mazanderan, par Mr Trezel Ingénieur', *Géographe*, November 26, 1808, no. 8, Ministère de la guerre, Archives historiques, Vincennes (Paris).
PRO, FO/248, vol. 325, Foreign Office, October 17, 1877[, which refers to a dispatch of Churchill from the British consulate in Anzali], no. 14, July 31, 1877.
PRO, FO/249, vol. 29, Tehran, October 10, 1833.
PRO, FO/284, vol. 325, [copie d'une note de la Légation Impériale au ministre persan en date du juillet 20, 1877], no. 88.
PRO, FO/369, vol. 32 (file 26543), August 3, 1906 (Esfahan, June 25, 1906), fol. 416.
PRO, FO/371, vol. 105 (file 5758), February 19, 1906.

PRO, FO/371, vol. 106 (file 8712), Foreign Office, March 19, 1906, fol. 237.
PRO, FO/371, vol. 107 (file 11106) Mach 31, 1906, fol. 264.
PRO, FO/371/114 (file 36259), October 29, 1906, fol. 500.
PRO, FO/60, vol. 20, no. 14, October 1821, fol. 108 and no. 22, December 16, 1821, fol. 117.
PRO, FO/60/90, Letter of Cl. Sheil, Tehran, September 6, 1842.
PRO, FO/60/90, Tehran, September 6, 1842.
United States Department of State (USDS), United States Department of State (USDS), online archive: http://digicoll.library.wisc.edu/cgi-bin/FRUS/FRUS-idx?type=article&did=FRUS.FRUS188788.i0031&id=FRUS.FRUS188788&isize= M, nos: 548–550, January 1887–August 1889.
USDS, Inclosure in no. 249, Legation of the US, May 28, 1896.
USDS, no 45, Legation of the US, Tehran, December 21, 1898.
USDS, no. 546, Legation of the United States, Tehran, November 29, 1886.
USDS, no. 63, American Legation, Tehran, July 23, 1904.
USDS, no. 912, Pratt to Bayard, Legation of the United States, Tehran, January 10, 1888.
USDS, no. 912, Tehran, January 10, 1888.
USDS, no. 914, Pratt to Bayard, Legation of the US, Tehran, May 27, 1888.

Published sources

Abbasi-Shavazi, Mohammad Jalal, Inhorn, Marcia C.; Razeghi-Nasrabad, Hajiieh Bibi and Toloo, Ghasem, 'THE "IRANIAN ART REVOLUTION"': Infertility, Assisted Reproductive Technology, and Third-Party Donation in the Islamic Republic of Iran', *Journal of Middle East Women's Studies*, vol. 4, no. 2, spring 2008: 1–28.
ᶜAbdol-Hossein-e Zenuzi-ye Tabrizi, *Matrah al-anzâr fi tarâjem an atebbâ al-aˈsâr va falâsefat al-amsâr*, edited by Mir Hâshem Mohaddeth (Tehran, Pejmân Publishers, 1388/2009).
Abrahamian, Ervand, 'The Causes of the Constitutional Revolution in Iran', *International Journal of Middle Eastern Studies*, 10 (1979), 381–414, p. 383.
Abu Nasr Asᶜad b. Ilyâs ibn Matrân, *Bustân al Atibbâ va rawzat al alibbâ*, edited with introduction of Mahdi Mohaghigh (Tehran, Centre for the Publication of Manuscripts, 1989).
Ackerknecht, Erwin H., *A Short History of Medicine*(Baltimore and London, Johns Hopkins University Press, 1982).
Ackerknecht, Erwin H., 'Broussais or a Forgotten Medical Revolution', *Bulletin of the History of Medicine*, 27 (4) (1953), 320–343.
Adamiyat, Fereidun, *Amir-Kabir va Iran* (Tehran, Khwârazmi, 1385/2006).
Adamiyyat, Fereidun & Nategh, Homa *afkâr-e ejtemâᶜi va siyâsi eqtesâdi dar âsâr-e montasher nashodeh-ye dowreh-ye Qâjâr* (*Socio-political and economic ideas in the unpublished literature of the Qâjâr period*) (Tehran, Agâh, 1357/1978).
Adamiyyat, F., *Andisheh-ye Taraqqi va Hokumat-e Qânun*, ᶜ*Asr-e Sepahsâlâr* (Tehran, Khwârazmi, 1385/2006).

Afkhami, Amir Arsalan, 'Health in Persia iii. Qâjâr Period', *Encyclopaedia Iranica*,
Afkhami, A.-A., 'Institut Pasteur', in *Encyclopaedia Iranica*, Vol. XIII, Fasc. 2, pp. 157–163.
Afkhami, 'The Sick Men of Persia: The Importance of Illness as a Factor in the Interpretation of Modern Iranian Diplomatic History', *Iranian Studies*, 36 (3), September (2003), 339–352.
Afkhami, A.-A., 'Public Health in Persia iii', *Encyclopaedia Iranica*, Vol. XIII, Fasc. 2, pp. 157–163.
Afkhami, Amir Arsalan, 'Compromised Constitutions: The Iranian Experience with the 1918 Influenza Pandemic', *Bulletin of the History of Medicine*, 77 (2) (2003), 367–393.
Algar, Hamid, 'Freemasonry ii In the Qâjâr Period', *Encyclopaedia Iranica*, Vol. XIII, Fasc. 2, pp. 157–163.
Alavi, Seema, *Islam and Healing: Loss and Recovery of an Indo-Muslim Medical Tradition, 1600–1900* (Palgrave Macmillan, 2008).
ᶜAli Akbar Dehkhoda, *Loghatnâmeh*, edited by M. Moᶜin & Jaᶜfar Shahidi, second edition, 1377/1998, mo'assesseh-ye enteshârât va chap-e dâneshgâh-e tehran.
ᶜAlinaqi Hakim al-mamâlek, *Ruznâmeh-ye safar-e Khorâssân*, Tehran, Enteshârât-e Farhang-e Iran Zamîn, 1977 (written in the second part of the nineteenth century).
ᶜAli-Qoli-Mirzâ Eᶜtezâd al-Saltaneh was a staunch opponent of traditional astrology and wrote a book refuting it: *Falak al-Saᶜâda*, lithograph edition, Tehran, 1278/1861.
Amanat, Abbas, *Pivot of the Universe* (London, New York, I.B. Tauris Publishers, 1997).
Amanat, Abbas, *Resurrection and renewal: the making of the Babi movement in Iran, 1844–1850* (Ithaca and London, Cornell University Press, 1989).
Amoli, Shams al-Din Mohammad b. Mahmud, *Nafâyes al-fonun fi ᶜarâyes al-ᶜoyun*, edited by Shaᶜrâni, 3 vols (Tehran, Ketâbforushi eslâmiyeh, 1377HQ/1958).
Andral, Gabriel, *Précis d'anatomie pathologique*, 5 volumes, vol. I (Paris, Chez Gabon, Libraire-éditeur, Rue de l'Ecole de Médecine, no. 10, 1829).
Ansari, Ali M., *Modern Iran* (London, New York, Bosto, San Francisco..., Pearson Lanman, second edition, 2007).
Ansari, Ali M., *The Politics of Nationalism in Modern Iran* (Cambridge, New York: CUP, 2012).
Arasteh, Reza, *Education and Social Awakening in Iran, 1850–1968* (Leiden, Brill, 1969).
Aramesh, Kiarash, 'Iran's Experience with Surrogate Motherhood: An Islamic View and Ethical Concerns', *Journal of Medical Ethics*, 35 (5) (2009), 320–322.
Arjomand, Kamran, 'The Emergence of Scientific Modernity in Iran: Controversies Surrounding Astrology and Modern Astronomy in the Mid-Nineteenth Century', *Iranian Studies*, 30 (1–2) (Winter–Spring 1997), 5–24.
Astarâbâdi, *Safina-ye Nuh*, text edited in H. Ebrahimnejad, 'Religion and Medicine in Qajar Iran', 2004, R. Gleave, ed., *Religion and Society in Qajar Iran* (London, Routledge, 2004), pp. 401–428.

Attewell, G., *Refiguring Uniani Tibb: Plural Healing in Late Colonial India* (New Delhi, Orient Longman, 2007).

Aubert, Théophile Edouard, *Coup d'œil sur la médecine envisagée sous le point de vue philosophique* (Paris, 1835).

Azizi, Mohammad-Hossein, & Azizi, Farzâneh, 'Government-sponsored Iranian Medical Students Abroad (1811–1935),' *Iranian Studies*, 43 (3) (2010), 349–363.

Badiʿi, Parviz (ed.), *Yâddâshthây-e ruzâneh-ye Nâser al-Din Shâh*, 1300–1303 H/1883–1885 (Tehran, Enteshârât-e Sâzmân-e asnâd-e melli-ye Iran, 1378/1999).

Bagley, F. R. C., 'Elgood, Cyril Lloyd', *Encyclopaedia Iranica*, 13.

Bamdad, Mhdi, *Sharh-e hâl-e rejâl-e Iran*, 6 vols. (Tehran, Zavvâr, 1378/1999).

Banani, Amin, *The Modernization of Iran 1921–1941* (Stanford, California: Stanford University Press, 1961).

Baron de Bode, *Travels in Luristan and Arabistan* (London, J. Madden & Co., 1845).

Basalla, George, 'The Spread of Western Science', *Science*, 156 (May 5, 1967), 611–22.

Bell, Evans, *Memoir of General John Briggs, of the Madras Army with Comments on Some of His Words and Work* (London, Chatto and Windus, 1885).

Bernard, Claude, *Principes de médecine expérimentale* (Paris, Quadrige, PUF, 2008).

Bonakdarian, Mansour, *Britain and the Iranian Constitutional Revolution of 1906–11: Foreign Policy, Imperialism and Dissent* (Syracuse: Syracuse University Press in association with the Iran Heritage Foundation, 2006).

Braunstein, Jean-François, *Broussais et le matérialisme: médecine et philosophie au XIXe siècle* (Paris, Méridiens Klincksieck, 1986).

Browne, Edward G., *Arabian Medicine* (Cambridge, Cambridge University Press, 1926).

Cabanis, Pierre-Jean-Georges, *Coup d'oeil sur les revolutions et la réforme de la médecine* (Paris, Crapart, Caille et Ravier, 1804).

Calmard, Jean, 'Feuvrier,' *Encyclopaedia Iranica*.

Canguilhem, Georges, *Le normal et le pathologique* (Paris, Quadrige, 2005, first Edition, 1934).

Chardin, Jean, *Voyages du chevalier Chardin en Perse et autres lieux de l'Orient*, vol. 2, La Perse et les Persans (Paris, M. Dreyfous, 1883).

Chomel, August François, *Elements de pathologie générale* (Paris, Crochard & Gabon, 1817).

Clot, Bey, Barthélémy, *Relation de l'épidémie de choléra qui a régné en 1831, en Arabie et en Egypte* (Paris, Victor Masson & Fils, undated).

Cohn, Bernard S., *Colonialism and its Forms of Knowledge: The British in India* (Princeton, Princeton University Press, 1996).

Collins, E. Treacher, *In the Kingdom of the Shah* (London, T. Fisher Unwin, 1896).

Comptes-Rendus des séances de la Société d'Ethnographie, tom premier (Paris, Challamel Ainé, Editeur, Librairie de la Société d'Ethnographie, 1860).

Conrad L. and Wujastyk D. (eds), *Contagion: Perspectives from Pre-Modern Societies* (Aldershot, Burlington USA, Singapore, Sydney, Ashgate, 2000).

Cooter, R., Harrison, M. and Sturdy S. (eds), *War, Medicine and Modernity* (Stroud, Sutton, 1998).

Daniel Balland, 'Cholera in Afghanistan', *Encyclopaedia Iranica* (online source, Access April 21, 2010).

de Croizier, Edme-Casimir, *La Perse et les Persans* (Paris, E. Dentu, 1873).

Dehkhoda, ʿAli Akbar, *Loghatnâmeh*, 15 vols., edited by M. Moʿin & Jaʿfar Shahidi, second edition (Tehran, mo'assesseh-ye enteshârât va chap-e dâneshgâh-e Tehran, 1377/1998).

de Khanikof, Nicolas, *Mémoire sur la partie méridionale de l'Asie Centrale* (Paris, de L. Martinet, 1861).

de Planhol, Xavier, 'Cholera in Persia', *Encyclopaedia Iranica* (online source, Access April 21, 2010).

Delaporte, François, *Le Savoir de la maladie: Essai sur le choléra de 1832 à Paris* (Paris, PUF, 1990).

Dieulafoy, Mme Jane, *La Perse, la Chaldée et la Susiane, Relation de voyage* (Paris, Librairie Hachette, 1887).

Djalal Khaleghi-Motlagh, 'Borzuya', *Encyclopedia Iranica*, online access: October 11, 2010.

Dölen, E. *Türkiye Üniversite Tarihi, 1: Osmanlı Döneminde Darülfünun (1863–1922)*, [*A History of the University in Turkey, 1, The Darülfunun in the Ottoman Era, 1863–1922*] Ýstanbul, Ýstanbul Bilgi Üniv. yay., 2009, pp. 7–21. On the influence of Istanbul's intellectual environment on Qâjâr reformists, see Mohammad Fazlhashemi, 'Istanbul's Intellectual Environment and Iranian Scholars of the Early Modern Period', in E. Özdagla et al. (eds), *Istanbul as Seen From a Distance: Centre and Provinces in the Ottoman Empire*, Istanbul, Swedish Research Institute Transactions, 2010, vol. 20.

Dols, Michael, *Medieval Islamic Medicine. Ibn Ridwân's Treatise 'On the Prevention of Bodily Ills in Egypt'* Translated by Dols M., Arabic Text edited by Adil S. Gamal (Berkeley, Los Angeles, London, University of California Press, 1984).

Ebn-e Sinâ, *Qânun dar tebb* (Canon of medicine), volume 1, translated by ʿAbdolrahmân Sharafkandi, edited by A. Pakdaman, M.-R. Ghaffâri, H. ʿErfâni, Tehran University Press, Tehran, 1357/1978.

Ebrahimnejad Hormoz, KâZem Rašti, *Encyclopaedia Iranica*, vol. 15.

Ebrahimnejad H., *Pouvoir et succession sous les premiers Qâjârs: 1726–1834* (Paris, L'Harmattan, Société d'Histoire de l'Orient, 1999).

Ebrahimnejad, H., 'Un traité d'épidémiologie de la médecine traditionnelle persane: *Mofaraq ol-heyze va'l-wabâ* (De la différence entre *heyze* and *wabâ*),' *Studia Iranica*, 27, (1998), 83–107.

Ebrahimnejad, Hormoz, 'Medicine in Islam or Islamic Medicine,' in Mark Jackson (ed.), *The Oxford Handbook of the History of Medicine* (Oxford, New York, OUP, 2011).

Ebrahimnejad, Hormoz, 'The waqf the state and medical education in the nineteenth century', in Ebrahimnejad, H. (ed.), *The Development of Medicine in Non-Western Countries* (London, New York, Royal Asiatic Society Books, Routledge, 2009).

Ebrahimnejad, Hormoz, *Medicine, Public Health and the Qâjâr State* (Leiden, Boston, Brill, 2004).

Ekhtiar, Maryam, *Modern Science, Education and Reform in Qâjâr Iran: the Dar Al-Funun* (London, Taylor & Francis Group, 2003).

Elgood, C., *Safavid Surgery* (Oxford, London, Edinburgh, New York, Pergamon Press, 1966).

Elgood, Cyril, 'Tibb-ul-Nabbi or Medicine of the Prophet', *Osiris*, 14 (1962), 33–192.

Elgood, Cyril, *A Medical History of Persia and the Eastern Caliphate from the earliest times until the year A.D. 1932* (Cambridge, CUP, 1951).

Eqbal-e Ashtiyâni, Abbas, *Mirzâ Taqi-Khân Amir-Kabir*, edited by Iraj Afshar (Tehran Enteshârât-e Tus, 1363/1884).

Eqbâl-e Ashtiyâni, ᶜAbbas, *Mirzâ Taqi-Khân Amir-Kabir*, edited by Iraj Afshar (Tehran, Tus, 1363/1984).

Eᶜtemâd al-Saltaneh, *Al-Mʾâser w'al-âsâr*, Tehran, Government Press, lithographed, 1888–1889).

Eᶜtemâd al-Saltaneh and Mohammad-Hassan-Khân, *Ruznâmeh ye Khâterât* (Diary), edited by Iran Afshar (Tehran, Amir-Kabir, 1385/2006).

Eᶜtemâd al-Saltaneh, *Tārikhe montazam-e nâsseri*, edited by Mohammad-Esmâᶜil Rezvâni, 3 vols, Donyâ-ye ketâb, 1363–1367/1984–1988).

Fahmy, Khaled, 'The Anatomy of Justice: Forensic Medicine and Criminal Law in Nineteenth-Century Egypt', *Islamic Law and Society*, 6 (1999), 224–271.

Fahmy, Khaled, 'The Anatomy of Justice: Forensic Medicine and Criminal Law in Nineteenth-Century Egypt', *Islamic Law and Society*, 6 (1999), 224–271.

Fahmy, Khaled, 'The Police and the People in Nineteenth-Century Egypt', *Die Welt des Islams*, 39 (3) (1999), 340–377.

Fahmy, Khaled, *All the Pasha's Men: Mehmed Ali, his Army and the Making of Modern Egypt* (Cambridge, Cambridge University Press, 1997).

Farmânfarmâiyân, Mehrmâh, *Zendeginâmeh-ye ᶜAbdol-Hossein Mirzâ Farmânfarmâ*, 2 volumes (Tehran, Toos, 1382/2003).

Farmânfarmâyân, Hâfez (ed.), *Khâterât-e siyâsi-ye Mirzâ ᶜAli-Khân Amin al-Dowleh* (Tehran, Amir-Kabir, 1355/1976).

Fassâyee, *Mirzâ Hassan Hosseini, Fârsnâmeh-ye Nâsseri* (*History and Geography of Fars*), edited by Mnsur Rastegâr Fassâyee, 2 volumes (Theran, Amir Kabir, 1282/2003).

Fereydoun-e Adamiyat, *Andisheh-ye taraqqi va hokumat-e qânun: ᶜasr-e Sepahsâlâr (The Politics of Reform in Iran: 1858–1880)* (Tehran, Khwârazmi, 1385/2006, first publication 1351/1972).

Feuvrier, Jean Baptist, *Trois ans à la cour de Perse* (Paris, F. Juven, second Edition, 1906).

Fleming, Donald, 'Science in Australia, Canada, and the United States: Some Comparative Remarks', *Proceedings of the 10th International Congress of the History of Science*, 1 (1962), 189–196.

Floor, Willem and Faghfoory, M. H., *The First Dutch-Persian Commercial Conflict* (California, Mazda, 2004).

Floor, Willem, 'The Art of Smoking in Iran and Other Uses of Tobacco', *Iranian Studies*, 35 (1–3), (Winter/Spring/Summer 2002), 47–85.

Floor, Willem, *Public Health in Qâjâr Iran* (Washington DC, Mage, 2004).

Floor, Willem, *The Persian Gulf: A Political and Economic History of Five Port Cities 1500–1730* (Washington DC, Mage, 2006).

Freyer, John A.,*New Account of East India and Persia in Eight Letters, begun 1672 and finished 1861* (London, R. Chiswell, 1698).

Gazette médicale d'orient, VIIIème année, no. 2, Mai 1864.

Gascoigne, John, *Science in the Service of Empire, Joseph Banks, the British State and the Uses of Science in the Age of Revolution* (Cambridge, CUP, 1998).

Ghaly, Mohammed, 'The Beginning of Human Life: Islamic Bioethical Perspectives,' *Zygon: Journal of Religion and Science*, 47 (1) (2012), 175–213.

Gheisari, Yousof, Baharvand, Hossein, Nayernia, Karim and Vasei, Mohammad, 'Stem Cell and Tissue Engineering Research in the Islamic Republic of Iran', *Stem Cell Rev and Rep*, 8, (2012), 629–639.

Gilmour, John, *Report on and Investigation into the Sanitary Conditions in Persia* (League of Nations, Geneva, 1925).

Gobineau, Arthur de, *Trois ans en Asie*, in *Œuvres*, II, edited by J. Gaulmier, P. Lésétieux, V. Monteil (Paris, Gallimard, 1983).

Gobineau, Comte Artur de, *Les Religions et les Philosophies dans l'Asie centrale*, in *Œuvres*, II. edited by J. Gaulmier, P. Lésétieux, V. Monteil (Paris, Gallimard, 1983).

Goddard, Hugh, *Muslim Perceptions of Christianity* (London, Grey Deal Books, 1996).

Golinski, Jan, *Making Natural Knowledge: Constructivism and the History of Science* (Cambridge, CUP, 1998).

Good Byron, Good, Mary-Jo DelVecchio, 'The Comparative Study of Greco-Islamic Medicine: The Integration of Medical Knowledge into Local Symbolic Contexts', in Leslie C. and Young A. (eds), *Paths to Asian Medical Knowledge* (Berkeley, Los Angeles, Oxford, University of California Press, 1992), pp. 257–271.

Goubert, Jean-Pierre, *La conquête de l'eau: L'avènement de la santé à l'âge industriel* (Paris, R. Laffont, 1986).

Grisolle, Augustin, *Traité élémentaire et pratique de pathologie interne* (3rd edition, 2 vols., Paris, V. Masson, 1848); idem, *Traité élémentaire et pratique de pathologie interne, cinquième édition, considérablement augmentée*, tom premier (Paris, Victor Masson, 1852).

Guerrini, Anita, *Experimenting with Humans and Animals From Galen to Animal Rights* (Baltimore and London, John Hopkins University Press, 2003).

Günergun, F. and Etker, S. '*Waqf* endowments and the emergence of modern charitable hospitals in the Ottoman Empire', in Ebrahimnejad, H. (ed.), *The Development of Modern Medicine in Non-Western Countries*, pp. 82–107.

Gurney, J. and Nabavi, G., 'Dâr al-Fonun', *Encyclopaedia Iranica* (online access, August 24, 2011).

Gutas, D., *Greek Thought, Arabic Culture* (London, New York, Routledge, 1998).

Hamlin, Christopher, 'Predisposing Causes and Public Health in Early Nineteenth-Century Medical Thought', *Social History of Medicine*, 5 (1992), 43–70.
Harley, David, 'Rhetoric and the Social Construction of Sickness and Healing', *Social History of Medicine*, 12, no. 3, pp. 407–435.
Harrison, Mark, 'Science and the British Empire', *ISIS*, 96 (2005), 56–63.
Harrison, Mark, *Climate and Constitutions: Health, Race, Environment and British Imperialism in India, 1600–1850* (Delhi, OUP, 1999).
Harrison, Mark, *Public Health in British India: Anglo-Indian Preventive Medicine 1859–1914* (Cambridge, Cambridge University Press, 1994).
Harrison, Mark, 'Medical experimentation in British India: The Case of Dr Helenus Scott,' in H. Ebrahimnejad (ed.), *The Development of Modern Medicine in Non-Western Countries*, 2009, pp. 23–41.
Haskell, Isaacs, 'European Influences in Islamic Medicine', *Mashriq: Proceedings of the Eastern Mediterranean Seminar*, University of Manchester 1977–1978 (Manchester, 1980).
Hawthorne, George Stuart, *The True Pathological Nature of Cholera and an Infallible Method of Treating It* (New York, W.H. Graham, 1849).
Heaman, E.A., 'The Rise and Fall of Anticontagionism in France', *CBMH/BCHM*, 12, (1995), 3–25.
Hippocratic Writings, Lloyd, G.E.R. (ed.) Chadwick, J., trans. (London, Penguin Books, 1987).
Hodge, Joseph M., 'Science and Empire: An Overview of the Historical Scholarship', in Bennett, Brett M. and Hodge, J. M. (eds), *Science and Empire: Knowledge and Networks of Science across the British Empire, 1800–1970 (Britain and the World)* (Basingstoke, Palgrave Macmillan, 2011).
Huff, Toby, *The Rise of Early Modern Science: Islam, China, and the West* (Cambridge, CUP, 1993).
Hume, John, 'Rival traditions: Western Medicine and Unani Tibb in the Punjab, 1849–1889,' *Bulletin of the History of Medicine*, 51, 2 (1977), 214–231.
Ibn Qayyim al-Jawziyya, *Medicine of the Prophet*, translated by P. Johnstone, see preface by Seyyed Hossein Nasr and Introduction of Johnstone.
Inclosure in No. 249, Legation of the US, May 28, 1896.
Iqbâl, Muzaffar, *Islam and Science* (Burlington, VT, Ashgate, 2002).
Iradj Amini, *Napoleon and Persia: Franco-Persian relations under the First Empire* (Washington, D.C.: Mage, 1999).
Iskandar, A.Z., *Arabic Manuscripts on Medicine and Science* (London, I.B. Tauris, 2000).
Issawi, C., *The Economic History of Iran 1800–1914* (Chicago and London, The University of Chicago Press, 1971).
Jackson, A. V. W., *From Constantinople to the Home of Omar Khayyam* (New York, Macmillan, 1911, AMS edition published in 1975).
Jalâlu'd-Din Abd'ur-Rahman As-Suyuti, *Tibb-ul-Nabbi* or *Medicine of the Prophet*, Translation and edition by C. Elgood, *Osiris*, 14 (1962), 33–192, p. 3;

Jalilvad, Hâmed 'Bozorgân-e musiqi-ye Iran: Aref-e Mashruteh,' online source, www.persianpersia.com, access: November 26, 2011.
Jaubert, Amédé, *Voyage en Arménie et en Perse faits dans les années 1805 et 1806* (Paris, E. Ducrocq, 1860).
Jordanova, Ludmilla, 'The Social Construction of Medical Knowledge', *Social History of Medicine*, 8 (3) (1995): 361–381.
Jorjâni, Esmâ'il b. Hassan b. Mohammad al-Hosseini, *Al-aghrâz al-tebbiyyeh v'al-mabâhes al-'alâiyyeh* (Goals of medicine and discussions of âlâ'i), edited by Hassan Tajbakhsh, 2 volumes (Tehran, Tehran University Press, 1388/2001).
Kapil Raj, *Relocating Modern Science: Circulation and the Construction of Knowledge in South Asia and Europe* (Basingstoke, New York, Palgrave Macmillan, 2007).
Kashani-Sabet, Firuzeh, 'The Politics of Reproduction: Maternalism and Women's Hygiene in *Iran*, 1896–1941', *International Journal of Middle East Studies*; 38 (2006): 1–29.
Kashani-Sabet, Firuzeh, *Conceiving Citizens: Women and the Politics of Motherhood in Iran* (Oxford, OUP, 2011).
Kashani-Sabet, Firuzeh, *Frontier Fictions, Shaping the Iranian Nation, 1804–1946* (Princeton: Princeton University Press, 1999).
Kasravi, A., *Târikh-e Mashruteh-ye Iran* (Tehran, Amir-Kabir, 1378/1999).
Katouzian, Homa, *State and Society in Iran. The Eclipse of the Qâjârs and the Emergence of the Pahlavis* (London, New York, I.B. Tauris, 2006)
Khârâbi, Fâruq, *Anjomanhâ-ye 'asr-e mashruteh*, Tehran, Mo'assesseh ye tahqiqât (Research Institute) 1386/2007).
Khodâdâdi, Jamshid, *Kelid-e vorud be tebb-e qadim* (Key to the traditional (ancient) medicine), Tehran: Shahr Press, 1389/2010).
Kuhn, Thomas S., *The Structure of Scientific Revolutions* (Chicago, London, University of Chicago Press, 1996).
Kumar, Deepak, 'Probing History of Medicine and Public Health in India', *Indian Historical Review*, 37 (2) (2010), 259–273.
Kumar, D., *Science and the Raj 1857–1905* (Delhi, OUP, 1995).
Laget, Pierre-Louis, 'Les Lazarets et l'émergence de nouvelles maladies pestilentielles au XIXe et au début du XXe siècle', online document (visited October 31, 2007).
Leiser, Gary, 'Medical Education in Islamic Lands from the Seventh to the Fourteenth Century', *The Journal of the History of Medicine and Allied Sciences*, 38 (1983), 48–75.
Léonard, Jacques, *La France médicale au XIXe siècle* (Paris, 2nd éditions, Gallimard/Julliard, 1978.).
Lester, Alan, 'Imperial Circuits and Networks: Geographies of the British Empire', *History Compass* 4 (1) (2006).
Lewis, B., *What Went Wrong? Western Impact and Middle Eastern Response* (London: Phoenix, 2003).
Libenskind, Claudia, 'Arguing Science: Unani Tibb, Hakims and Biomedicine in India, 1900–50', in Ernst, W. (ed.), *Plural Medicine: Tradition and Modernity, 1800–2000* (London, New York, Routledge, 2002), pp. 58–75.

MacLeod Roy, Lewis M. (eds), *Disease, Medicine and Empire* (London, Routledge, 1988).

MacLeod, R., 'On Visiting the Moving Metropolis: Reflections on the Architecture of Imperial Science', *Historical Records of Australian Science*, 5 (1982), 1–16.

MacLeod, R., *Technologies and the Raj: Western Technical Transfers to India 1700–1947* (New Delhi, Sage, 1995).

MacLeod, R., 'Scientific Advice for British India: Imperial Perceptions and Administrative Goals, 1898–1923', *Modern Asian Studies*, 9 (3), (1975), 344–358.

MacLeod, Roy, 'On Visiting the Moving Metropolis: Reflections on the Architecture of Imperial Science', Reingold, Nathan, Rothenberg, M. (eds), *Scientific Colonialism: A Cross-Cultural Comparison* (Washington, Smithsonian Institution Press, 1987).

Magner, Lois I., *A History of Medicine* (Boca Raton, FL, London, New York, Singapore: Taylor & Francis, 2005).

Mahbubi Ardakâni, Hossein, *Târikh-e mo'assessât-e tamaddoni-ye jadid dar Iran* (*History of Modern Institutions in Iran*), 2 vols. (Tehran, Tehran University Press, 1354/1975).

Mahdavi, Shirin, 'Andrew Jukes, British East India Company Surgeon and Political Agent (1774–1821)', *Encyclopaedia Iranica*, Vol. XV, Fasc. 2: 216–217.

Mahdavi, Shirin, 'Introduction of Western Medicine to Iran', *Encyclopaedia Iranica*. Online source.

Mahdavi, Shirin, 'Shahs, Doctors, Diplomats and Missionaries in 19th Century Iran', *British Journal of Middle Eastern Studies*, 32 (2) (2005), 169–191.

Majd, Mohammad-Gholi, *The Great Famine and Genocide in Persia, 1917–1919* (Lanham, Maryland: University Press of America, 2003).

Makdisi, George, *The Rise of Islamic College; Institutions of Learning in Islam and the West* (Edinburgh, Edinburgh University Press, 1981).

Malekzâdeh, Elhâm., *Negâhi be omur-e kheiriyeh*, pp. 53–54.

Malekzâdeh, Elhâm, *Negâhi be omur-e kheiriyeh dar dowreh-ye Qâjâr* (Tehran, Dâneshgâh-e Azâd, Shahr-e Rey, 1385/2006).

Manton, John, 'Making Modernity with Medicine: Mission, State and Community in Leprosy Control, Ogoja, Nigeria, 1945–50,' in H. Ebrahimnejad (ed.), *The Development of Modern Medicine in Non-Western Countries*, pp. 160–183.

Masnawi-ye Ma^cnawi, Book I. p. 1. Digitized source.

Matthee, Rudolph, *Persia in crisis: Safavid decline and the fall of Ispahan* (London, I.B. Tauris, 2010.).

Mehdi-Qoli Hedâyat Mokhber al-Saltaneh, *Khâterât va Khatarât* (Tehran, Zawwâr Publishers, 1385/2006) Mir, Mohammad Taqi, *Pezeshkân-e nâmi-ye fârs* (Shirâz, enteshârat-e dâneshgâh-e Shirâz, 2nd edition, 1263/1684).

Mirsepassi, Ali, *Democracy in Modern Iran: Islam, Culture, and Political Change* (New York, NYU Press, 2010).

Mirzâ Aqâ-Khân-e Kermâni, 'Ey Jalâl al-Dowleh', in *Seh maktub (Three writings)*, edited by Bahram-e Chubineh (Lindenallee, Essen (Germany), Nima Verlag, 2000).

Mirzâ Qahramân Amin-Lashkar, *Ruznâmeh-ye Safar-e Khorâssân*, edited by Iraj Afshar & M.-Rasul Daryâgasht (Tehran, Asâtir, 1374/1995).

Mohammad-Pâdeshâh (Shâd), *Farhang-e Anendrâj*, 1882, edited by Mohammad Dabir Siyâqi, 1335/1956, 7, p. 4471.

Mohit-e Tabâtabâ'i, M. 'Dâr al-Fonun va amir-Kabir', in Rowshani Zaᶜferânlu, Qodratollah, *Amir-Kabir va Dâr al-Fonun* (Tehran, Enteshârât-e ketâbkhâneh-ye markazi, 1354/1975).

Mollâ Rezâ Hamedâni, 'Resâleh-ye jahâdiyyeh-e Hâji Mollâ Rezâ-ye Hamedâni (died 1247/1831),' see http://www.hawzah.net/en/ArticleView.html?ArticleID= 81882 (access date, 19/04/2012).

Moradi, Mohammad 'Ganj-anduzi va habs-e mâl,' *Faslnâmeh-ye waqf mirâs-e jâvidân*, no. 71.

Morier, James, *A Journey Through Persia, Armenia, and Asia Minor, to Constantinople in the years 1808 and 1809; in which is included, some account of the proceedings of His Majesty's Mission under Sir Harford Jones, Bart. K.C. to the court of the King of Persia* (Philadelphia, Boston, C. Palmer, 1816).

Morier, James, *Sargozasht-e hâji bâbâ esfahâni*[translation of *The adventures of Hajji Baba*, by Mirzâ Habib Esfahâni] (Tehran, Mo'asseseh-ye fahangi, honari, sinemâyee Elset fardâ, 1378/1999).

Morier, James, *The Adventures of Hajji Baba of Ispahan*, edited by C.J. Wills (London, Lawrence and Bullen, 1897).

Mostowfi, ᶜAbdollâh, *Sharh-e zendegâni-ye man yâ târikh-e ejtemâ ᶜI va edâri-ye dowreh-ye qâjâriyeh*, 3 vols. (Tehrân, Ketâbforushi MohammadᶜAli ᶜElmi, 1324/1945).

Moulin, Anne Marie, 'Disease transmission in nineteenth-century Egypt,' in Ebrahimnejad, H. (ed.), *The Development of Modern Medicine*, op. cit.

Moulin, Anne Marie, and Chuvin, Pierre, *Lady May Montagu L'Islam au peril des femmes*, second edition (Paris, La Découvert, 2000).

Moulin, Anne Marie, *Le médecin du prince: voyage à travers les cultures* (Paris, Odile Jacob, 2010).

Moulin, Anne Marie, 'L'esprit et la lettre de la modernité égyptienne. L'enseignement médical de Clot bey', in Panzac, D. and Raymond, A. (eds), *La France et l'Egypte à l'époque des vice-rois 1805–1882* (Paris, IFAO, Cahier des Annales islamologiques, 22 (2002), 119–134.

Moulin, Anne Marie, 'Révolutions médicales et politiques en Egypte (1865–1917)', *Revue de l'Occident musulman et de la Méditerranée*, 52 (1989), 111–123.

Moulin, Anne Marie, 'The Construction of Disease Transmission in Nineteenth-Century Egypt', in Ebrahimnejad H. (ed.), *The Development of Modern Medicine in Non-Western Countries*, pp. 42–58.

Moulin, Ann Marie, 'Introduction: Se repérer dans le labyrinthe du corps et de l'Histoire' in Moulin, A.M. (ed.), *Le Labyrinth du corps* (Paris, Editions Karthala, 2012).

Nafisi, Saᶜid, 'Doktor ᶜAli-Akbar-Khân-e Nafisi Nâzem al-Atebbâ (1263–1342/1847–1924)', *Majalleh-ye Yâdegâr*, (4): 52–53.

Naficy, Abbas, *La médecine en Perse* (Paris, Les éditions Véga, 1933).

Najmabadi, Mahmud, 'Les relations médicales entre la Grand Bretagne et l'Iran et les médecins anglais serviteurs de la médecine contemporaine de l'Iran', *Proceedings of the XXIII International Congress of the History of Medicine*, Wellcome Institute, London, 1972: 704–708, p. 705.

Najmi, Nâsser, *Terhrân-e ᶜahd-e nâsseri* (Tehran, ᶜAttâr Publishers, 1364/1985)

Nâsser al-Din Shah, *Ruznâmeh-ye khâterât dar safari avval-e farangestân*, edited by Fatemeh Qâzihâ (Tehran, Sâzmâne asnâd, 1377/1998)

Nâsser al-Din Shah, *Ruznâmeh-ye khâterât-e Nâsser al-Din-Shah dar safar-e dovvome farangestân*, edited by Fâtemeh Qâzihâ (Terhan, Sâzmâne asnâd, 1379/2000)

Nâsser al-Din Shah, *Ruznâmeh-ye khâterât*, edited by Abdolhossein Navâyee & Elhâm Malekzâdeh (Tehran, Sâzmân-e asnâd Publishers, 1384/2005)

Nâsseri, Mohsen, 'Dr Ahmadiyeh and clinical traditional medicine in Iran', in *Yâdnâmeh ye Hakim Doktor Abdollâh-Khân-e Ahmadiyeh*, published by the National Museum of the History of Medical Sciences, 1381/2002)

Natchkebia, Irène, 'Jean-Baptist-Louis-Jacques Rousseau's Manuscript: 'Tableau general de la Perse modern (1807)': Military possibilities and Trades of Persia in the Context of the Indian Expedition' (unpublished manuscript).

Nategh, Homa, *Iran dar râhyâbi-ye farhangi 1834–1848 (Iran and cultural enlightenment)* (London, Payamco, 1988).

Nategh, Homa, *Kârnâmeh-ye farhangi-ye farangi dar Iran (Les français en Perse. Les écoles religieuses et séculières (1837–1921)* (Paris, Khâvarân, 1375/1996).

Nikpour, M., Ghaffârinejad, A., *Pishineh-ye Pezeshki-ye Kerman dar sadeh-ye akhir (1280–1377)* (Tehran, Sâzmân-e Asnâd-e Melli-ye Iran, 1377/1998).

Nutton, Vivian, 'Did the Greeks Have a Word For It? Contagion and Contagion Theory in Classical Antiquity', in Conrad, L. I., Wujastyk, D. (eds), *Contagion. Perspectives from Pre-Modern Societies* (Aldershot, Ashgate, 2000), pp. 137–162.

Oberlin, Charles, *Barnâmeh-ye Behdâsht bara-ye Iran (A program for health preservation [public health] in Iran)*, translated from French (Tehran, Dânesh Publishers, 1320/1941).

Olivier, G.A., *Voyage dans l'empire Ottoman, l'Egypte et la Perse, fait par ordre du gouvernement, pendant les six premières années de la République [1792–98]* (3 vols. Paris, H. Agasse, 1807).

Omidiyâni, S. Hossein, 'Gâhanbâr dar Iran-e bâstân', *Waqf Mirâs-e Jâvidân*, no. 3, Tehran Autumn 1372/1993.

ᶜOnsor al-Maᶜâli Kaykâvus b. Eskandar, *Gozideh-ye Qâbus nâma*, ed. Gholâmhossein Yusofi, Tehran, enteshârât-e ᶜelmi, 1382/2003).

Pahlavi, Farah, *Mémoires* (Paris, Edition J'Ai Lu, 2003).

Parsons, Talcott, *Structure of Social Action* (New York, Free Press, 1937).

Parsons, Talcott, 'The Place of Ultimate Values in Sociological Theory', *International Journal of Ethics*, 45 (3) (April 1935), 282–316.

Plessis, Jean-Louis and Théodoridès, Jean, 'Tholozan: médecin militaire à compétence étendue', *Histoire des sciences médicales*, tom XXXII (3) (1998), 279–286.

Polak, J-E., *Safarnâmeh-ye Polak, Iran va Iranian*, Persian translation of *Persien, das Land und Seine Bewohner*, by Keykâviis Jahângiri (Tehran, Khwârazmi, 1361/1982).

Pormann, P. E., Savage-Smith, E., *Medieval Islamic Medicine* (Edinburgh, Edinburgh University Press, 2007).

Porter, R., *The Greatest Benefit to Mankind: A Medical History of Humanity from Antiquity to the Present* (London, Harper Collins, 1997), pp. 313–314.

Qâem-Maqâmi, Jahângir, 'Ravâbet-e nezâmi-ye iran va faranseh dar dowreh-ye Safaviyeh' (The Military relationship between Iran and France under the Safavids), *Majalleh-ye barrasihâye târikhi, Journal of the Army*, nos. 1–2 (1345/1966), 106–124.

Qezel Ayâq, Dr Hossein ʿAli, *Hefz al-sehheh ya kelid-e sehhat (Preservation of health or key to health)* (Tehran, Ferdowsi, 1307/1928).

Qodrat ollâh Rowshani Zaʿferânlu (ed.), *Amir-Kabir va Dâr al-Fonun*, the proceedings of a conference at the University of Tehran (*Enteshârât-e ketâbkhâneh-ye markazi va markaz- asnâd*, 1354/1975).

Ragab, A., 'Le Bimâristân al-Mansuri: L'institution dans le contexte politique et social mamelouk (1285–1390)', PhD thesis, Paris, 2012.

Rémond, René, *L'anticléricalisme en France* (Paris, Editions Complexe, 1985).

Rezâ-Qoli-Khân-e Hedâyat, *Rowzat al-Safâ-ye Nâsseri*, vol. 10

Rezâyee, Omid, 'Bimârestân-e Najmiyeh', *Mirâs-e Jâvidân*, year 7, no. 4.

Rezwâni, Mohammad-Esmâʿil (ed.), *Gozideh-ye farmânhâ* (Royal decrees) from the collections of the National Library, Iran, with the collaboration of H Saidi, Z. Haddâd, M. Vedâdi (Tehran, National Library Publications, 1375/1996)

Richard, Francis, *Raphaël du Mans missionnaire en Perse au XVIIe s.* 2 vol. (Paris, Socieìteì d'Histoire de l'Orient, Editions l'Harmattan, 1995).

Ringer, M., *Education, Religion, and the Discourse of Cultural Reform in Qâjâr Iran* (California, Mazda, 2001).

Robert E. Speer, *The Hakim Sahib, The Foreign Doctor, A Biography of Joseph Plumb Cochran* (New York, London, Flemming H. Revell Company, 1911).

Rochard, Philippe, 'Le sport antique des zurkhâne de Téhéran: formes et significations d'une pratique contemporaine', PhD thesis, Université Aix/Marseille, 2000).

Rochard, Philippe, Chehabi, H. E., 'The Identities of the Iranian Zurkhanah,' *Iranian Studies*, 35 (4), 313–340

Rostam al-Hokamâ, *Rostam al-Tavârikh*, edited by Mitra Mehrâbâdi (Tehran, Donyâ-ye Ketâb, 1382/2003).

Rousseau, Jean-Baptiste, *Lettres sur la contagion du choléra-morbus de l'Inde* (Épernay, Noël-Boucart, 1866). [Lettres au journal L'Union médicale, août 28, 1849, et au Ministre du Commerce, octobre 19, 1865 'A Son Excellence M. Le Ministre du Commerce et des Travaux Publics'].

Rustâyee, Mohsen, *Târikh-e teb va tebâbat dar Iran*, 2 volumes (Tehran, Sâzmân-e asnâd-e ketâbkhâneh-ye melli, 1382/2003).

Rustâyee, M., 'Aʿlam al-Dowleh, Khalil-Khân-e Saqafi, tabib va nâji-ye mashrutiyyat', *Ganjineh-ye asnâd*, (44) (Winter 1380/2001), 58–79.

Ruznâmeh-ye khâterât-e (the Diary of) *Eʿtemâd al-Saltaneh* (Tehran, Amir Kabir, 2006).

Sabra, Adam, *Poverty and charity in medieval Islam: Mamluk Egypt, 1250–1517* (Cambridge, Cambridge University Press, 1999).

Sadiq, Isâ, *Modern Persia and Her Educational System* (New York, Teachers College, Columbia University, 1931).

Sajjadi, Dentistry, *Encyclopaedia Iranica*, 7, 292–296.

Saniei, M., and Vries, R. D., 'Embryo in Stem Cell Research in Iran; Status and Ethics', *Indian Journal of Medical Ethics*, 5 (2008), 181–184.

Saniei, Mansureh, 'Human Embryonic Stem Cell Research in Iran: The Role of the Islamic Context', *Scripted*, 7 (2) (August 2010), 324–334, p. 330.

Sardâriniâ, Samad, *Dâr al-Fonun-e Tabriz* (Tabriz, Nedâ-ye shams 1382/2003).

Savâqeb, Jahanbakhsh, 'Abʿâd-e farhangi-ye emtiyâzenâmeh-ye tanbâku, *Meshmât* (62–65), (1978/1999), 339–360.

Savage-Smith, Emilie, 'Attitude Towards Dissection in Medieval Islam', *Journal of the History of Medicine and Allied Sciences*, 50, 67–110.

Schacht J., Meyerhof, M., *The Medico-Philosophical Controversy Between Ibn Butlan of Baghdad and Ibn Ridwan of Cairo. A Contribution to the History of Greek Learning Amongst the Arabs* (Cairo, 1937).

Schayegh, Cyrus, *Who Is Knowledgeable Is Strong, Science, Class, and the Formation of Modern Iranian Society, 1900–1950* (Berkeley, Los Angeles, London, University of California Press, 2009).

Schneider, J. 'La médicine persane. Les médecins français en Perse: leur influence', *Bulletin de l'Union Franco-Persane*, deuxième année, no. 5, décembre 1910–janvier 1911.

Schneider, J., 'La médecine Persane', *Bulletin de l'Union Franco-Persan*, première année, no. 4, octobre–novembre 1910.

Scoutetten, Henry, *Histoire chronologique, topographique et étymologique du choléra depuis la haute antiquité jusqu'à son invasion en France en 1832* (Paris, Victor Masson et Fils, 1869).

Sepehr, Mohammad-Taqi Lesân al-Molk, *Nâssekh al-Tavârikh* (NT), 3 vols., edited by Jamshid Kiyânfar (Tehran, Asâtir, 1377/1998).

Seyf al-din Mohammad Jaʿfar Astarâbâdi, *Safineh-ye Nuh*, translated and edited in H. Ebrahimnejad, 'Religion and Medicine in Qâjâr Iran', in: R. Gleave (ed.), *Religion and Society in Qâjâr Iran* (London, Routledge, 2004), pp. 401–428.

Seyyed Mostafa Matbaʿeh-chi Esfehâni, 'Negâhi be waqf-shenâsi-ye faqihân (2)' *Faslnâmeh-ye waqf mirâs-e jâvidân*, no. 72.

Sheikholeslami, A-Reza, *The Structure of Central Authority in Qâjâr Iran 1871–1896* (Atlanta, Georgia, Scholar Press, 1997).

Shiffoleau, Silvia, 'Genèse de la coopération sanitaire internationale. L'Europe et l'Orient face aux épidémies (1825–1938)', HDR dissertation, 2010.

Snow, J. *Snow on Cholera, being a Reprint of Two Papers* (London, Oxford University Press, 1936).

Snow, John, *Snow on Cholera: Being a Reprint of Two Papers by John Snow, M.D., Together With a Biographical Memoir by B.W. Richardson and an Introduction by*

Wade Hampton Frost (New York, The Commonwealth Fund; London, H. Milford; Oxford, UOP, 1936.)

Spronk, Maurice, 'La question Persane', *Bulletin de l'Union Franco-Persane*, 2e année, no. 5, décembre 1910–janvier 1911.

Stearns, J. K., *Infectious Ideas: Contagion in Premodern Islamic and Christian Thought in the Western Mediterranean* (Baltimore, The John Hopkins University Press, 2011).

Stenberg, Leif, *The Islamization of science: four Muslim positions developing an Islamic modernity* (Lund: Religionshistorika avdelningen, Lunds universitet, 1996).

Tajbakhsh, Hassan, *Tâtikh-e bimârestânhâ-ye Iran, History of hospitals in Iran* (Tehran, Pajuheshgâh-e ʿolum-e ensâni va motâleʿât-e farhangi, 1379/2000).

Tâjbakhsh, Hassan, *Tărikh-e bimârestânhâ-ye Iran*, pajuheshgâh-e ʿolum-e ensâni va motâleʿât-e farhangi, Tehran, 1379/2000).

Tancoigne, J. M., *Lettres sur la Perse et la Turquie d'Asie* (Paris, Nepveu, Libraire, Passage des Panoramas, 1819), 2 vols.

Tancoigne, J. M., *A Narrative journey into Persia, and residence at Teheran*, translated from the French (London, William Wright, 1820).

Tavakoli-Targhi, Mohamad, *Refashioning Iran: Orientalism, Occidentalism and Historiography* (Houndmills, Basingstoke, New York, Palgrave, 2001).

Tavakoli-Targhi, Mohamad, 'The Homeless Texts of Persianate Modernity', in Ramin Jahanbegloo (ed.), *Iran: Between Tradition and Modernity* (Lanham, MD: Lexington Books, 2004), pp. 129–160.

Tavakoli-Targhi, Mohamad, *Bahâyeesetizi va eslâmgarâyee dar Iran* (Anti-Babism and Islamism in Iran), *Iran-Nâmeh*, Winter 1379/2000–Spring 1380/2001, (73–74), pp. 79–124.

Temkin, Owsei *Galenism: Rise and Decline of a Medical Philosophy* (Ithaca, London, Cornell University Press, 1973).

The Koran, translated from Arabic by Rodwell, J. M. (London, Phoenix, 2009, first published 1909).

Théodoridès, Jean, 'Un grand épidémiologiste franco-mauricien: Joseph Désiré Tholozan (1820–1897)', *Bull. Soc. Path. Ex.* (*Bulletin of the Exotic Pathology Society*), 1 (1998): 104–108.

Théodoridès, Jean, 'Tholozan et la Perse,' *Histoire des sciences médicales*, tome XXXII, (3) (1998).

Théodoridès, Jean, 'Tholozan et la Perse', *Histoire des sciences médicales*, tome XXXII, (3) (1998), 287–296.

The Illustrated London News, June 14, 1873.

Tholozan, Joseph-Désiré, *Histoire de la peste bubonique en Mésopotamie* (Mémoire lu, en août 1873, à la Société Impériale de médecine de Constantinople et imprimé dans la *Gazette médicale d'Orient*), p. 51.

Tholozan, J.-D., *Histoire de la peste bubonique en Perse* (Paris, G. Masson, 1874).

Tholozan, J.-D., *L'histoire de la peste bubonique en Mésopotamie* (Paris, 1874).

Tholozan, J.-D., *Prophylaxie du choléra en Orient* (Paris, Victor Masson et Fils, 1869).

Tremayne, Soraya, 'The Dilemma of Assisted Reproduction in Iran', FVV in ObGyn, 2012, Monograph, pp. 70–74.
Twort, F. W., 'The Discovery of the "Bacteriophage"', *The Lancet*, 205 (5303), (1925), 803–852.
Wasti, Nayyar, 'Iranian Physicians in the Indian Sub-continent,' *Studies in History of Medicine*, 2 (4), (December 1978), 264–283 (Department of History of Medicine, Institute of History of Medicine and Medical Research, New Delhi).
Werner, Christoph, *An Iranian Town in Transition: A Social and Economic History of the Elites of Tabriz, 1747–1848* (Wiesbaden: Harrassowitz, 2000).
Worboys, Michael, 'Science and Colonial Empire, 1895–1940', in Kumar, D. (ed.), *Science and Empire, Essays in Indian Context, 1700–1947* (Delhi, Anamika Prakashan, 1991), pp. 13–27.
Wright, Denis, *The Persian Amongst the English* (London, I.B. Tauris, 1985).
Wright, Denis, *The English amongst the Persians: Imperial Lives in Nineteenth-Century Iran* (London, New York, IB Tauris, 1977).
Yarshater, Ehsan (ed.), *Encyclopaedia Iranica* (New York, Encyclopaedia Iranica Foundation Inc., published since 1975).
Yildirim, Nuran, *A History of Healthcare in Istanbul* (Istanbul, ajansfa, 2010).
Zahedi, Farzaneh & Larijani, Bagher, 'National Bioethical Legislation and Guidelines for Biomedical Research in the Islamic Republic of Iran', *Bulletin of the World Health Organisation*, 86 (8) August 2008: 630–634.
Zakeri, Mohsen, *Sâsânid Soldiers in Early Muslim Society: the origins of Ayyârân and Futuwwa* (Wiesbaden, Harrassowitz verlag, 1995).
Zenuzi Tabrizi, ᶜAbdol-Hossein (Filsuf al-Dowleh), *Matrah al-anzâr fi tarâjem al-atebbâ al-aᶜsâr va falâsefat al-amsâr* (Biography of the physicians and philosophers across centuries and regions), written in 1324/1906, edited by Mir-Hâshem Mohaddes (Tehran, Enteshârât-e Hoquqi, 1388/2009).

Author Index

ʿAbdol-Hossein Khân, 91
ʿAbdol-Hossein Zenuzi Tabrizi (Filsuf al-Dowleh), 3, 169, 174
ʿAbdollâh Mostowfi, 15, 16, 134, 170, 173
ʿAbdol-Sabur-e Khoi, 56, 175, 183, 185– 186, 188
ʿAbdul-Rahmân al-Suyuti, 51, 183
ʿAlinaqi Hakim al-Mamâlek, 23, 47, 97, 176
ʿAlinaqi Khân-e Mokhber al-Dowleh, 98
ʿAliqoli Khân-e Mokhber al-Dowleh, 90, 99, 194, 199
ʿAqili Khorâsâni, 16, 32–3, 50–1, 174, 179, 183
ʿAref-e Qazvini, 208
ʿAziz-Khân, 160
ʿAziz Khân Ajudan Bâshi, 108, 188
ʿAzod al-Dowleh Daylami, 133
Abbasid Caliphate, 19, 50, 138, 174
Abdulhaq Molla, 169
Abol Hassan (Khân-e) Tafreshi, 82, 127, 175, 193
Achaemenid, 212
Ackerknecht, Erwin, 65
Adib al-Saltaneh, 208
Agha Mohammad Khân, 22, 58
Ahmadiyeh hospital, 159, 216
Alexandrian school, 54
Alexandria (port of), 181

Alloy, Sprenger, 2
Amanat, Abbas, 10
Amânollâh Khân-e Tabib, 105
Amin al-Dowleh, Mirzâ ʿAli-Khân (prime minister of Mozaffar al-Din Shah), 134, 185
Amin al-Soltân, Ebrahim Khân, 124, 128, 185, 206
Amin al-Zarb, Hâj Mohammad Hassan, 164, 215
Amir-Aslân Khân, 70
Amir Assadollâh Khân, 95
Anatolia, 56
Ancient Greece, 166
Andral, Gabreil, 82, 193
Anglo-Iranian Oil Company, 160
Anglo-Persian Convention, 158
Anjoman-e Moqaddas-e Melli-ye Esfahân, 204, 213
Anjoman-e Dabestâniyân, 135
Anjoman-e Farhang, 146
Anjoman-e hefz al-Sehheh-ye Melli, 213
Anjoman-e kheiriyyeh, 152
Anjoman-e Maʿâref, 134
Anjoman-e Maʿâref-e Guilân, 214
Anjoman-e melli, 147
Anjoman-e mohasselin, 145
Ansari, Ali, 152, 172, 216–17
Antimon (antimony), 71
Anzali, 70, 130, 178, 216

Aqâ Mirzâ Mohammad-e Tehrâni, 96, 196
Ardeshir-Mirzâ Rokn al-Dowleh, 90
Armenian church, 90
Armenian plateau, 27, 177
Askar Khân-e Afshâr, 89, 98
Astara, 130
Astarâbâdi, Seyf al-din Mohammad Jaʿfar, 19, 71, 188–9
Austria, 63–5, 123, 192, 220
Ayatollah Khamenei, xi, 7, 166
Ayatollah Khomeini, xi, 166
Ayubids, 105
Azerbaijan, 21, 25, 27, 56, 57, 59, 90–1, 178

Bâb, ʿAli-Mohammad, 173
Babur, 50
Bahâ al-Dowleh, 49, 52
Baqer-e Larijani, 7, 171
Baron de Bourgogne, 98
Bernard, Claude, 77, 80, 191, 228
Bernard, Jean, 188
Bimârestân-e Bistun, 152
Bistun, 126
Bohler (instructor of mathematics), 64
Boqrât al-Molk, 112, 202, 213
Borhân al-Din Nafis b. ʿAvaz b. Tabib-e Kermâni, 117
Borowski, 64, 187
Briggs, George, 55, 184
Britain, 8, 21, 55, 58–66, 83, 87, 97–8, 121, 130–1, 158, 161, 172
British Church Missionary Society, 131, 159
British India, 55, 168, 171, 172, 232, 234
British Legation, 56, 57, 61, 64, 116, 124, 154, 156, 205, 217
Broussais, François-Joseph-Victor, 77, 82, 179, 180, 226
Browne, Edward G., 126, 204, 215
Bruguière, Jean-Guillaume, 57
Brujerd, 27, 105, 200
Bushehr (Bushir), 26, 56–7, 99

Cabanis, Pierre-Jean-Georges, 12, 106, 133, 172, 200
Cairo, 33, 181
Calcutta, 56
Caliph Hârun al-Rashid, 84
Calomel, 70, 84
Campbell, James, 56
Campbell, John, 27, 70, 74
Canon (of Avicenna), 15–16, 34, 73, 114, 183, 187, 191, 229
Captain Charnota, 64
Champagny, Comte de (Napoleon's minister of Foreign Affairs), 62
China, 6, 26, 55, 84, 170, 172, 232
Chomel, August-François, 74, 82, 191, 228
Christoval Acosta, 45
Church Mission Hospitals, 131
Clerics (Shiite), 7, 19, 92, 96, 134, 152, 165, 189, 208
Cloquet, Ernest, 61–5, 71–4, 92, 189–91, 198
Cloquet, Jules, 190
Clot, Barthélémy, 33, 38–9, 179–81
Colombe, Philip, 169
Colonel Mahmud Khân, 109
Colonel Shiel, 61
Comte des Alleurs, 186
Comte de Sartiges, 63, 186–7
Constantinople, 180, 186, 202
Cormick, Dr (Junior), 64, 83, 154
Cormick, John, 27, 43, 56, 57, 67, 70, 74, 77, 188
Corps de santé militaire, 125–6
Crimean War, 180, 212
Cte de Croizier, 212

Dâmghân, 91
Dâr al-Fonun, xiii, 3–4, 9–10, 12, 15, 23–4, 33, 44, 47, 63–4, 69, 71–4, 76–9, 81, 88–99, 101–2, 104–5, 107–8, 113, 117–19, 125, 127, 140–1, 144, 147, 150, 154, 156, 158, 165, 170, 175–6, 187, 191, 194–5
Dâr al-shafâ, 20, 106, 152, 175, 200

Author Index • 243

Dar-e Gaz, 130
Davalu, (clan), 201
de Balloy, Marie-René-Davy de
    Chavigné, 124–6, 130, 204–6
De Khânikof, Nicolas, 126, 206, 229
Delaporte, François, 181
Delhi, 49
Descartes, René, 196, 198
Dieulafoy, Jane, 71, 229
    couple, 138
Dieulafoy, Marcel, 126, 205–6
Diphtheria, 144, 146, 149
Dissection, 5, 38, 44, 51, 53–4, 63,
    71–3, 78, 165, 174, 179, 189
Douvillier, 125
Dr. ͨAli-Khân-e Falâti, 144
Dr. Ahmadiyeh, 141, 210
Dr. Amir Aͨlam, 157
Dr. Arno, 104
Dr. Babayev, 83
Dr. Bahrâmi, 161
Dr. Bâqer Khân, 154
Dr. Barrachin, 61
Dr. Barthelemy, 65
Dr. Basil, 28, 82–3, 178, 182, 202
Dr. Begmez, 104
Dr. Bongrand, 154
Dr. Carr, 159
Dr. Castaldi, 83
Dr. Charles Bell, 60–1
Dr. Cherebrin, 123
Dr. Dâvood Khân-e Afshâr, 155
Dr. Dickson, 64–5, 104, 121–3, 183,
    188
Dr. d'Obermayer, 154
Dr. Dolmage, 64–5
Dr. Ernest Cloquet, 61, 64–5, 71–4, 92
Dr. Feuvrier, 124, 174, 204–5, 224
Dr. Focceti, 191, 195
Dr. Gachet, 154, 156
Dr. Galley, 154
Dr. Georges, 154
Dr. Gustav Frank, 144
Dr. Haaz, 144
Dr. Hakim al-Molk, 154
Dr. Ilberg, 158

Dr. Isidor Albo, 64, 78–80, 78–80, 82,
    125, 156, 158, 192, 213, 216, 220
Dr. Joseph P. Cochran, 129–30, 206
Dr. Kazollani, 65, 75, 188
Dr. Kelley, 213
Dr. Labat, 61, 186
Dr. Lattes, 156
Dr. Le Blanc, 156
Dr. Lindley, 154
Dr. Mahmud-e Moͨtamed, 153
Dr. McNeill, 57, 60
Dr. Mesnard, Joseph, 153, 161
Dr. Minas, 169
Dr. Mirzâ ͨAli Moͨtamed al-Atebbâ, 117
Dr. Neligan, 154
Dr. Oberlin, Charles, 161, 217
Dr. Odling, 57, 154
Dr. Regling, 154
Dr. René Legroux, 161
Dr. Riach, 60, 186
Dr. Sadowsky, 154
Dr. Salvatori, 55–6
Dr. Schtrong, 144
Dr. Stump, 156
Dr. Trafus, 141
Dr. Treacher Collins, 57, 185
Dr. Wibier, 154
Dr. Wishard, 154, 176
Dr. W. Torrence, 128
Dual, 125
Dust-Mohammad Khân-eMoͨayyer
    al-Mamâlek, 198

Eͨtemâd al-Saltaneh,
    Mohammad-Hassan Khân, 35, 72,
    91, 125, 173–4, 189, 190, 199,
    201
Eͨtezâd al-Saltaneh, ͨAli-Qoli Mirzâ,
    90–1, 93–4, 96, 99, 101, 169, 196
East India Company, 56, 60, 184, 234
Ebrâhim-Khân-e Zahir al-Dowleh, 152
Edinburgh, 22, 56, 230, 234, 237
Egypt, 33, 38, 105, 165, 168, 171, 172,
    173, 179, 180, 181, 185, 207, 228,
    229, 230, 235, 236, 238

Elgood, Cyril, 65, 126, 154, 174, 176, 177, 178, 182, 183, 184, 185, 186, 187, 188, 189, 200, 206, 215, 216, 228, 230, 232
Elyâs, 71–2
Emâd al-Din Mahmud-e Shirâzi, 20–1, 49, 175, 183, 191
Emâd al-Din Mahmud-e Shirâzi, 49, 175, 183, 191, 220
Erevan, 186, 188
Esfahan, 26–7, 57, 90, 118, 124, 156, 159, 177, 204, 213–15
Eskandar Mirzâ, 117
Esmâʿil Jorjâni, 15, 72, 178, 223, 233

Fahmy, Khaled, 168, 171, 179, 230
Farâhân, 210
Farâmushkhâneh, 98, 100, 171, 197
Farrokh Khân-e Amin al-Dowleh, 97
Fathʿ Ali Tabib, 213
Fathʿ Ali Shah, 22, 23, 24, 28, 35, 55, 56, 57, 59, 60, 62, 67, 70, 71, 73, 74, 89, 94, 117, 178, 185, 200
Firuz Mirzâ Nosrat al-Dowleh, 153, 207
Flandrin, Georges, 188
Foccetti, 74
France, 8, 56–9, 61–4, 66, 73–4, 77, 84, 87, 89, 97–9, 117, 123–30, 144, 153, 160–1, 169, 173, 180–1, 186, 188, 197, 205–6, 211–12, 216, 222, 232–3, 235, 237–8

Galen, 3, 15, 17, 18, 36, 46, 50, 53, 54, 56, 65, 71, 77, 81, 169, 191, 231
Geertz, Clifford, 10, 171
General Gardanne, 55, 56, 58, 61, 62, 185
General Nazar-Aqâ, 127, 147, 205
Germany, 64, 66, 123, 130, 141, 160
Gobineau, Comte Arthur de, 75, 99, 126, 146, 191, 198, 231
Goubert, Jean-Pierre, 209, 231
Grand Orient de France, 98
Grisolle, Augustin, 3, 32, 33, 77, 81, 82, 176, 179, 192, 193, 220, 231
Guilân, 25, 27, 70, 75, 95, 118, 152, 177, 185, 214

Guillaume-Antoine Olivier, 57, 58, 185, 193, 236
Guizot, François, 61, 63
Gvarret, Jules, 82

Hâfez, 98
Hâjeb al-Dowleh, 107, 215
Hâj Esmâʿil Hushmand, 140
Hâji Bâbâ Afshar (student to England), 59
Hâji Bashir, 132
Hâji Mirzâ Abolfazl-e Sâvaji, 101
Hâji Mirzâ Âghâssi, 22, 61, 62, 65, 66, 74, 88, 110, 188, 190, 202
Hâji Mohammad Jaʿfar Esfahâni, 215
Hâji Seyyed Ali Qazvini, 215
Hâj Mohammad Ebrahim-e Esfahâni, 215
Hâj Mohammad Karim Khân-e Rashti, 135
Hâj Seyyed ʿAli Aqâ Fumani, 152
Hakim ʿAlavi Khân, 49, 183
Hakim Mohammad, 53, 55, 112, 184, 220
Hakim Qoboli, 175, 185, 221
Harvey, William, 65, 79
Hassan-ʿAli Khân-e Eʿtezâd al-Molk, 136
Health Council (*Dâr al-Showrâ-ye Tebbi or Tebbiyyeh*), 100
Hejâz, 49
Hippocrates, 15, 34, 36, 37, 50, 71, 81, 113, 179, 180
Hojjat al-Eslâm Sadr'al-Ulama, 152
Horr ibn Yazid-e Riyâhi, 212
Hospital ʿAzodi, 133
Hospital Amini, 132, 159, 160
Hospital Najmiyeh, 132, 207, 237
Hospital Nuriyeh, 152
Hosseinʿ Ali-Mirzâ Farmânfarma, 132
Hunain b. Isʾhâq, 53
Hussen Qouli Aga [Hossein Khân-e Qâjâr], 62

Ibn Mâsawayh, 53
Ibn Nafis, 15, 92, 191

Ibn Qayyim al Jawziyya, 7, 19, 167, 171, 174, 232
Ibn Ridwân, 53, 177, 229, 238
Imam-Rezâ, 100
Imperial Military Medical Academy (Ottoman Empire), 63
India, 1–2, 5–6, 13, 18, 26, 28, 33, 42, 45, 49–51, 55–6, 58–60, 84, 106, 119, 130, 165, 167–8, 170–2, 177, 183–4, 206, 208
Indo-European Telegraph, 65
Iraj-Mirzâ (Ra'is al-Atebbâ), 97, 101
Iraq, 71, 102, 153, 214
Islamic Republic of Iran, 171, 218, 226, 231, 240

Jackson, A. V. Williams, 139, 210, 232
Jaubert, Amédé, 55, 186
Jones, Harford, 56, 59, 175, 235
Jordanova, Ludmilla, 169, 233
Jouannin, 56
Jukes, Andrew, 43, 56, 57, 184, 234
Julfa, 57

Kâmrân Mirzâ Nâyeb al-Saltaneh, 124
Karbala, 19, 71, 95, 153
Karbalâyee Khur-Mohammad-e Abeleh-kub, 95
Karim Khân-e Zand, xiii, 49, 106
Kârun River, 66
Katuzian, Homayun, 116
Kâvoos, Kay, 145
Kazullani, 190
Kerman, 130–1, 135, 152, 159, 214
Kermânshâh, 27, 105, 153, 159, 177, 214
Khalil ebn-e Mirzâ ᶜAbdol-Bâqi Eᶜtezâd al-Atebbâ, 78
Khalkhâl, 91, 190
Khorâssân, 23, 28, 91, 108, 176, 197, 213, 227, 235
Koran, 18, 54, 132, 137, 167, 204, 208

Kuhn, Thomas, 165, 217, 233
Kumar, Deepak, 129, 172
Kurdistan, 105

Lahore, 167
Lajard, 56
Lamy, captain of engineering, 58
Lavoisier, 80, 81
Lebanon, 161
Lemaire, 125
Lieutenant Bernard, 56
Lieutenant Colonel D'Arcy, 59
Lieutenant Colonel T. W. Mercer, 167
Lord Minto, 56
Louis Philip, 63
Louvre, 138

Madrasa-ye Soltâni, 134
Madresseh-ye Nâsseri, 125, 126
Mahd-e ᶜOlyâ, 21, 103
Majd al-Dawla, 70
Majles (parliament), 135, 154, 161, 182, 188–9
Majusi, 15–16, 36, 46, 112–13
Makdisi, George, 172, 234
Malâyer, 90–1
Malcolm, John, 55, 56, 184
Malek al-Motekallemin, 137
Malek al-shoᶜarâ Mirzâ Fathᶜali, 175
Malek-Tâj, 207
Marseilles, 186
Maryam-e Mozayyen al-Saltaneh, 145
Mashhad, 20, 25, 28, 100, 108, 118, 152, 159, 182, 197, 199, 217
Mashhad Zolfâbâd, 210
Mausoleum of ᶜAbdol-ᶜAzim, 135
Mawlavi Jalâl al-Din Mohammad Balkhi, 190
Mâzandarân, 25, 27, 92, 95, 107, 118, 177
McNeill, John, 27, 56, 60
Mecca, 49, 65, 186, 198
Mehdi-ye Ghodsi, 161
Mirzâ ᶜAbbâsᶜAli Khân, 93
Mirzâ ᶜAbbas Khân, 208
Mirzâ ᶜAbdol Ali Seif al-Atebbâ, 83, 193

Mirzâ ʿAbdol-Ghaffâr, 203
Mirzâ ʿAbdol-Karim-e Tabib-e Tehrâni, 76
Mirzâ ʿAbdollâh, 92, 94, 105
Mirzâ ʿAbdollâh Loqmân, 150, 156
Mirzâ ʿAbdollâh Shirâzi, 94
Mirzâ ʿAbdol-Wahhâb, 92, 193, 220
Mirzâ ʿAli-Akbar, 110
Mirzâ ʿAli-Akbar-e Kermâni (Nâzem al Atebbâ), 101, 117, 158, 203
Mirzâ ʿAli-Akbar Khân-e Nâzem al-Atebbâ, 117, 158
Mirzâ ʿAli-Akbar Khân-e Shirâzi, 105
Mirzâ ʿAli-Asghar Khân-e Moʾaddab al-Dowleh, 117
Mirzâ ʿAli Doktor, 92, 104, 196
Mirzâ ʿAli (Dr, Moʿtamed al-Atebbâ), 23, 33, 82–3, 85, 92, 104, 117, 176, 179, 192, 206
Mirzâ ʿAli Khân-e Hâjeb al-Dowleh, 107
Mirzâ ʿAlinaqi (Hakim al-Mamâlek), 23, 47, 97, 117, 176, 203, 227
Mirzâ ʿAli Tabib-e Tehrâni, 117
Mirzâ ʿIsâ Vazir, 132
Mirzâ Abolfazl Tabib-e Kâshâni, 114
Mirzâ Abol-Hasan Khân, 60
Mirzâ Abolhasan Khân-e Ilchi, 98
Mirzâ Abol-Hassan Khân-e Tafreshi, 82
Mirzâ Abol-Qâsem, 92, 95, 101, 104
Mirzâ Abolqâsem Asʿadol-Hokamâ, 213
Mirzâ Abolqâssem, 117
Mirzâ Ahmad-e Hakimbâshi, 23, 69, 92, 93, 117
Mirzâ Ahmad-e Hakimbâshi-ye Kâshâni, 69, 92–3, 97, 117, 175
Mirzâ Ahmad-e Hakimbâshi-ye Tonekâboni, 23, 67, 74, 117
Mirzâ Ahmad-e Kâshâni Hakimbâshi, 97
Mirzâ Ahmad-e Tabib-e Tonekâboni, 23, 67
Mirzâ Amir Khân, 148
Mirzâ Aqâ Khân-e Kermâni, 146, 234
Mirzâ Aqâ Khân-e Nuri, 96, 97
Mirzâ Askar Khân-e Afshâr, 98

Mirzâ Assadollâh Esfahâni, 144
Mirzâ Bâbâ Hakimbâshi, 200
Mirzâ Dâvood Khân, 63
Mirzâ Ebrahim Khân, 154
Mirzâ Habibollâh, 114, 117
Mirzâ Hakimbâshi, 83
Mirzâ HassanʿAli Khân (ambassador), 98
Mirzâ Hassan-e Roshdiyeh, 134
Mirzâ Hedâyat, 102, 107, 210
Mirzâ Hessâm al-Din, 105
Mirzâ Hossein-e Afshâr, 78
Mirzâ Hossein-e Doktor, 69, 81, 92
Mirzâ Hossein Khân-e Sepahsâlâr, 98, 99, 100, 107, 108, 109, 110, 124, 128, 135, 152, 199, 213
Mirzâ Jaʿfar Khân-e Moshir al-Dowleh, 201
Mirzâ Jaʿfar (student of medicine in Britain), 60
Mirzâ Jafar-e Tabib, 83
Mirzâ Kâzem, 47, 92, 93, 101, 104, 114, 215
Mirzâ Mahmoud Khân, 154
Mirzâ Mahmud-e Kalântar, 95
Mirzâ Malkam Khân (Nâzem al-Molk), 98, 100
Mirzâ Massih, 117
Mirzâ Mehdi Khân, 208
Mirzâ MohammadʿAli Khân-e Shirâzi, 62, 63
Mirzâ Mohammad, 104, 220
Mirzâ Mohammad-e Râzi-ye Kani Fakhr al-Atebbâ, 189
Mirzâ Mohammad-e Tabib, 110
Mirzâ Mohammad-Hossein, 94
Mirzâ Mohammad Hossein-e Afshâr, 77, 78, 222
Mirzâ Mohammad-Kâzem-e Rashti, 21, 70, 78, 80, 103, 189, 220
Mirzâ Mohammad-Khân, 127, 224
Mirzâ Mohammad-Taqi Kâshâni, 192
Mirzâ Mohammad Taqi Sepehr, 96

Author Index • 247

Mirzâ Mohammad-Taqi Shirâzi (Malek al-Atebbâ), 16, 27, 28, 43, 45, 47, 67, 68, 97, 103, 173, 178, 183, 185, 221
Mirzâ Moqim-e Mostowfi, 97
Mirzâ Mostafâ, 118
Mirzâ Mozaffar Khân, 148
Mirzâ Musâ Sâvaji, 96, 178, 183
Mirzâ-Musâ Vazir, 96
Mirzâ Nasrollâh Khân-e Moshir al-Dowleh, 89
Mirzâ Nosrat-e Quchâni, 79
Mirzâ Qahramân-e Amin-Lashkar, 198
Mirzâ Qawâm al-Din, 148
Mirzâ Razi Hakimbâshi, 107
Mirzâ Rezâ Doktor, 33, 81, 92, 94, 101, 118
Mirzâ Rezâ Jarrâhbâshi, 94
Mirzâ Rezâ-Qoli Jarrâhbâshi, 79
Mirzâ Saʿid-e Sharif-e Kermâni, 117
Mirzâ Saʿid Khân, 98
Mirzâ Seyyed ʿAli, 92, 104
Mirzâ Seyyed ʿAli Boqrât al-Molk, 213
Mirzâ Seyyed Hossein Khân-e Nezâm al-Hokamâ, 93
Mirzâ Seyyed Razi Raʾis al-Atebbâ, 94, 101, 104, 145, 196, 212
Mirzâ Seyyed Razi-ye Hakimbâshi, 94
Mirzâ Soleimân, 114
Mirzâ Taqi Kâshâni, 81
Mirzâ Taqi Khân-e Lashkarnevis, 208
Mirzâ Yahyâ Khân-e ʿEmâd al-Hokamâ, 150, 151
Mirzâ Zaki, 62
Mirzâ Zeyn al-ʿAbedin-e Kâshâni, 101
Moʿin al-Atebbâ, 127, 206, 224
Moʾazzen Jâmi, Mohammad-Hâdi, 167
Mohammad-ʿAli Amini, 159
Mohammad ʿAli Mirzâ, 21, 135
Mohammad-Ali Pasha (ruler of Egypt), 171
Mohammad b. Mahmud Chaghmini, 92
Mohammad Hossein, 182, 200, 209
Mohammad-Hossein b. ʿAbdolwahhâb-e Tehrâni, 209

Mohammad Hossein-e Zakâʾul-Molk-e Forughi, 133
Mohammad-Hossein ibn Mohammad-Hâdi-ye ʿAqili-ye ʿAlavi-ye Khorâssâni-ye Shirâzi, 183
Mohammad-Kâzem (student of painting in England), 59
Mohammad-Reza Beg, 186
Mohammad-Reza Shah, 188
Mohammad-Sâdeq Khân, 91
Mokhber al-Dowleh, ʿAliqoli Khân, 64, 90, 105, 110, 194, 216
Mollah Bâshi, 96
Mollâh Zein al-Din-e Lâri, 169
Molla Lâlazâr-e Hamadani, 99
Mollâ Mohammad-e Qoboli, 97
Mollâ Rezâ Hamedâni, 213, 235
Montesquieu, 106
Morier, James, 59, 70, 74, 89, 172, 173, 174, 175, 184, 186, 191, 194, 235
Morshed Morâdi, 211
Mossadegh, Mohammad, 207
Mostafâ Falâti, 144
Moulin, Anne Marie, 24, 168, 170, 172, 174, 176, 179, 180, 199, 235
Muʿâwiya ibn Abi Syfyân, 212
Mughal Empire, 50
Mustafâ Rashid Pasha, 61

Nafisi ʿAbbâs, 117
Nafisi Abolqâsem, 117
Nafisi, Saʿid, 117, 203, 235
Najm al-Saltaneh, 132
Napoleon Bonaparte, 58, 59, 61, 62, 66, 87, 97, 184, 185, 186, 232
Nasrollâh Mirzâ Qâjâr, 3, 170, 221
Nâsser al-Molk, 152
Nategh, Homa, 226, 236
National Assembly, 139, 147
Nayyer al-Molk, the Minister of Education, 134, 194
Nazar Aqâ, 91, 127, 147, 205, 206, 224
Neishabur, 207
Netherlands, 64–5, 70
Newton, 11, 80

Nezâm al-Din (Sufi saint), 49
Nezâm al-Hokamâ, 93, 154
Njoman-e Dabestâniyân (Society for Primary Education), 135
North Africa, 5
Nurollâh-Khân-e Zahir al-Mamâlek, 152
Nurses, 110, 131

Ologh-Beg Timurid, 117
Onsor al-Maʿâli Kaykâvus b. Eskandar, 113, 202
Ottoman Empire, 55, 58, 63, 95, 134, 164, 169, 171, 186–7, 201, 207, 229

Paracelsus, 56, 57, 67, 77, 185, 191, 222
Parliament, 7, 135–6, 139, 147, 213
Parsons, Talcott, 129, 206, 215, 236
Pasteur Institute, 153, 161, 164, 207
Pasteur, Louis, 3, 65
Persian Gulf, 6, 25–6, 55, 58
Pharaoh Ramses III, 173
Pietro Della Valle, 169
Pinel, Philip, 82
Pishdâdiyân, 133
Poland, 64
Pratt, Spencer, 128, 176, 226
Prince Yamin al-Dowleh, 118
Prussia, 63–4
Punjab, 167–8, 232

Qâzân, 133
Qâzi ibn Kâshef al-Din Mohammad Yazdi, 55
Qom, 22, 25, 27, 105, 200, 203
Quarantine, 25, 28, 83, 95, 102, 116, 156, 169, 181, 202
Quchân, 150

Rashid al-Din Fazlollâh, 133, 186
Rasht, 27, 43, 67, 95, 114, 130, 152, 159, 189
Rawlinson, Henry, 126
Râzi, 15, 16, 19, 25, 45, 52, 71, 169, 174, 176, 191, 220

Relief council (*majles-e eʿânât-e foqarâ*), 110
Renan, Ernest, 138, 146
Rey, 135, 234
Rezâ Qoli Khân, 96, 154, 223, 237
Rezâ Qoli Khân-e Hedâyat, 97, 98, 99, 170
Richard, Joseph, 64, 125
Robʿ-e Rashidi, 133
Rome, 84
Roshdiyyeh School, 208
Rostam, 49, 145
Rostam al-Hokamâ, 183, 237
Rousseau, Jean-Baptist-Louis-Jacques, 42, 177, 181, 193, 236, 237
Rousseau, J.-J., 66
Royan Institute, 166
Russia, 8, 27, 55, 58–63, 66, 87–8, 98, 110, 130, 164, 172, 197, 213

Saʿadi, 98
Safavids, 3, 19, 49, 55, 106, 169, 184, 237
St Petersburg, 22, 169, 208, 219–20
Sâleh ibn Nasrollâh al-Halabi, 169, 222
Samad-Khân (Mozaffar al-Din Shah's minister in Paris), 58
Sanitary Council (*Majles-e Hefz al-Sehheh*), 9, 10, 26, 41, 70, 71, 88, 95, 101, 102, 103, 104, 105, 114, 118, 119, 122, 140, 142, 147, 152, 154, 155, 158, 159, 164, 202
Sartiges Comte de, 63, 186, 187
Sassanid, 138, 146
Sauvage, 82
Sâvaji, Mirzâ Mussâ, 24, 31, 32, 36, 37, 39, 42, 44, 96, 172, 178, 183, 192, 199, 211, 221
Schlimmer, Johannes, 64, 65, 69, 70, 79, 82, 85, 89, 95, 114, 178, 179, 189, 192, 193, 194, 198, 222
Schneider, Jean-Etienne Justin, 16, 29, 46, 93, 102, 125, 126, 127, 128, 130, 140, 142, 147, 153, 154, 173, 174, 182, 187, 190, 196, 199, 203, 205, 206, 213, 215, 224, 238

School of political science (*madrassa-ye siyâsi*), 89
Scoutetten, Henry, 34, 179, 180, 238
Seljuk, 207
Semnân, 27, 91
Seyyed ʿAli Boqrât al-Molk, 112, 202, 213
Shah ʿAbbas the Great, 117, 209
Shâhnâmeh, 145
Shah Safi I Safavid, 55, 184
Shah Soltan Hossein Safavid, 169
Shah Tahmasp Safavid, 175
Shaqâqi (tribe), 107
Shâyegh, Cyrus, 212
Sheikh-e Bahâyee, 203, 204
Sheikh Mohammad-e Tehrâni, 93
Sheikh Mohammad-Sâleh, 92
Sheikholeslâmi, A.-R., 116, 198, 203, 238
Shemran (Shemiran), 176
Shirâzi, Mirzâ Kâzem-e Rashti, 16, 21, 47, 70–3, 77–8, 80, 101, 103–4, 114–15, 135
Shokrollâh Khân-e Qâjâr-e Qovânlu, 210
Showrâ-ye baladiyyeh (City council), 138
Shush, 126, 138
Sir Harford Jones, 56, 59, 175, 235
Sistan, 130, 131, 156, 159, 213, 215
Snow, John, 41, 83, 181, 193, 238
South Persian Rifles, 131
Spronk, Maurice, 58, 185, 239
State Council (*Dâr al-Showrâ-ye Kobrâ*), 96, 108
Syria, 161

Tabriz, 15, 20–2, 60, 64, 90–1, 105–6, 133, 135, 144, 159, 195, 208, 216
Talbot, Major Gerard, 186
Tancoigne, J. M., 168, 185, 239

Tavakoli-Targhi, Mohammad, 1, 7, 168, 169, 171, 194, 211, 239
Tehran, 15, 21–3, 27–8, 35, 41, 52, 55–7, 60–1, 63–5, 74, 90–1, 93, 95–6, 98, 102, 104, 107–8, 110, 114, 117, 124–8, 130, 132, 134, 139–40, 144, 150, 153, 156, 158–61, 166–7, 175, 178–9, 186, 188, 201, 203, 208, 216, 219
Timurids, 49
Toyserkân, 90, 91
Transcaucasia, 87
Treacher Collins, Dr, 57, 185
Trebizond, 178
Tytler, John, 2

United Kingdom, 6
United States, 6, 24, 123, 129, 168, 176, 188
Urumia, 129
Usuli Shiʿa, 194

Vale de Grace (military school), 98
Verdier, captain of infantry, 58
Voltaire, 106

Waterloo, 59
Weqâr al-Molk, 134
Wosuq al-Dowleh, 216
Wright, Denis, 59, 184, 186, 239, 240

Yamut Turkmen, 107
Yazd, 26–7, 95, 105, 140, 159, 166
Yazd Research and Clinical Centre for Infertility, 166
Yazid (Horr ibn Yazid-e Riyâhi), 212

Zakhirah-ye Kâmela, 184, 202
Zand dynasty, 50, 53
Zell al-Soltân, 57, 124, 214
Zeyn al-'Abedin Khân-e Doktor, 210
Zoroastrian Parsis, 208
Zurkhâneh, 144

Subject Index

allopathic medicine, 167
anal fistula, 57
anatomical pathology, 3–4, 9, 12, 33–4, 38–40, 44, 48, 73, 77, 78, 81–2, 103, 112, 119
anatomy, 4, 8, 23, 51–3, 72, 73, 77–9, 80–3, 112, 118, 119, 122, 127, 148, 165, 171, 175, 176, 179, 185, 190, 193, 196, 206, 220, 222, 230
anti-Semitic ideas, 146
applied science, 5, 165
Arab medicine, 154, 215
artificial insemination, 7, 166
Assisted Reproductive Technique (ART), 166, 218, 226
Assyrian, 212
auscultation, 99, 127

biomedicine, 2, 3, 5, 10, 12, 121, 122, 153–4, 161–2, 166, 234
British hospital in Esfahan, 159

chemical drugs, 47, 67, 85, 94, 102, 155, 185, 221
cholera, 17, 18, 24–8, 31–2, 34–6, 38–46, 52, 67, 71, 75, 77, 83, 96, 99, 100–1, 104, 114–15, 131, 137–8, 142, 150, 152, 173, 176–83, 193, 198, 202, 220–2, 228–9, 232, 237–9
choleric fever, 39, 42

Christianity, 54, 204, 231
circular fever, 4
cirrhosis, 57
cloning, 6–7, 166
colonial context, 5, 8, 11, 123
colonial country, 5, 6
colonial medicine, 3, 5
colonial science, 1
conceptual transformation, 163, 165
Constitutional Revolution, 3, 10, 12, 66, 121, 134–5, 138–9, 144, 152, 154, 171–2, 205, 211, 226, 228
continuous fever, 4, 81
Crusades, 54

dissection, 5, 38, 44, 51, 53–4, 63, 72–3, 78, 165, 174, 179, 189, 238
divine science, 18

ejtehâd, xii, 6, 18, 71
embryo stem cell research, 166
enlightenment, 106, 137, 164, 166, 188, 217, 236
epidemic fever, 39, 45, 78, 178

faith (religious) healing, 17–18, 20, 167, 173
fatwa, 166
feqh, xii, 18
fever (s), 3–4, 27, 29–39, 45, 47–8, 53, 55, 67, 71, 75, 78, 80–3, 210

Subject Index

francophily, 66
freemasonry, 10, 98, 197, 227
French Revolution, 66
Fundamental Law, 139

Galenic medicine, 7–8, 40, 79
Galenico-Islamic medicine, 80
Gendarmerie, 159
Gilmour, John, 16, 155, 159, 161, 173, 214, 216, 231
Greece, 84, 166
Greek medicine, 19, 50, 167, 174
Greek science, 8, 54

hadith, 18–19, 51, 54
hasba (typhoid), 179, 192
Hippocratic aphorisms, 52
homeland, 147, 157, 211–12
hospital Ahmadiyeh, 216
human rights, 136, 139
humoral theory, 4, 15, 19, 29, 31, 34, 36–7, 40, 45–6, 48, 75, 83
hygiene, 100, 137, 139, 144–6, 216, 233
hygienist movement, 138

ideology of progress and modernity, 137
imperial science, 168, 172
Industrial Revolution, 55
infectious fever, 78
inoculation (variolisation), 56, 104, 153, 185, 221
institutional modernization, 114
intelligentsia, 5, 10, 12, 22, 66, 85, 88, 98–9, 116, 118, 121–2, 130, 139, 146, 164–5
intermittent fever, 4, 27, 32, 46–7, 74–5, 81, 104, 114, 191, 198, 200, 222
Islamic science, 8, 19, 80, 193

Judaism, 54

Kashani-Sabet, Firuzeh, 157, 211
Kérandel, Jean, 161
Khâdeʿa fever, 35
koranic medicine, 167

leishmaniasis, 99
Leukemia, 188

magic and faith healing, 17–18, 173
magnesium sulfate, 68–70
measles, 40, 42, 83
medical science, 61, 78, 181, 200, 210
medical transformation, 4, 6–9, 25, 122, 131, 163–5
Medicine of the Prophet, 7, 19, 51, 167, 174, 183
mercury (*zanbiq*), 70
miasma, 34–5, 37–8, 40–3, 71, 83–4, 137–8, 179, 181
miasma theory, 38, 83, 116
midwifery, 148, 157
midwives, 140
military medicine, 72, 110, 112, 125
military science, 112
mission hospitals, 85, 123, 129, 131, 133, 153
modernity, 1, 5, 7, 10, 12–13, 121, 133, 135, 137, 149, 156, 164, 167–9, 171–2, 202, 227, 229, 233–4, 239
modern anatomy, 23, 122, 165, 176, 196
modern science, 1, 5, 7, 9–13, 57–9, 62–3, 66, 84, 88–9, 92, 94, 97–8, 116, 118–19, 121, 126, 129, 134–7, 139–41, 147, 152, 157, 160, 164–5, 167–9, 172, 176, 204, 217
modern surgery, 78, 93
mohreqa, 31–3, 42, 107, 118
Mothbeqa, 31–3, 37, 42, 74, 81, 107, 118, 173, 190, 192, 200, 222
Mothbeqa fever, 81, 192
Movement for progress and modernity, 133

natural science, 15, 18, 51
neo-Hippocratic medicine, 3, 50, 75
neo-Lamarkian genetics, 146, 212
non-Islamic science, 8, 19, 51

ophthalmology, 16, 53, 77, 112, 117, 127, 148, 192, 222
orient, 1, 123

Subject Index • 253

oriental culture, 123
orientalism, 1, 8, 168
orientalist, 123
orientalist discourse, conception, 1

Paracelsian iatrochemistry, 3
paradigm of modern science, 164
patriotism, 139, 146, 157, 211
percussion, 99, 127
pestilential fever, 34, 35, 39
political science, 89, 137
professionalization, 85, 112, 114, 116, 118–19, 164
public health, 5, 17, 24, 95–7, 100–4, 108, 116, 119, 137, 139–41, 144, 145–6, 150, 154, 156–61, 167–8, 170, 173–4, 181, 209, 212, 217
putrid fever, 36, 178

quarantine, 25, 28, 83, 95, 102, 116, 156, 169, 181, 202
quinine, 47, 71, 74, 155, 182–3

rational science (*maʿqul*), 19
recurrent fever, 82
religion, 17, 20, 50–1, 54, 58, 73, 88, 90, 92, 99, 106, 112, 128, 165, 172–4, 179, 183, 189, 191, 194–5, 198, 200
religious science (manqul), 19–20, 51, 92, 94
reuter concession, 66, 183

sarsâm, 29, 114
Science of divination, 19
semicolonial, 5, 170
Shariʿa, 6–7, 92, 100, 204
smallpox, 32, 42–3, 55–7, 77, 83, 95, 104, 145, 150, 174–5, 185, 198–9
smallpox vaccination, 95, 122
smallpox vaccine, 142
social policy, 166
space science, 6
spirit of enquiry, 166
Stem Cell Research, 7, 166, 171

Sufism, 88
sunna, 6, 18, 172, 174
Surgery, 4, 16, 18, 24, 52–3, 63, 71–3, 77–80, 90, 92–4, 101, 110, 112–13, 117, 119, 122, 127, 148, 150, 165, 183, 184, 191–2, 200
suyursât, 96, 197
syphilis, 55, 183, 192, 220

Tanzimât (reform), 63, 187, 201
taqiyya (dissimulation), 146
tashrih-e ʿamali (practical dissection), 51
Tâʾun, 27, 40, 43, 67, 173, 182, 185, 202, 221
tazkerat al-atebbâ (biography of physicians), 3
tebb-e ʿamali, (manual medicine), 53, 112
tebb-e nazari (theoretical medicine), 53, 72
tebb-e qorʾâni, xi, 167
tonsillitis, 71, 102
traditional science, 85, 92, 118, 147
translation movement, 7, 80, 84
typhoid, 28, 32, 81, 107, 154, 176, 191
typhus, 32, 74, 107, 114, 179, 192

Unani medicine, 5, 51, 167, 168, 170
unnatural fever, 45

vaccination, 43, 56, 67, 95, 104, 122, 150, 165, 185
Vandidad, 54
vatan, 147, 157, 208, 211, 214, 223
Vivisection, 51, 53, 78

wabâ (*wabâyee, wabâiyyeh*), 17, 25–8, 31, 34–6, 38–45, 52, 96, 137, 172–3, 177–9, 182–3, 189–90, 192–3, 196, 202, 207, 212, 219
waqf (plural owqâf), 21, 89, 131–5, 150, 152–3, 159, 194, 207–8, 214, 230
western imperialism, 87, 127

western influence, xii, 2, 8,
    12–13, 84, 122–3, 130,
    204

western knowledge, 1, 6

western medicine, xiii, 1–5, 8–9, 24, 41,
    48–50, 57, 61, 63, 67–8, 70, 84,
    88, 92–4, 122, 139, 168

western science, 3, 5, 22, 63, 84, 89,
    122, 126, 145, 164, 168

GPSR Compliance

The European Union's (EU) General Product Safety Regulation (GPSR) is a set of rules that requires consumer products to be safe and our obligations to ensure this.

If you have any concerns about our products, you can contact us on

ProductSafety@springernature.com

In case Publisher is established outside the EU, the EU authorized representative is:

Springer Nature Customer Service Center GmbH
Europaplatz 3
69115 Heidelberg, Germany

www.ingramcontent.com/pod-product-compliance
Lightning Source LLC
Chambersburg PA
CBHW071615100426
42873CB00004B/50